T0332334

Recent Advances in Management and Engineering

First published 2024
by Routledge
4 Park Square, Milton Park, Abingdon, Oxon OX14 4RN

and by Routledge
605 Third Avenue, New York, NY 10158

Routledge is an imprint of the Taylor & Francis Group, an informa business

British Library Cataloguing-in-Publication Data
A catalogue record for this book is available from the British Library

ISBN: 9781032873183 (pbk)
ISBN: 9781003532026 (ebk)

DOI: 10.1201/9781003532026

Typeset in Sabon LT Std
by Ozone Publishing Services

Recent Advances in Management and Engineering

Dr. Narendra Kumar

Contents

List of Figures

Chapter 1

Chapter 2

Chapter 3

Chapter 4

Chapter 5

Chapter 6

Chapter 7

Chapter 8

Chapter 9

Chapter 10

Chapter 11

Chapter 12

Chapter 13

Chapter 14

Chapter 15

Chapter 16

Chapter 17

Chapter 18

Chapter 19

Chapter 20

Chapter 21

Chapter 22

Chapter 23

Chapter 24

Chapter 25

Chapter 26

Chapter 28

Chapter 29

Chapter 31

Chapter 32

Chapter 33

Chapter 34

Chapter 36

Chapter 37

Chapter 39

Chapter 40

Chapter 41

Chapter 42

Chapter 43

Chapter 44

Chapter 45

Chapter 46

Chapter 47

Chapter 49

Chapter 50

Chapter 51

List of Tables

Chapter 19

Chapter 20

Chapter 21

Chapter 22

Chapter 23

Chapter 24

Chapter 25

Chapter 26

Chapter 27

Chapter 29

Chapter 30

Chapter 31

Chapter 32

Chapter 33

Chapter 48

Chapter 49

Chapter 50

Chapter 51

Foreword

It is with great pleasure that I present to you the proceedings of our Recent Advances in Management and Engineering held on November 24 – 27, 2023 in Male. Maldives. This conference represents a milestone in our ongoing journey towards academic excellence where we aspire to become a renowned platform for the exchange of ideas, collaboration, networking, and learning.

These proceedings contain contributions that are very amazing in innovations in management. It covers a wide range of issues, ranging from the most recent trends in business to innovations in fundamentals of management. A broad collection of scholars, practitioners, and thought leaders from four continents across the world worked together to produce these results, which are a reflection of their combined efforts.

I would like to express my most sincere gratitude to all of the writers for the high-quality contributions they made, to the reviewers for the insightful criticism and constructive criticism they provided, and to the organizing team for the hard work they put forth to ensuring that this conference was a positive experience for everyone involved. In the course of your exploration of these proceedings, I strongly recommend that you approach each paper with an open understanding. The ideas and conclusions that have been presented in these articles have the potential to inspire inventive solutions, create interesting discussions, and spark new thoughts. In spite of the difficulties that we have encountered over the course of the past year, our community has demonstrated resiliency and adaptability in its efforts to pursue the advancement of our collective knowledge. There is no question in my mind that the discoveries and concepts that were discussed in these proceedings will play a significant part in determining the direction that management innovations and digitalization will take in the future.

I am sure about proceedings of this conference, which stands as a tribute to the intellectual curiosity, rigorous scholarship, and innovative spirit that characterizes our academic community. The purpose of this compilation is to present a roadmap for future research and development, as well as a picture of the enormous strides that we are making in the domains on Science, Technology and Management.

Your participation in this adventure is greatly appreciated. The lively and enriching experience that is ICOSTEM is a direct result of your engagement, your inquisitiveness, and your commitment to pushing the boundaries of human understanding.

Dr. Rajeev Lochan Pareek
Visiting Professor: SRM University Amaravati, India
Ex Vice Chancellor: The ICFAI University, Tripura, India
December 18, 2023

Preface

Dear Participants,

It is with immense pleasure that we extend a warm welcome to all of you to the recently concluded conference, international conference on Advances in Science, technology and Management (ICOSTEM 2023) which took place from November 24 – 27, 2023, in the picturesque Maldives, Male. This significant event focused on the "Recent Advances in Management and Engineering" with special sessions on Applied Sciences, Management and Engineering.

In an epoch characterized by swift technological progress and digital ingenuity, the dynamics of management continually undergo transformation. The conference served as a congregation point for thought leaders, industry professionals, academics, and innovators from across the globe, fostering the exchange of ideas and serving as a catalyst for transformative change. The era of digital transformation has not only revolutionized personal lives but has also profoundly impacted the business landscape. It has emerged as a strategic imperative, compelling companies to reevaluate their business models, reinvent strategies, and redefine value propositions. Amidst this evolution, the central theme became ensuring sustainability, constructing resilient, adaptable, and future-proof businesses.

Throughout the conference, we delved into the ways in which digital transformation is reshaping the management landscape and explored innovative strategies to ensure sustainability. Discussions encompassed fostering a culture of continuous learning and adaptation, leveraging digital technologies for sustainable growth, and striking a balance between innovation and ethical, responsible conduct.

The conference featured a diverse program, including keynote addresses by esteemed Indian industrialists and businessmen, panel discussions, workshops, and networking sessions. A dynamic cohort of speakers shared real-world experiences, providing practical insights into the challenges and opportunities of managing in the digital era. Topics such as the role of artificial intelligence in sustainable management, the impact of blockchain technology on supply chain management, and the significance of data analytics in decision-making were thoroughly examined.

We express my sincere gratitude to the dedicated members of the conference committee, who tirelessly worked over the past three months to bring this conference and proceedings to fruition. Kind blessings of Dr. Rajeev Lochan Pareek, Ex Vice chancellor The ICFAI University Tripura, Dr. Sunder Lal, Ex Vice Chancellor Purvanchal University Jaunpur, India. Unconditional support from Dr. Sanjeev Kumar, Dr Bhimrao Ambedkar University, Agra, India, Dr. Alok Aggarwal, University of Petroleum and Energy Studies, Dehradun, Dr. Anil Bhardwaj, Rajasthan University Jaipur, India, and Dr. M. S. Gill, Executive Director, GGI Khanna, India.

As we continue to navigate the exhilarating realm of digital transformation, my hope is that the insights gained from this conference will not only inspire and challenge your thinking but also equip you with the knowledge and tools necessary to drive sustainable innovation in your respective fields. Thank you for your active participation and valuable contributions, which have undoubtedly contributed to the success of this conference. We eagerly anticipate our continued collaboration in the pursuit of sustainable management innovations.

Warm regards,

Prof. (Dr.) Ilona Paweloszek, Conference Chair
Faculty of Management
Czestochowa University of Technology, Poland

Prof. (Dr.) Sanjeev Kumar, Conference Chair
Department of Mathematics
Dr. Bhimrao Ambedkar University, Agra, India

Acknowledgement

Dear Attendees,

We would like to extend our deepest gratitude to each and every one of you for your participation in our international conference. We understand the time and effort required to attend such events, especially amidst your busy schedules, and we are immensely grateful for your commitment and contribution.

A special thanks goes to SERF India for successful completion of this conference. A huge gratitude to all the contributors/authors who displayed their exceptional patience and commitment to research and developments. Your contributions have significantly enriched knowledge base and sparked important discussions that will undoubtedly lead to progress in research domains.

We would also like to specifically acknowledge our keynote speakers: Prof. Dorota Jelonek, Dr Tomasz Turek and Dr. Ilona Paweloszek from Poland, Dr R K Chaurasia and Dr Narendra Kumar from India, Dr. Dalia Younis from Egypt. Your insightful presentations were a highlight of the conference, providing invaluable perspectives and stimulating thought-provoking dialogue.

Our heartfelt thanks go out to the organizing committee. Your tireless efforts, dedication, and meticulous planning have made this conference a resounding success. Dr Sunil Kha, Dr Satyabhan Kulshestha, Dr. Aanya Chaudhary, Dr Mahesh Joshi, Dr Amit Kumar Sharma, Dr Preeti Narooka, Dr Naveen K Sharma, Dr Mamta Chahar, Dr Mukesh Jangir - without your efforts, this event would not have been possible.

Last, but certainly not least, we want to thank all attendees. Your active participation, insightful questions, and shared experiences truly enriched the event. We hope that you found the conference informative and worthwhile, and we look forward to seeing you at our next event.

Once again, thank you for your valuable contribution to this successful conference.

Best Regards,

Prof. (Dr.) Dorota Jelonek, Mentor
(Czestochowa University of Technology, Poland)
Shanti Educational Research Foundations,
Jaipur, India

Prof. (Dr.) Anirudh Pradhan, Mentor
GLA University, Mathura India)
Shanti Educational Research
Foundations, Jaipur, India

About the Editors

Prof. (Dr.) Ilona Paweloszek is a Doctor of Economics in the discipline of Management and Quality Sciences, a graduate of the Faculty of Management of the Czestochowa University of Technology, specializing in "development management and consulting." She also completed pedagogical studies in training teachers of technical subjects. She has been a researcher and didactic employee at the Faculty of Management of the CUT since 1999, currently as an assistant professor at the Department of Management Information Systems. From 2002-2006, she researched mobile technologies in knowledge management, the usability of mobile solutions, and their integration with enterprise information systems. Her scientific and research interests focus on using semantic web technology, big data, and data mining to support managerial decisions. She is the author of several dozen publications, including three original monographs. She is a member of the Scientific Association of Business Informatics and the Polish Association for Production Management.

Prof. (Dr.) Dorota Jelonek is a full professor of Economics at the Faculty of Management of the Czestochowa University of Technology in Poland. Her scientific and research interests focus on solving problems related to the implementation of management information systems in enterprises and improving management information processes. She is the author of 6 books and the editor of 10. Additionally, she has authored or co-authored 250 articles in Polish and foreign journals and book chapters. She previously held the positions of Associate Dean for Science and Dean of the Faculty. Since 2019, Prof. Dorota Jelonek has been the President of the Scientific Society for Economic Informatics and is a member of many societies, including the Polish Association for Innovation Management and the Informing Science Institute.

Prof. (Dr.) Munish Sabharwal is an accomplished academician, researcher, and proactive decision-maker with a diverse background and a passion for education and innovation. Possesses dual PhD qualifications in Computer Science (CS) and Management with a focus on Management Information Systems (MIS). He is a seasoned professional with over 25+ years of extensive experience in Teaching, Education Management, Research, and Software Development. He is currently working as Executive Director & Professor (CSE) at IILM University, Greater Noida, India. His present research interests include Data Science, AI & ML, Biometrics, and E-Banking,

Prof. (Dr.) Narendra Kumar is doctors of Computer Science & Engineering and Mathematics. He is working in NIMS Institute of Engineering and Technology, NIMS University Rajasthan, Jaipur. He has completed his M.Phil.(1994) with gold medal and Ph.D. (2003, 2012) from Dr. Bhimrao Ambedkar University, Agra. He has an academic experience for more than 28 years. He worked as Dean, Joint Director and Director in various universities. He has published more than two dozen books in the domain of mathematics, statistics and computer science and engineering, more than 70 research papers in national/ international journals. He has guided more than a dozen students for research degree. He has many patents in his credit and member of Board of studies in many universities. His key areas of research work are Mathematical modeling, Theory of relativity, Data science, Big data and brand management.

Prof. (Dr.) Karlibaeva Raya is a Full Professor of Economics at the Faculty of Finance of Tashkent State University of Economics. Her research interests are focused on solving problems related to the management of financial resources in enterprises, improving the management of cash flows and investment processes of enterprises. She is the author of 5 books and the editor of 9 monographs. In addition, she is the author or co-author of 137 articles in Uzbek and foreign journals and book chapters. She is the head of the Finance Department. Since 2019, Karlibaeva Raya has been a member of the Scientific Council for the defense of her PhD thesis and a member of many societies, including the Association of Corporate Governance under the Government of Uzbekistan

1. The Application of the Internet of Things (IoT) in Creating Enhanced Industry 4.0 for Sustainable Growth and Development

Pradeep Kumar Bharadwaj[1] and G. Vinodini Devi[2]

[1]Research Scholar, Department of Law, Koneru Lakshmaiah Education Foundation, Green Fields, Vaddeswaram, Guntur, Andhra Pradesh, 522302

[2]Assistant Professor, Department of Law, Koneru Lakshmaiah Education Foundation, Green Fields, Vaddeswaram, Guntur, Andhra Pradesh

ABSTRACT: The aim of this chapter is to demonstrate how IoT has been revolutionising the manufacturing sector. In this context, a thorough explanation of the IoT's historical development and contemporary dynamics has been provided. This chapter also highlighted the importance of IoT in various industrial processes. To understand how people view IoT, a study of 150 individuals working in the industrial sector was undertaken. For this, three closed-ended questions are created, and 150 respondents' binary responses are gathered to conduct the questionnaire with the help of an online survey. This research chapter also provides examples of contemporary IoT system problems and the role they are performing in the current predicament. This research's final section includes a concluding summary that may be used to acquire a general understanding of it. This chapter also examined critically the role of technology in the industrial sector.

KEYWORDS: Internet of Things (IoT), sustainability growth, industry 4.0, industry revolution

1. INTRODUCTION

In the present circumstances, the world is switching from conventional mode to digital mode in the context of executing most of its work processes, which is popularly known as digitalisation. As a consequence, technological advancement is evolving at an enormous pace and several digital devices are emerging to make this switching process more convenient. Besides the invention of the internet has remained one of the most pioneering steps that have helped immensely in making digitalisation more organised. Moreover, these devices are going through an evolutionary process which reduces their size and upgrades their storage and processing power continuously (Rahman *et al.*, 2020).

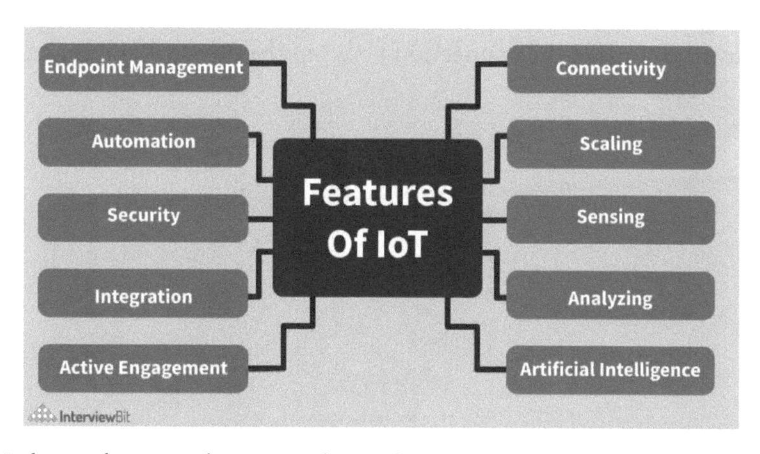

Figure 1. Relevant features of IoT regarding industry 4.0. Source: Shahbazi and Byun (2021).

DOI: 10.1201/9781003532026-1

Internet of Things (IoT) is a relatively new concept in this field which is proving to be immensely beneficial in various fields of daily needs. The main aim of introducing IoT is to create a bridge between the conjectural world and the virtual world with the help of the internet (Paweloszek, 2015). It is expected that IoT in the industrial sector can bring a revolution in the field of industry with the help of artificial intelligence and autonomous systems. The people are pretty much excited about this aspect of IIoT and coined this upcoming industrial revolution as the ***fourth industrial revolution*** (Industry 4.0). It is also expected that the features of IIoT like inter-machinery communication on a large scale; systems regarding cyber security, cyber-physical systems (CPs) can become extremely fruitful in this context.

1.1. Objectives and Hypothesis

The key objectives of this study are:

- To evaluate the current state of adoption of the latest technological advancements in the context of Industry 4.0
- To explore the significant role of IoT technologies in promoting sustainable growth and development to enhance Industry 4.0
- To identify some potential barriers or challenges of IoT implementation in Industry 4.0
- To recommend some suitable solutions and strategies to implement IoT effectively for sustainable development and growth

Hypothesis 1:*With the help of IoT industrial processes can be enhanced and sustainable growth can be achieved.*
Hypothesis 2: *There is a significant relationship between IoT adoption and improved industrial processes*
Hypothesis 3: *There are certain challenges associated with IoT implementation for sustainable growth and development*

2. LITERATURE REVIEW

The IoT has brought several promising aspects to change the dynamics of the industrial sector in the last few years. It is pretty much evident that Industry 4.0 will become integrated, automated and digitised. These traits of the upcoming industrial revolution can only be possible by the proper implementation of IoT. There are four parameters that will become extremely important in the future for driving the industry and its related work process. They are:

- Connectivity and data computation ability
- Intelligence and analysis ability
- Interaction between human and nonhuman beings in the industrial work process
- Conversion between the conjectural and virtual world (Mercan *et al.*, 2020).

The simultaneous helping hand of cutting-edge machinery advancement and the conventional manufacturing power of existing industry can make the connection between the physical and digital world more efficient (Lampropoulos *et al.*, 2018).

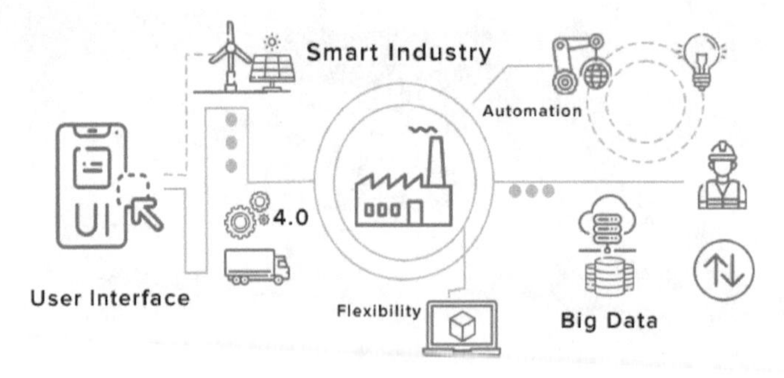

Figure 2. Internet of Things in manufacturing. Source: Ghosh et al. (2020).

These three things are extremely vital in the context of improving the digital business process and its reach among the customers. They are:

- Interconnecting the horizontal value chain system with its vertical component.
- Making products and services more digitalised.
- Creating an organised system to make the interaction process between human and non-human beings easier.

Modern industries must also focus on switching their working process from a centralised to a decentralised model. Resultantly, the IIoT will become a major contributor in fulfilling this demand.

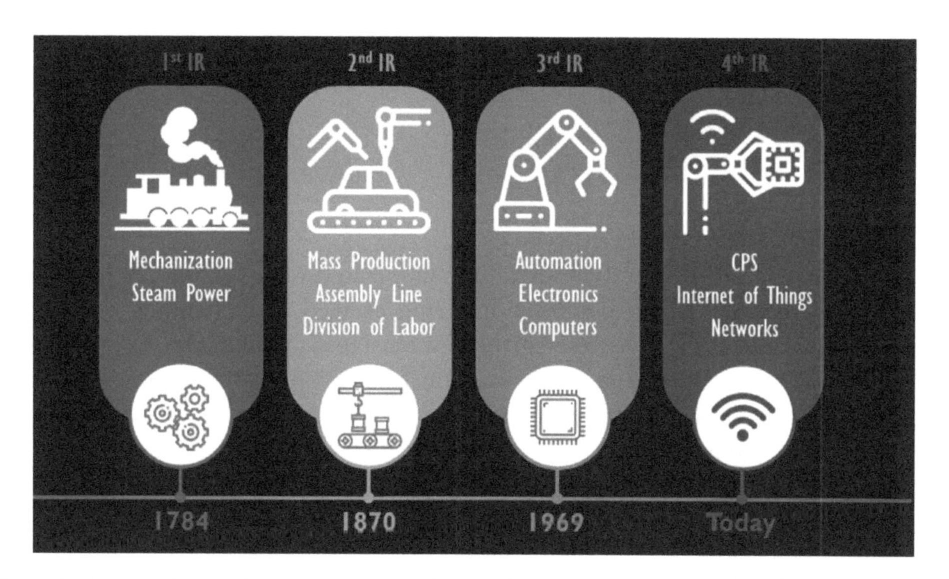

Figure 3. *Four industrial revolutions and their main contributors. Source: Ghosh et al. (2020).*

It is worth mentioning here that the IIoT and its role are somewhat different from the traditional IoT as IIoTs are specially designed to improve the function of industry-related works. There are several key components that work in parallel to organise the IIoT system and help it to work to its fullest potential. IIoT enhances the efficiency of software, storage systems, sensors and different applications to provide an overall authority on the process and assets of industries. As a consequence, more organised planning regarding time scheduling and manufacturing priorities can be seen nowadays in the industries (Nimbalkar *et al.*, 2020). This aspect has largely upgraded the growth of the industrial sector and also brought sustainability to its work procedure.

The emergence of ***cloud computing technology***, ***big data analysis*** and ***cyber-physical systems (CPS)*** due to the advancement of IoT-based technology are also becoming immensely beneficial for industries in recent times. The feature of CPS to analyse and interpret large data sets and its transdisciplinary methods are helpful for interconnecting physical assets and virtual network systems (Caro and Sadr, 2019). The different segments of cloud computing like SaaS (***software as a service***), PaaS (***platform as a service***) and IaaS (***infrastructure as a service***) are important for assisting in various segments of virtualisation and managing solution stacks in an efficient manner. Thus modern industries are taking advantage of structured monetisation and scaling features of cloud computing nowadays to improve and sustain the future endeavour (Cicconi and Raffaeli, 2020).

IoT has empowered the concept of smart manufacturing as it uses ***service-oriented architecture*** (SOA) to provide holistic solutions in modern manufacturing mechanisms (Li and Liu, 2020). It is pretty much evident that the IoT has used the economic information of demand dynamics which are largely dependent on customers, the public and partners to provide an overall revolutionary advantage in the modern industrial sector (Bersani *et al.*, 2022).

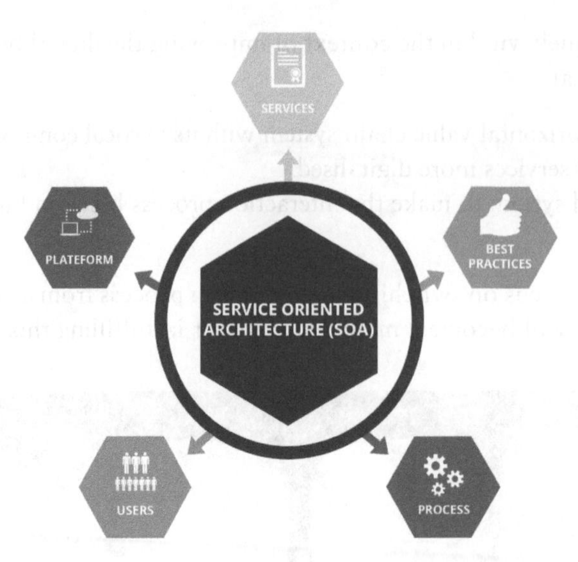

Figure 4. *Service-oriented architecture. Source: Li and Liu (2020).*

However, there are certain challenges and opportunities associated with the application of IoT to develop industrial processes. It also helps in real-time monitoring as well as controlling industrial assets. IoT also encourages innovation and the development of new business models. On the other hand, there are some major challenges (Javaid *et al.*, 2022). Apart from that, scalability is another crucial concern for IoT systems which need flexible architecture to grow effectively. Thus, it can be summarised that IoT has the capabilities to improve industrial processes and ensure sustainable growth. However, it is essential to resolve certain challenges such as data privacy and security, interoperability, and scalability and utilize opportunities such as promoting sustainable practices, fostering innovation and optimising resource allocations to ensure an inclusive and sustainable future.

3. RESEARCH METHODOLOGY

A systematic way and implementation of a proper method help immensely in conducting a research project in an organised manner. Therefore, this research project has been conducted by incorporating such relevant research methods to gain an authentic and precise idea about the respective topic. Primary as well as secondary research have been considered to gain an overall idea about the relevance of IoT in bringing the fourth industrial revolution. The secondary research method has remained immensely helpful regarding quantitative analysis as the data collected in the primary method remained much lesser in amount for the purpose of conducting quantitative analysis. In terms of primary data collection, an online survey will be conducted with 150 individuals working in the industrial sector. The sample size of the population is 150 and some open-ended questionaries are asked to gather relevant information from the research study.

4. ANALYSIS AND INTERPRETATION

Based on an analysis of the data, it can be said that IoT is one of the most innovative technologies that has been snowballing. It has various applications and services in different markets and industries. Although it can be identified that the recent developments of IoT have made many applications feasible, there are few that are currently available in the market. The application of IoT plays a very crucial role in driving sustainable growth and development across different sectors. Through the definition of IoT, it can be interpreted that it is the process of connecting various physical devices and systems through the Internet by enabling data collection, analysis, resource management, and productivity.

It has been estimated that the global market size of IoT will increase exponentially and reach a valuation of $994.01 billion by the end of 2025 with a CAGR of 26.9% . These can effectively reduce energy consumption by more than 15–20% and lower greenhouse gas emissions. In agriculture and farming, IoT can be used for smart farming, resource management and sustainable farming practices. Within the next few years, the smart agricultural market will reach $22.1 billion with a CAGR of 9.8% from 2021 to 2025. Within 2026, the number of IoT-connected devices in smart cities will be around 1.3 billion. On the other hand, IoT can enhance supply chain management.

5. DISCUSSION AND FINDINGS

By going through several relevant journals and articles it is evident that the usage of IoT in several aspects of industrial work has increased rapidly in the recent past and there is no sign of backward integration of this process (Li *et al.*, 2020). The needs and requirements of the consumers are changing day to day. Besides, competency has increased in a drastic manner in the global business market. In parallel , IoT is providing promising features that can help in a significant manner in the sustainability aspects of an organisation. Industry 4.0 will bring different levels of challenges in the upcoming days. IoT may alter the process of value chain management through its autonomous feature to bring self-reliability in this sector (itpro.co, 2022). It is also pretty much capable of handling smart manufacturing processes and data optimisation (Liu *et al.*, 2018).

According to a recently conducted survey, the IoT has influenced the operational efficiency of the industrial sector by approximately 47% (Elnagar *et al.*, 2020). The study also suggests that its influence has impacted several other segments of the industry. The productivity has been raised by approximately 31%, the asset utilisation has increased by 22% (Ramanathan, 2018). Moreover, the new business opportunities, ease in selling services and products, the safety of employees and other similar factors have also improved by a noticeable amount in the last couple of years (assets.publishing, 2022).

The IIoT is also going through a similar situation as there are several recent incidents where an IIoT-based gadget has created problems in the industrial work process (Malina *et al.*, 2019). The capital expenditure in incorporating IIoT in the industrial sector has also a major hindrance in recent days. The encryption of data is not that rigid, and piracy of data is a regular incident in this regard (raeng.org, 2021). Therefore, IIoT has not managed to address issues regarding the privacy and confidentiality of data and information. Resultantly, people in the industrial arena are showing unwillingness to use this technology in the present scenario (Bal and Badurdeen, 2019). Although the scalability factor of IIoT is good, nonetheless it has still not achieved a standard level in this context (ofcom.org, 2022).

6. CONCLUSION

It is discernible that IoT will be immensely helpful in the context of upgrading the industrial sector in the upcoming days. This technological aspect provides holistic solutions and connects the conjectural world with the virtual world. IoT is a growing technological aspect that is keeping its mark in most of the segments of modern-day lifestyle. It is evident that the proper implementation of IoT can reduce the complexities of modern life in an organised manner.

REFERENCES

Bal, A., and F. Badurdeen (2019). "A business model to implement closed-loop material flow in IoT-enabled environments," *Procedia Manufacturing*, 38, 1284–1291.

Bersani, C., C. Ruggiero, R. Sacile, A. Soussi, and E. Zero (2022). "Internet of things approaches for monitoring and control of smart greenhouses in industry 4.0," *Energies*, 15(10), 3834.

Caro, F., and R. Sadr (2019). "The Internet of Things (IoT) in retail: bridging supply and demand," *Business Horizons*, 62(1), 47–54.

Cicconi, P., and R. Raffaeli (2020). "An industry 4.0 framework for the quality inspection in gearboxes production."

Elnagar, S., and M. A. Thomas (2020). "Federated deep learning: a conceptual model and applied framework for industry 4.0."

Ghosh, A., D. J. Edwards, and M. R. Hosseini (2020). "Patterns and trends in Internet of Things (IoT) research: future applications in the construction industry," *Engineering, Construction and Architectural Management*, https://doi.org/10.1016/j.matpr.2021.07.288.

itpro.co.uk (2022). Retrieved on 7 February from: https://www.itpro.co.uk/network-internet/internet-of-things-iot/357637/smart-factories-key-to-growth-in-industrial-iot.

Javaid, M., A. Haleem, R. P. Singh, R. Suman, and E. S. Gonzalez (2022). "Understanding the adoption of Industry 4.0 technologies in improving environmental sustainability," *Sustainable Operations and Computers*, 3, 203–217.

Lampropoulos, G., K. Siakas, and T. Anastasiadis (2018). "Internet of Things (IoT) in Industry: contemporary application domains, innovative technologies and intelligent manufacturing," *People*, 6(7).

Li, W., and L. Liu (2020). "Using IoT to enhance knowledge management: acase study from the insurance industry."

Li, Y., X. Su, A. Y. Ding, A. Lindgren, X. Liu, C. Prehofer, J. Riekki, R. Rahmani, S. Tarkoma, and P. Hui (2020). "Enhancing the internet of things with knowledge-driven software-defined networking technology: future perspectives," *Sensors*, 20(12), 3459.

Liu, X., S. Tamminen, X. Su, P. Siirtola, J. Röning, J. Riekki, J. Kiljander, and J. P. Soininen (2018). "Enhancing veracity of IoT generated big data in decision making," *2018 IEEE International Conference on Pervasive Computing and Communications Workshops (PerCom Workshops)* (pp. 149–154). IEEE.

Malina, L., G. Srivastava, P. Dzurenda, J. Hajny, and S. Ricci (2019). "A privacy-enhancing framework for internet of things services," *International Conference on Network and System Security* (pp. 77–97). Springer, Cham.

Mercan, S., L. Cain, K. Akkaya, M. Cebe, S. Uluagac, M. Alonso, and C. Cobanoglu (2020). "Improving the service industry with hyper-connectivity: IoT in hospitality," *International Journal of Contemporary Hospitality Management*.

Nimbalkar, S., S. D. Supekar, W. Meadows, T. Wenning, W. Guo, and J. Cresko (2020). "Enhancing operational performance and productivity benefits in breweries through smart manufacturing technologies," *Journal of Advanced Manufacturing and Processing*, 2(4), e10064.

ofcom.org.uk (2022). Retrieved on 7 February from: https://www.ofcom.org.uk/__data/assets/pdf_file/0025/38275/iotstatement.pdf.

Paweloszek, I. (2015). "Approach to analysis and assessment of ERP system. A software vendor\'s perspective," *Proceedings of the 2015 Federated Conference on Computer Science and Information Systems*, M. Ganzha, L. Maciaszek, M. Paprzycki, Annals of Computer Science and Information Systems, pp. 1415–1426, IEEE, doi: 10.15439/2015F251.

Rahman, A., M. J. Islam, M. S. I. Khan, S. Kabir, A. I. Pritom, and M. R. Karim (2020). "Block-SDoTCloud: enhancing security of cloud storage through blockchain-based SDN in IoT network," *2020 2nd International Conference on Sustainable Technologies for Industry 4.0 (STI)* (pp. 1–6). IEEE.

Ramanathan, K. (2018). "Enhancing regional architecture for innovation to promote the transformation to industry 4.0," *Industry*, 4, 361–402.

Shahbazi, Z., and Y. C. Byun (2021). "Integration of blockchain, IoT and machine learning for multistage quality control and enhancing security in smart manufacturing," *Sensors*, 21(4), 1467.

2. An Investigation of Machine Learning Approaches and their Impact on Resource Management & Sustainable Development

Pradeep Kumar Bharadwaj[1] and G. Vinodini Devi[2]

[1]Research Scholar, Department of Law, Koneru Lakshmaiah Education Foundation, Green Fields, Vaddeswaram, Guntur, Andhra Pradesh, 522302

[2]Assistant Professor, Department of Law, Koneru Lakshmaiah Education Foundation, Green Fields, Vaddeswaram, Guntur, Andhra Pradesh

ABSTRACT: Earth science and statistical fields are becoming more intertwined as a result of the wide usage of descriptive statistical processes to give evaluations of ecology, farming, and ecological sustainability. In light of this, they researched the empirical methods for machine learning commonly employed to analyse remote sensing information in the literature. Opportunities for remote sensing data are being researched as 'machine learning' methods have advanced and more high-quality sensor data are becoming available to the public. Now, information on land use, land cover change, crop segmentation, degrading and flood control are derived from machine learning data processing. Survey data collecting is taken into consideration to gather statistical information concerning the usage of ML methods and their influence on environmental sustainability.

KEYWORDS: Machine learning (MI), artificial intelligence (AI), development, sustainability, survey, as well as resource management

1. INTRODUCTION

'Machine Learning' (ML) is a part of science that involves 'data analysis' for creating analytical models. It is also a part of artificial intelligence to showcase that machines are capable of understanding, identifying patterns and trends, and making judgments without the need for human intervention. This machine-learning strategy is computer software that performs a job using raw data rather than being deliberately trained to achieve a certain outcome. Such algorithms are designed in such a way that they may readily alter their design as a consequence of practice such as experience to improve their performance (El Naqa and Murphy, 2015).

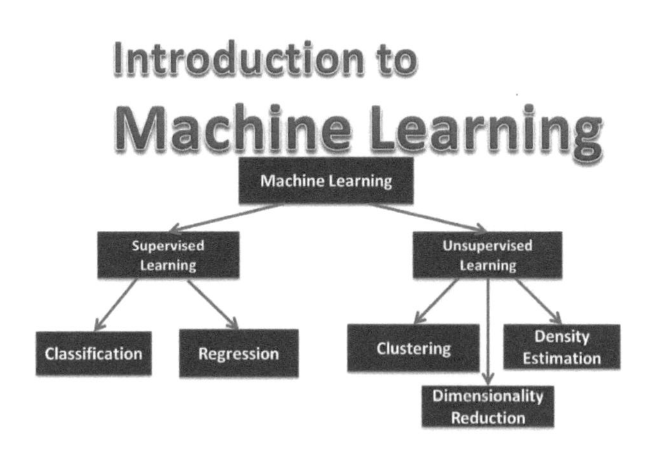

Figure 1. Machine learning or ML introduction. Source: El Naqa and Murphy, 2015.

DOI: 10.1201/9781003532026-2

ML has made great strides in the previous two decades, from a research project to a game-changing technology with widespread commercial application. As stated by Jordan and Mitchell (2015), ML has grown in importance as a method for developing application protocols in fields such as computer imaging, voice recognition, computational linguistics, adaptive control, resource management, environmental sustainability and other fields. ML is reconfiguring the financial system by modernising a diverse number of sectors, including healthcare, education, aviation, agriculture, hospitality and a variety of industrial lines, to mention a few. It will have an impact on nearly every aspect of people's lives, which include their homes, automotive, shopping and even food placing an order (Singh, 2020). Many AI system experts now agree that, for a variety of reasons. Workflow estimate, task scheduling, resource optimisation, VM consolidation and energy optimisation are just a few of the resource management activities that ML is mostly used for. Human resources may use ML to help them manage the recruitment process from beginning to end.

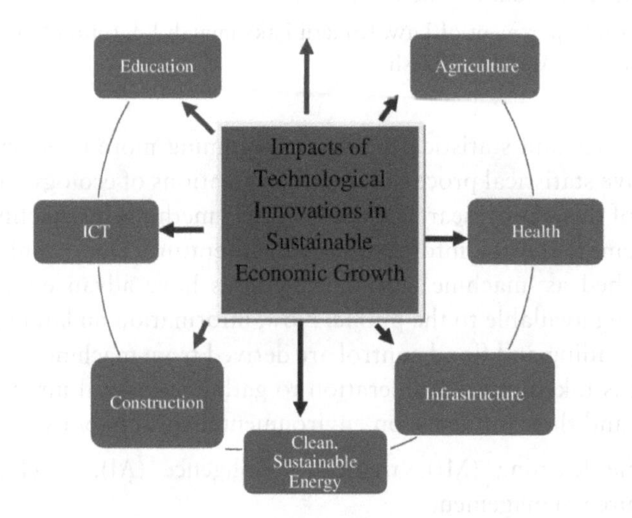

Figure 2. Growth of technologies in sustainable development. Source: Pathak, 2021.

As per the above figure, it has been observed that technological upgradation such as AI development, MI implementation and Deep Learning adaptation can help to build sustainable development in this era. Approaches to ML and its impact on resource management and sustainable development will be examined in this research study.

1.1. Objectives and Hypotheses

- To identify various ML processes and their importance regarding sustainability and resource management.
- To evaluate the effectiveness of current ML algorithms regarding efficient allocation, conversation and utilisation of resources
- To identify the challenges and opportunities associated with the implementation of machine learning for sustainable development and resource management
- To propose some recommendations to support sustainable development goals and effective resource management using machine learning

Hypothesis 1: *ML approaches can significantly improve resource management and ensure sustainable development*

Hypothesis 2: *There is a significant relationship between machine learning approaches and improved resource allocation and utilization*

Hypothesis 3: *The implementation of machine learning approaches can significantly contribute to sustainable development*

2. LITERATURE REVIEW

Machine Learning (ML) seems to be the study of enabling machines to think and act like humans, and to improve their understanding over the period in a self-contained approach, integrating knowledge/analysis from observations and real-world experiences. This kind of learning feedback process is used to classify the best ML techniques. The bulk of ML problems can be solved using one of these techniques. The four types of techniques include 'supervised learning', 'semi-supervised learning', 'unsupervised learning' and 'reinforcement learning' (Sharma *et al.*, 2020).

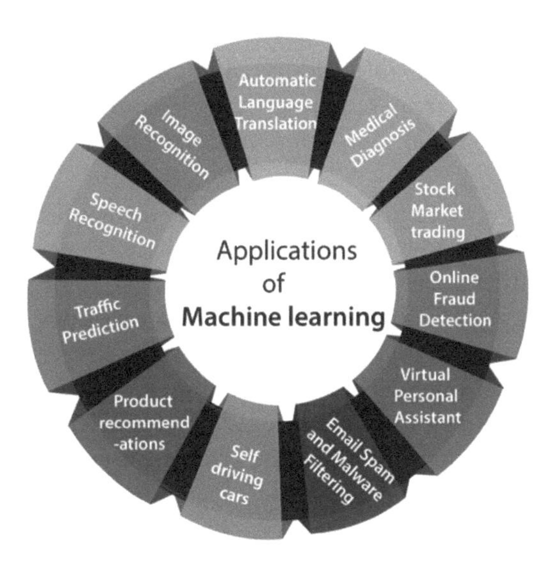

Figure 3. *Different applications of Machine Learning. Source: Sharma et al. (2020).*

In this present era, Machine Learning or ML is one of the latest and advanced technologies that can help to create sustainable development in various sectors such as house pricing prediction, healthcare, education and so on. Random forest algorithms, Bayesian networks, decision trees and regression analysis are examples of supervised learning techniques. According to Sharma *et al.* (2020), unsupervised learning approaches operate with information that is unlabelled and has uncertain input and output variables. Unsupervised learning, which is commonly used for data pre-processing as well as information extraction, uses an unlabelled dataset to identify hidden patterns.

Figure 4. *Approaches of ML. Source: Stanford.edu (2021).*

In contrast to using previously obtained data, first, it is essential to investigate new as well as unknown activities to obtain further data. Whereas Reinforcement learning methods are generally utilised in 'autonomous navigation', 'machine instructional strategies', and 'real-time decision-making'. 'Reinforcement

ML' methods include 'Q-learning (SARSA max)', 'Deep Q-learning (DQL)' and 'Dataset Aggregation (dAgger)'. Traditionally, resource allocation problems have been tackled using optimisation algorithms that take into account the users' immediate CSI and QoS needs. Because resource allocation issues are seldom convex, the solutions achieved using standard methods are rarely globally optimum. Furthermore, the answers might not be available in real time. More effective (both in terms of computation and performance) systems are necessary for widespread implementation. Researchers may not have been able to formulate the optimal control issue in some circumstances where the resource provisioning issue is not well-defined analytically (e.g., considering the nature of the modulation scheme, and users' mobility characteristics). It encourages researchers to look at novel resource allocation systems (Martinez and Ipek, 2009).

To maximise the performance achievable with fixed amounts of total resources, supervised learning systems that execute in huge search areas require effective resource allocation algorithms. Performance measurements, which provide feedback information during the search process, influence strategy (Tamburino *et al.*, 1989).

HR, on the other side, has a variety of challenges that distinguish it from many other industries where AI has been implemented. The first is that human resource results are complicated. Consider what being a 'good employee' entails. Labour accounts for around 60% of total spending in the economy. These database management duties, according to practitioners, constitute a major barrier to studying HR practises and outcomes. ML refers to a collection of methodologies that use data to create algorithms that are frequently used to predict outcomes. The most common use of machine learning algorithms in organisations has been 'supervised application,' in which a data scientist 'trains' an ML algorithm on a subset of relevant details and determines the best metric to measure the algorithm's effectiveness (Sharma *et al.*, 2020).

In the domain of natural resource management, the approach offers a significant paradigm change in statistics, wildlife management and conservation. ML can be employed in a wide range of scenarios. Humphries *et al.* (2018) have mentioned that when data is complicated, not regularly distributed (statistically), and messy' (as is usual in ecological and environmental data).

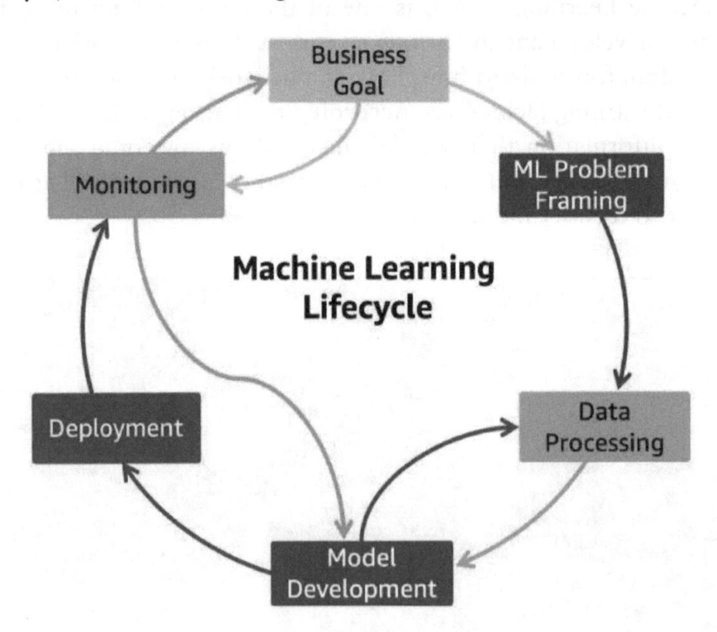

Figure 5. *Business sustainability through Machine Learning. Source: Sharma et al. (2020).*

In this scenario, data-driven machine learning (ML)-based allocation of resources algorithms would be feasible options in such circumstances, and they should be flexible to the rapid evolution of IoT networks. The purpose of applying machine learning algorithms to agricultural data is to create effective ASCs. ML algorithms' ability to estimate timely rainfall during the phase of pre-production and assist in water planning and effective resource management (Sharma *et al.*, 2020). Various models of ML help in weather predictions, assisting decision-making in various areas. Irrigation management that is efficient improves environmental performance while also increasing output and production.

3. RESEARCH METHODOLOGY

To produce sustainable solutions in integrated 'human–environment ecosystems', advanced technologies such as Machine Learning or ML approaches are required. This research chapter has considered a secondary method of data collection to gather essential information and relevant outcomes regarding the topic. A secondary qualitative data is collected from different sources such as journals, articles, newspapers and company websites regarding the progression of ML approaches in enhancing sustainable development in this era.

4. ANALYSIS AND INTERPRETATION

Through the data analysis and interpretation, it can be said that machine learning approaches have several contributions to sustainable development and resource management. It helps in data-driven decision-making, predictive analytics, and optimisation. Concerning the working of machine learning includes some algorithms that analyze large data sets and identify patterns to make accurate decisions that lead to sustainable development, conservation, and resource utilization. Machine learning approaches have a wide range of applications in waste management, renewable energy forecasting, energy management and sustainable transportation.

Regarding machine learning approaches and energy management, the algorithms of machine learning usually optimise energy consumption by identifying patterns to enhance energy efficiency. Furthermore, it can be interpreted that the use of machine learning approaches to improve grid stability, real-time monitoring, and reduce peak load can efficiently reduce energy consumption by more than 20%. In terms of renewable energy forecasting, machine learning approaches can also be used. In comparison with traditional methods, machine learning approaches are able to improve forecasting accuracy by 40%. Moreover, machine learning algorithms are often used in recycling, landfill management, and waste collection. Hence, it can achieve a recycling rate of more than 90% compared with 40% of traditional methods. Regarding sustainable transportation, machine learning algorithms are also used to optimise transportation, reduce fuel consumption and improve traffic management. Along with these, it can also be useful in congestion prediction, routing optimisation and greenhouse gas reduction. By implementing machine learning approaches in traffic management, harmful gas emissions and fuel consumption can be reduced by 15%.

Thus, based on the analysis and interpretation, it has been found that machine learning has immense potentials such as reinforcement learning, supervised learning and machine learning algorithms to improve decision-making, optimise resource utilisation and promote sustainable practices. This research study has also analyzed the challenges associated with machine learning which include data integration and transformation, adoption of evolving IoT networks, and complexities in database management. On the other hand, machine learning focuses on various opportunities to achieve social, economic and environmental goals.

5. DISCUSSION AND FINDINGS

Growth begins by streamlining the procedure, improving efficiency, and streamlining the process. ML has a broad range of applications in the natural resource management sector, financial resource management sectors and so on. Current issues include achieving sustainable development in industrial systems and preserving essential resources for future generations can be mitigated by implementing ML approaches. Sustainable development has been increasingly crucial for organisations as well as the industry in recent history (Pathak, 2021). ML assists everyone in achieving this sustainability. Moreover, academics and professionals have successfully used machine learning methods and approaches to critical real-world difficulties such as allocation of resources, rescheduling, resource provisioning, and design space exploration. While allocating resources and rescheduling decisions in soft real-time systems instantaneously, a reinforcement learning technique for adaptive work scheduling is used. Conventional resource distribution strategies in wireless IoT networks mostly depend on optimisation algorithms. Whenever the percentage of subscribers grows high and/or a variety of cellular situations are addressed, these solutions confront difficulties.

6. CONCLUSION

Based on the detailed discussion, it can be concluded that to create long-term sustainable development, three machine learning techniques are used such as 'supervised, unsupervised and reinforcement learning'. The study's important contribution is the ML function application platform, which will help academicians and practitioners better grasp the current condition of the field's literature. The framework proposed in this chapter is supported by findings from a review article that has not been examined. As a result, additional research may be done to empirically validate this approach. Future research can also look into the scope of machine learning's deployment to sustainable development in different parts of the world and compare the results. Thus, this research significantly highlighted the immense capabilities of machine learning for sustainable development as well as resource management. Organisations can effectively harness the power of machine learning algorithms to reduce waste, contribute to economic growth and optimise resources. However, further research and collaborations are needed to exploit the full capabilities of machine learning to drive sustainable growth and build a better future.

7. FUTURE SCOPE

In the previous decade, global accomplishments in productivity expansion, reduced poverty and greater welfare have been offset by a rising strain on the biosphere. 'Big data' would become the 'dominant scientific perspective' as technology improves, improving prediction strength and transparency. Machine learning, for instance, may be used to forecast environmental dangers and dangers as an extra set of modelling techniques. Thus, it can be stated that ML approaches can help to develop sustainable development in this era by eliminating business issues.

REFERENCES

El Naqa I., and M. J. Murphy (2015). "What is machine learning?" *Machine Learning in Radiation Oncology*. I. El Naqa, R. Li, and M. Murphy (eds.). Springer, Cham. https://doi.org/10.1007/978-3-319-18305-3_1.

Humphries, G. R., D. R. Magness, and F. Huettmann (Eds.). (2018). *Machine Learning for Ecology and Sustainable Natural Resource Management* (p. 3). Switzerland: Springer.

Jordan, M. I., and T. M. Mitchell (2015). *Machine Learning: Trends, Perspectives, and Prospects*, Retrieved from: https://www.science.org/doi/abs/10.1126/science.aaa841, Retrieved on: 25/02/2022.

Martinez, J. F., and E. Ipek (2009). "Dynamic multicore resource management: a machine learning approach," *IEEE Micro*, 29(5), 8–17.

Pathak, R. (2021). *Types of Machine Learning*, Retrieved from: https://www.analyticssteps.com/blogs/types-machine-learning, Retrieved on: 25/02/2022.

Sharma, R., S. S. Kamble, A. Gunasekaran, V. Kumar, and A. Kumar (2020). "A systematic literature review on machine learning applications for sustainable agriculture supply chain performance," *Computers & Operations Research*, 119, 104926.

Singh, V. (2020). *How Machine Learning is Changing the World*, Retrieved from: https://www.datasciencecentral.com/how-machine-learning-is-changing-the-world/, Retrieved on: 25/02/2022.

Stanford.edu (2021). *Machine Learning and Decision Making for Sustainability*, Retrieved from: https://forum.stanford.edu/events/2016/slides/plenary/Stefano.pdf, Retrieved on: 25/02/2022.

Tamburino, L. A., M. M. Rizki, and M. Zmuda (1989). "Computational resource management in supervised learning systems," *Proceedings of the IEEE National Aerospace and Electronics Conference* (pp. 1074–1079). IEEE.

3. Critical Determinants of the Internet of Things (IoT) in Supporting Business Performance among Small and Medium-Sized Enterprises

Rajesh Deb Barman[1], Pravin D Sawant[2], Vijayalakshmi P[3], Mohit Tiwari[4], Melanie Lourens[5], Ashim Bora[6], and Leszek Ziora[7]

[1]Assistant Professor, Department of Commerce, Bodoland University Kokrajhar BTR, Assam, India 783370

[2]Associate Professor, Department of Commerce, Narayan Zantye College of Commerce Bicholim Goa

[3]Associate Professor, Faculty of Management Studies, CMS Business School, Jain (Deemed to-be University), Karnataka, Bengaluru 560009

[4]Assistant Professor, Department of Computer Science and Engineering, Bharati Vidyapeeth's College of Engineering, Delhi A-4, Rohtak Road, Paschim Vihar, Delhi

[5]Deputy Dean Faculty of Management Sciences, Durban University of Technology South Africa

[6]Principal, Kampur College, Kampur, Assam, India

[7]Faculty of Management, Czestochowa University of Technology, Poland

ABSTRACT: In the modern, 'digitalised world', businesses are spending money to put cutting-edge approaches and technology into practice to create effective, connected business models. Additionally, SMEs use advanced business models and IoT applications to tackle their company issues because they lack significant resources and cutting-edge concepts. The study's objective is to analyse IoT services and assess how IoT enhances SME company's profitability. To acquire factual and private information, this research employs both primary and secondary research approaches. The responses of 50 employees from various SME businesses are taken into consideration in this connection. They are asked three closed-ended questions. The responses are examined using various methods such as charts and graphs. For doing secondary analysis, journals, publications and newspapers are also taken into consideration.

KEYWORDS: Internet of Things (IoT), Industry 4.0, SMEs, sustainability, technology

1. INTRODUCTION

In this current era, the application of the latest technological advancements such as '***Artificial Intelligence***' (AI) and '***Internet of Things***' (IoT), a form of Industry 4.0 revolution, is increasing in several industries. Industry 4.0 has been studied for years by both research as well as industry, with several corporations and academic institutes attempting to define the technologies and techniques of the 4.0 industry revolution. IoT is one of the most advanced technologies and can increase business performance, challenges and financial issues among small and medium enterprises.

- Improve business efficiency
- Develop smart workplaces among organisations
- Improve business security
- Connect billions of customers
- Share updated news among customers at a time

DOI: 10.1201/9781003532026-3

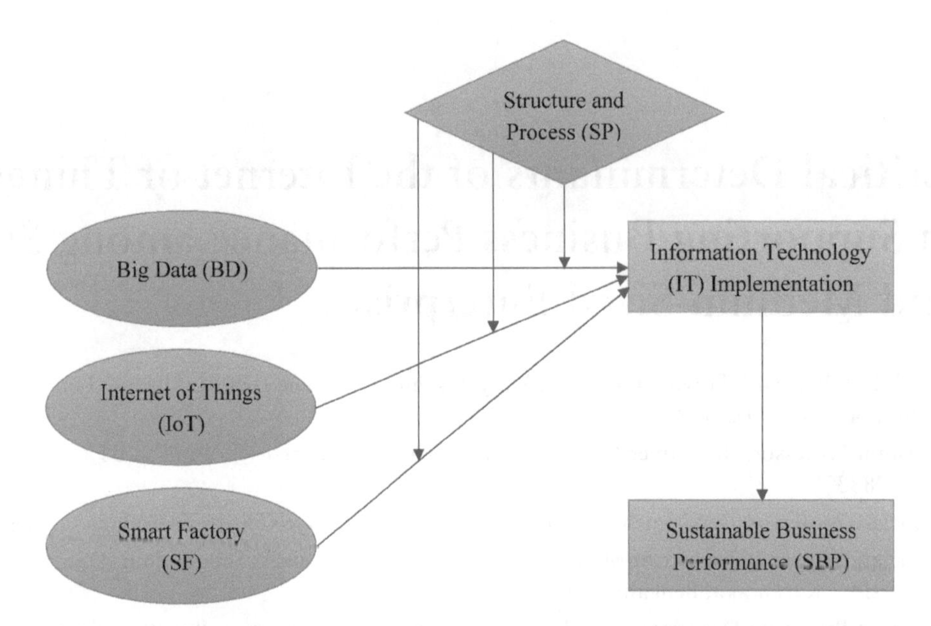

Figure 1. *Industry 4.0 effect on productivity and business performance among SMEs. Source: Haseeb et al. (2019).*

According to observations, there will be '4.9 billion gadgets' connected to the Internet by 2015, and that number will continue to grow every day (Hamidi and Jahanshahifard, 2018). Small- and medium-sized businesses are therefore concentrating on embracing IoT as a revenue stream. Additionally, it seems as though the IoT is about to permeate business: companies have started implementing IoT technology rapidly, and 'IoT devices connected' are anticipated to expand by '43 billion by the end of 2023' (Sestino *et al.*, 2020).

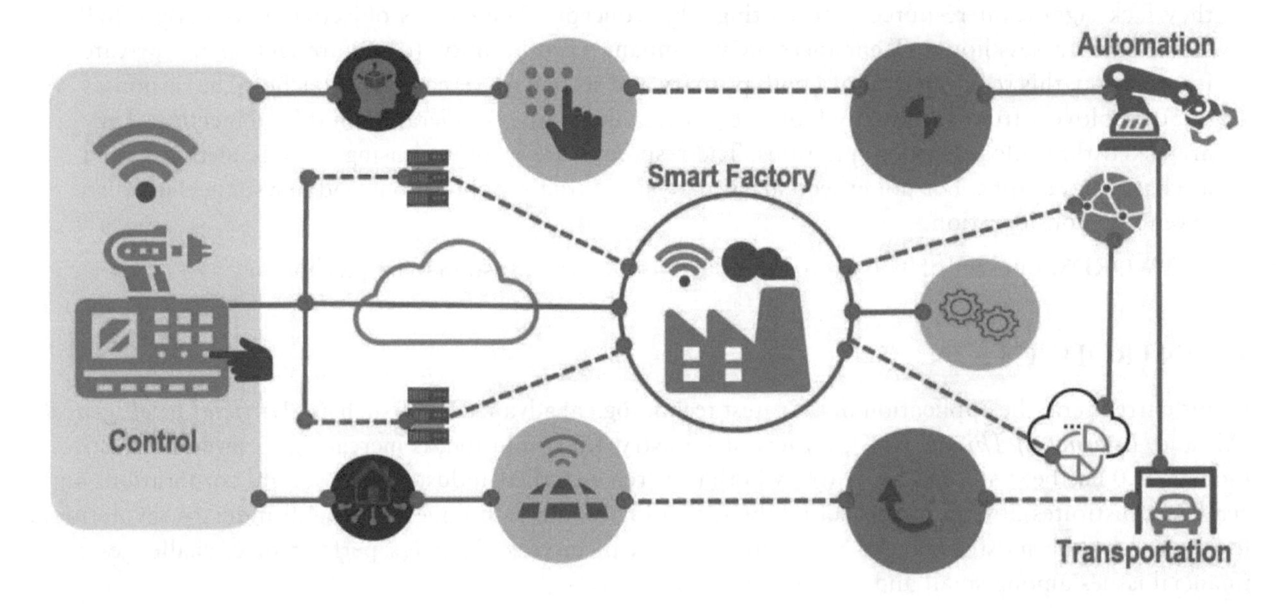

Figure 2. *Business growth and performance increases due to adaptation of IoT in small and medium enterprises. Source: Sestino et al. (2020).*

Based on the above figure, it has been observed that the business profit rate has increased by almost 5–15% in recent years which will expect to reach 53% in the next year due to the use of IoT technology in business. The 'Internet of Things (IoT)' thus illustrates the growing trend toward physical items with communication and computing capabilities that may gather data in real time.

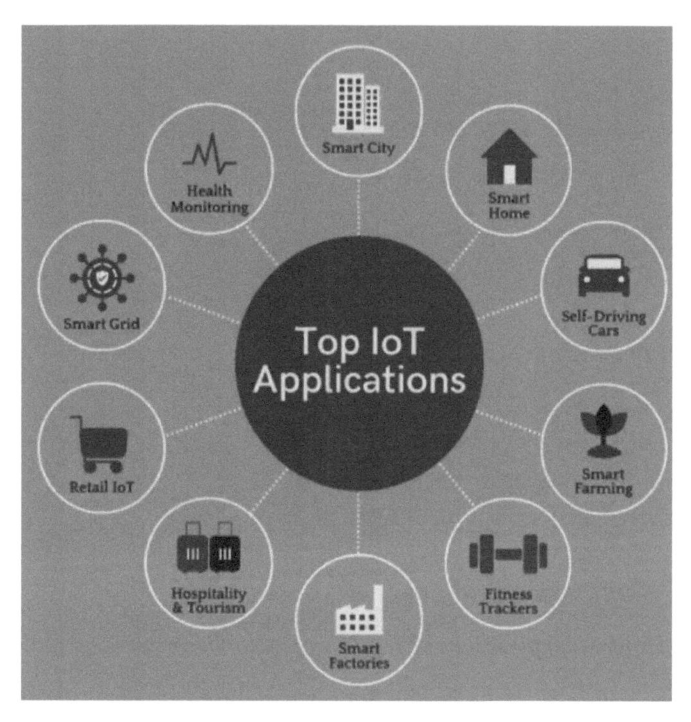

Figure 3. *Application of IoT in enhancing business performance. Source: Haseeb et al. (2019).*

IoT instruments can be used to study consumer behaviour, and also help to make customers' decisions during purchasing, which have implications for marketing research. Haseeb *et al.* (2019) have identified that IoT could be a '***crucial enabler of business digitalisation***', allowing maintenance strategies and daily routines to be improved which has a huge impact on small and medium enterprises. Therefore, financial efficiency might be regarded as a major aspect of fully utilising the IoT in the workplace (Chatterjee, 2021).

This research chapter sheds light on the applications of IoT in enhancing business performance among enterprises in different industries. Additionally, this chapter also highlights how IoT technology has helped to increase overall productivity among medium and small size enterprises in the recent era.

1.1. Objectives and Hypotheses

- To investigate the critical determinants associated with adopting IoT in SMEs
- To evaluate the impact of IoT on various business performances in SMEs
- To examine the challenges and opportunities associated with IoT adoption in SMEs
- To provide suitable recommendations and strategies for SMEs to integrate IoT within businesses

Some hypotheses have been developed at the start of the study,

Hypothesis 1: *IoT applications have the ability to improve decision-making and business performance*

Hypothesis 2: *IoT adoption positively influences customer satisfaction, revenue growth, and cost reduction, and improves business processes*

Hypothesis 3: *There are some barriers and challenges for SMEs to integrate IoT into organisations*

2. LITERATURE REVIEW

'Small and medium enterprises', or SMEs, are amongst the most significant foundations of the development of a country's economy, and they play a significant role in enhancing economies on their route to long-term growth. SMEs mainly represent 99% of all businesses, and offer more than 50 to 60% of job offers along with almost 65.8% of the total population being offered jobs by SMEs.

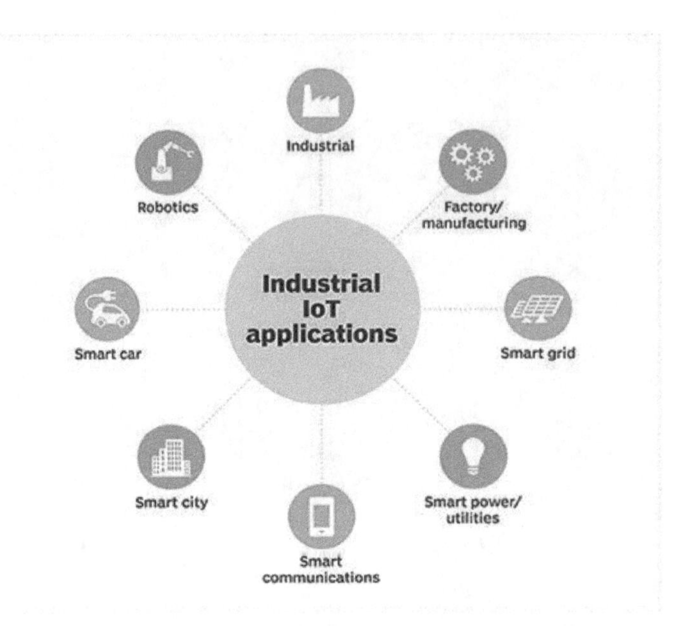

Figure 4. Different applications of IoT in enterprise business. Source: Lee (2019).

Generally, SMEs do not have more opportunities to achieve success, and sustainable growth in their business due to lack of 'technological advancements' and 'lower expectation profitability'. Moreover, from different studies, it has been observed that small or medium organisations do not use digitised technologies, IoT and cloud computing. As a result of this, SMEs cannot achieve success and sustainable business growth. The main obstacle is a lack of knowledge, poor-strategic planning related to business behind less development. SMEs with a stronger innovation focus may profit through 'external knowledge sourcing', which is nothing but an open innovation. For example, 'Romanian SMEs' are focusing on transitioning to digitalisation 4.0 with a specific focus on industries, and while they have the ambition and ability to execute change in their businesses, they identified that lack of resources may hamper their business success and future development. Thus, it can be stated that IoT technologies are significantly beneficial for business enterprises, mainly SMEs in Romania, to manage business issues, financial challenges in a successful manner.

World population and number of connected devices			
No.	Year	Number of Connected Devices	World population
1	2003	6.3 billion	500 million
2	2010	6.8 billion	12.5 billion
3	2015	7.2 billion	25 billion
4	2020	7.6 billion	50 billion

Figure 5. World population and connected devices.

The primary goal of implementing IoT is to create a smart system that can be helpful for connecting millions of devices around the world. From the above image, it has been seen that almost 25 billion devices were connected through IoT by 2015 which will be enhanced by 50 billion by 2020. In the context of other business aspects for example '***nature of entry threats***', '***power of buyer***', '***power of suppliers***', '***threats from alternative businesses***', and '***rivalry among existing enterprises***,' are linked to these technologies. Investing alone in IoT does not ensure a company's long-term performance; it must be combined with other physical, administrative and organisational resources. Because of the introduction and growth of new innovations, businesses must reconsider and adapt their marketing strategies, transforming them into sustainable alternatives through service-oriented digitised approaches.

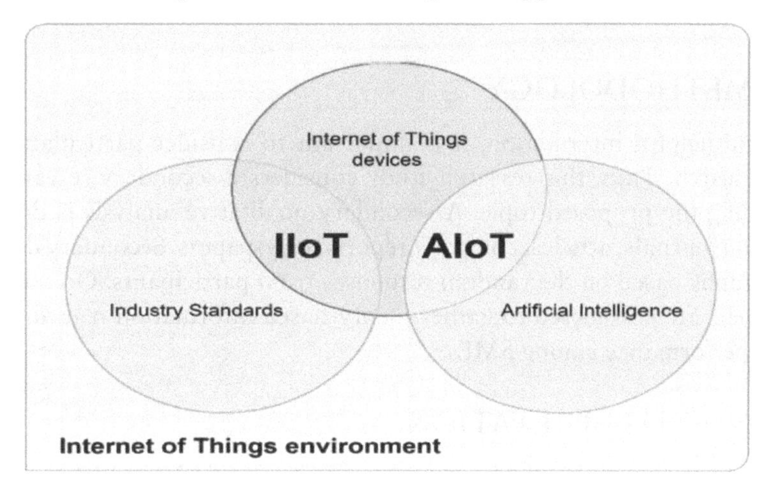

Figure 6. *Relations between AI, IoT with SMEs.*

As a consequence of digitalisation, also known as the 4.0 Industrial Revolution, which raises the bar for organisational efficiency, IoT and Big Data are changing marketing and operational strategies. According to Lee (2019), the IoT has undergone a change in thinking, enabling companies to create efficient and effective offerings via a ***'network of machines and devices,' enhance 'service business models' and boost 'enterprise stability'***.

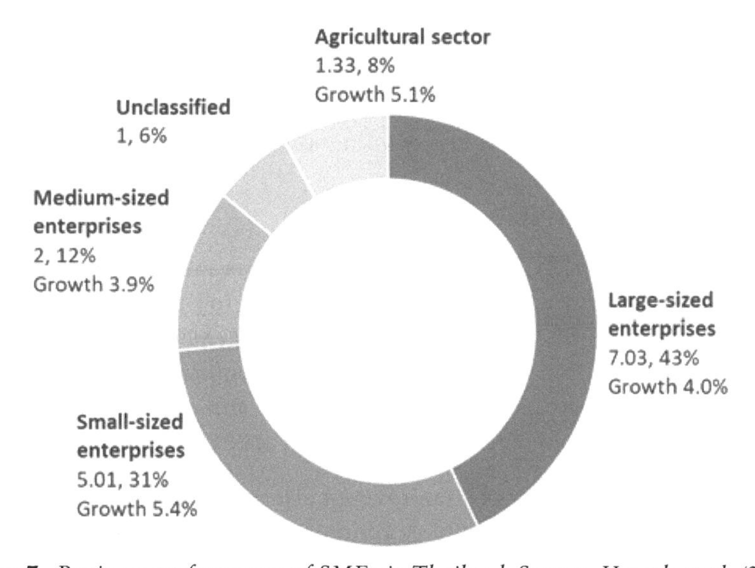

Figure 7. *Business performance of SMEs in Thailand. Source: Haseeb et al. (2019).*

Based on the above figure, it has been identified that Industry 4.0 evolution helps to prevent business challenges, financial issues among SMEs. This above figure has emphasised business performance and overall productivity growth among SMEs in Thailand. The annual revenue growth of SMEs in Thailand has increased by 44% by 2017 (Haseeb *et al.*, 2019). SME businesses contribute a higher economic contribution towards the country's economy. However, there are some limitations in SMEs business due to lack of technological adaptation. Thus, they are unable to achieve long-term success, constant growth along with sustainable development. In this context, implementing advanced technological development and adaptation of IoT can prevent business issues.

IoT software provides different exciting services such as '*Machine to Machine*' or M2M, '*data and device management*', '*protocol translation*', '*security and storage*', as well as '*programming frameworks*'. Many researchers have suggested that IoT simplifies corporate procedures, such as inventory tracking along with administration.

3. RESEARCH METHODOLOGY

To gather relevant and helpful information, it is important to consider particular research methods to manage systematic research. Thus, this research study considers a secondary research method to collect authentic data regarding the proposed topic. A secondary qualitative analysis is done to gather theory-based information from journals, articles, company reports, newspapers. Secondary data has been analysed through charts and graphs based on the random responses from participants. On the other hand, articles, journals and newspapers are considered to gather theory-based information regarding how IoT supports business growth and performance among SMEs.

4. ANALYSIS AND INTERPRETATION

In the current era, IoT is revolutionary and has many significant potentials which can subsequently lead to greater business performance for SMEs. There are many determinants of IoT that provide support to SMEs by enhancing their operational efficiency, improving decision-making ability and saving costs. However, several critical determinants have been identified that directly or indirectly affect the adoption of IoT in SMEs such as cost, security, data analytics, flexibility and scalability and skillset. Concerning the adoption and utilisation of IoT in SMEs, the cost is the most significant determinant. High capital investment is required for SMEs to implement IoT. It has been found that SMEs that have adopted IoT witnessed a significant increase in revenue by 15% and a reduction of operations costs by 20%. Apart from that, it is also essential for SMEs to interpret the scalability and adaptability of changing business needs in the context of IoT. For IoT deployment scalability is an important determinant. It can be seen that, by adopting IoT, more than 70% of organisations have increased scalability and flexibility. On the other hand, security and privacy are other crucial determinants of IoT adoption in SMEs. It can be interpreted that more than 90% of SMEs claimed that security vulnerabilities affect IoT deployment.

5. DISCUSSION AND FINDINGS

As per the views of Tang *et al.* (2018), for businesses, the IoT is more than just a term; it's a growing trend, a tried-and-true strategy, and a cutting-edge technology. IoT-enabled businesses reach a tipping point, but they may also face a slew of technological and management obstacles. Agrawal *et al.* (2021) have stated that through digitalisation, businesses are currently embracing the '4.0 industrial revolution'. Enhancement of technology can create continuous development among organisations and most SMEs can develop a sustainable business environment due to accept advance technology.

- IoT apps expand corporate prospects by improving business processes and service quality.
- It improves business performance by enabling training for employees which ensures employees' work efficiency and eliminates skill mismatches. For that, employees can be up to date on their responsivities, work activities with updated organisational culture.
- IoT applications reduce overall corporate costs by improving business modules, equipment monitoring, asset usage and employee training services.

6. CONCLUSION

Based on the analysis and discussion it can be said that IoT helps small and medium size enterprises to quickly separate themselves from their competition. Some businesses are utilising IoT to gain data about specific products that are widely discussed on social media. Due to the increase in the competition level in this era, organisations switched from conventional mode to digitalised mode. Thus, it can be stated that IoT is the main contributor to advanced strategies that help to achieve ultimate success and long-term growth among SMEs. IoT would allow users to customise services, products, and offers based on customers' preferences. 'With more data about consumer behaviour' available to marketers, it becomes much easier to entice customers and even impact their own buying decisions. To get proper data, this research article has considered mixed methods such as primary quantitative and secondary qualitative for developing empirical analysis.

7. FUTURE SCOPE

In this modern age, IoT is one of most growing technological aspects that help to provide advanced ideas, quick responses among businesses. Implementation of IoT can also be beneficial for reducing higher complexities among businesses in a significant manner. As IoT is the most growing technology, SMEs can get maximum benefits in future from IoT services. Not only the business sector, applications of IoT are used in many other sectors such as healthcare, agriculture, education and many more.

REFERENCES

Agrawal, R., V. A. Wankhede, A. Kumar, A. Upadhyay, and J. A. Garza-Reyes (2021). "Nexus of circular economy and sustainable business performance in the era of digitalization," *International Journal of Productivity and Performance Management*.

Chatterjee, S. (2021). "Antecedence of attitude towards IoT usage: a proposed unified model for IT Professionals and its validation," *International Journal of Human Capital and Information Technology Professionals (IJHCITP)*, 12(2), 13–34.

Hamidi, H., and M. Jahanshahifard (2018). "The role of the Internet of Things in the improvement and expansion of business," *Journal of Organizational and End User Computing (JOEUC)*, 30(3), 24–44.

Haseeb, M., H. I. Hussain, B. Ślusarczyk, and K. Jermsittiparsert (2019). "Industry 4.0: a solution towards technology challenges of sustainable business performance," *Social Sciences*, 8(5), 154.

Lee, I. (2019). "The Internet of Things for enterprises: an ecosystem, architecture, and IoT service business model," *Internet of Things*, 7, 100078.

Sestino, A., M. I. Prete, L. Piper, and G. Guido (2020). "Internet of Things and big dData as enablers for business digitalization strategies," *Technovation*, 98, 102173.

Tang, C. P., T. C. K. Huang, and S. T. Wang (2018). "The impact of Internet of Things implementation on firm performance," *Telematics and Informatics*, 35(7), 2038–2053.

4. Design and Empirical Analysis of a Machine Learning-Based Human Resource Management Processing Systems for Detecting Personal Stress

Chhaya Nayak[1], Dr. Radha Raman Chandan[2], Shaik Rehana Banu[3], Mohit Tiwari[4], Melanie Lourens[5], Bratati Kundu[6], and Rania Mohy ElDin Nafea[7]

[1]Assistant Professor, Pimpri Chinchwad College of Engineering, Pune

[2]Associate Professor, Department of Computer Science, School of Management Sciences, Varanasi, India

[3]Post Doctoral Fellowship, Department of Business Management, Lincoln University College Malaysia

[4]Assistant Professor, Department of Computer Science and Engineering, Bharati Vidyapeeth's College of Engineering, Delhi A-4, Rohtak Road, Paschim Vihar, Delhi

[5]Deputy Dean Faculty of Management Sciences, Durban University of Technology, South Africa

[6]Assistant Professor, Department of Management, ABS Academy of Science Technology and Management

[7]University of technology, Bahrain

ABSTRACT: 'Machine learning (ML)' has previously been the subject of in-depth study in a number of disciplines. Almost all businesses are now integrating intelligent technology and ML into their operating divisions to increase employee performance. ML is used in human resources for anything from hiring employees to reviewing employee performance. Additionally, ML exhibits a strong correlation with creativity, ingenuity and usability, suggesting that it affects HR regarding innovations and usability. Stress can be defined as the effort of a body to exert self-control in the face of environmental alterations through mental, bodily and emotional states. Worldwide, people of all ages are experiencing psychological stress at higher rates. Users might be able to effectively organise and manage their stress by using a reliable, affordable acute stress detection method, which would lessen the negative effects on their health in the long run. In this study, we review and discuss studies on ML-based methods for emotion recognition. Researchers have known for a long time that mental stress negatively affects human health.

KEYWORDS: Machine learning, management, sensors, human resource, physiological, stress detection techniques, technology

1. INTRODUCTION

Machine Learning (ML) is the technology used to undertake a job that needs some amount of expertise to complete. It refers to a technology that has been taught to function in the same way as a person can. The elements of ML distinguish it from typical software because they include high-speed processing, advanced algorithms and a vast volume of elevated information. ML employs algorithms that combine high-quality data with rapid computing services, leading to Core ML providing correctness and reliability to common activities (Picard, 2016). ML technologies provide several chances to enhance HR services like recruiting, payroll and self-service operations, as well as accessibility rules and processes (England *et al.*, 2012).

According to the American Psychological Association, pressures from the recent past and the immediate future are causes of acute stress. Sporting difficulties, test distress or worry while interacting with new individuals can all cause acute stress (Nashef *et al.*, 2013).

DOI: 10.1201/9781003532026-4

This chapter discusses about stress detection method by utilising psychological information collected from sensors. Individuals may also use our approach in their everyday lives. Individuals' movements are free in real-life environments, and artefacts develop as a result. We used various innovative artefact identification and elimination algorithms to make our system useful in these environments (Rosch, 1996). There are various methods or techniques available for machine reading and stress reading. These methods generally include various models and computational algorithms to understand and extract information. By combining and integrating these methods into comprehensive stress reading or machine reading, it is possible to extract information. However, the choice of appropriate method or technique is directly dependent on data availability, the complexity of text analysis and the objectives of the tasks.

1.1. Objectives and Hypotheses

- To analyse and design a ML-based HRM system to detect the personal stress of employees
- To identify, understand and analyse various factors of employee stress
- To design and implement various ML-based systems to collect accurate data on the personal stress levels of employees
- To evaluate the effectiveness of ML-based systems to detect personal stress

Some hypotheses have been developed,

Hypothesis 1: *ML-based HRM systems can successfully detect personal stress levels of employees*

Hypothesis 2: *A correlation between work–life balance and personal stress levels of employees is apparent*

Hypothesis 3: *Negative interpersonal relationships in workplaces have a direct impact on the personal stress levels of employees*

2. LITERATURE REVIEW

2.1. What Exactly is Stress?

Stress can be defined as a person's response to adversity in an environment that tests his or her usual adaptation capabilities (Daugherty, 2011). Additionally, it has been shown that psychological stress impairs physiological functions which leads to affecting daily job performance and impacts economic growth (Rosch, 1997).

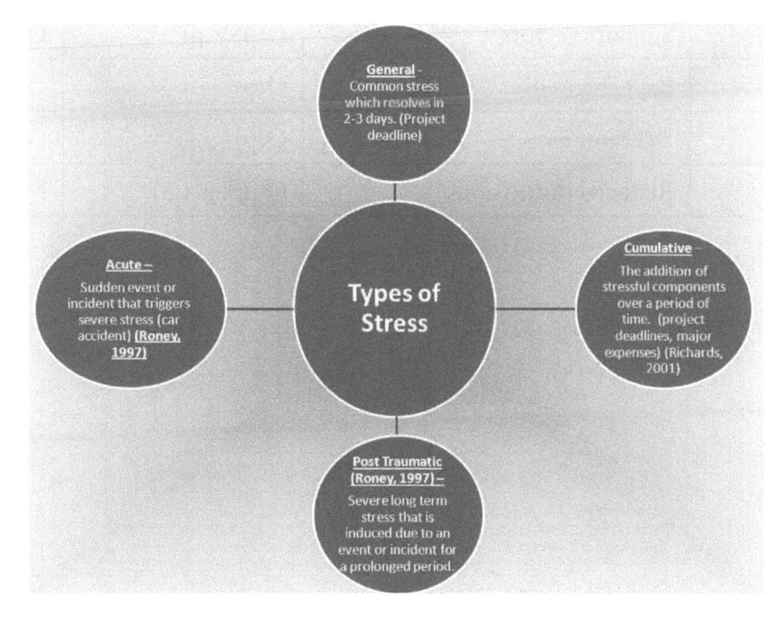

Figure 1. *Depicts instances of objective and subjective stress measurements.*

Negative stress has two unique impacts:

(i) Physiological or Objective Stress
(ii) Psychological or Perceived Stress

Changes in parameters associated with physiology such as an increase in blood pressure and rising pulse indicate objective stress. Subjective stress is an individual's assessment of whether or not a circumstance is distressing. Some questionaries associated with stress such as the 'DASS 21 (Depression, Anxiety, and Stress Scale – 21 Items)', 'STAI (State-Trait Anxiety Inventory)', and 'POMS (Perceived Anxiety and Stress Scale)' are the most common way to assess perceived stress ('Profile of Mood States'). There are two main stress markers: adrenal and evaluation for physiological signals such as GSR ('Galvanic Skin Response'), ECG ('Electrocardiogram') and EEG ('Electroencephalogram'). In the study by Baumer *et al.* (2013), we looked at a variety of stress measurements concerning physiological issues and the technology associated with stress-measuring systems.

Using facial cues, identifying stress and anxiety in movies.

2.2. Stress Tracking in the Workplace Using ML

The link between physiological indicators and accompanying stress levels may be represented by characteristics derived from these physiological indicators. ECG signals generally contain the following characteristics: 'SDNN (standard divergence of the R–R peak interval)', 'RMSD (RMS valuation of the progressive variations between R–R peak intervals)', 'pNN20 and pNN50 (percentage of the progressive sinus rhythm R–R intervals that are greater than 20 ms or 50 ms in both)', are the standard variations of the Variation graphs in the longitudinal and across directions, respectively. For the most part, GSR is composed of two elements: 'skin conductance level (SCL)', and an extremely fast-changing phased element known as the 'skin conductance response (SCR) (SCR)'. Showcases such as SCL, the length of time or extent of SCR as well as SCRR (SCR Rate), and the timeframes during which the effects of the stimulation continue, also known as OPD are among the characteristics that have been investigated extensively in the sense of stress detection and identification (Kiranashree *et al.*, 2021). Feature characteristics of common ML algorithms are listed in Table 1.

Table 1.

Algorithm	Rule of Classification	Testing	Training
SVM	Support vectors	O(nSV p)	O(n 2p + n 3)
NB	Bays theorem	O(p)	O(np)
DT	Decision tree	O(p)	O(n 2p)
LDA	Reduced dimensions	O(npt + t 3)	NA

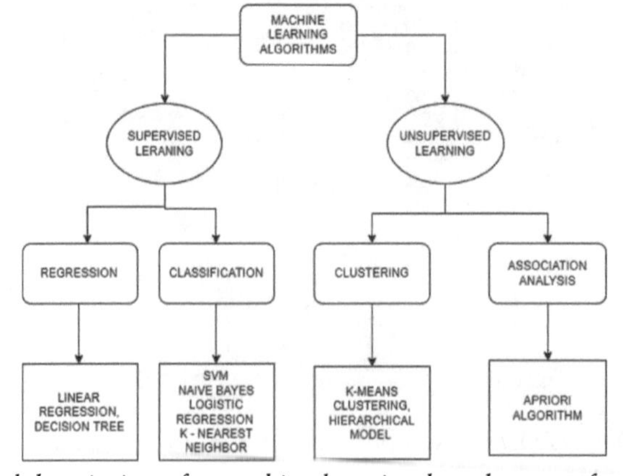

Figure 2. *A high-level description of a machine learning-based system for stress categorisation.*

2.3. ML Equipment

Our categorisation of information was carried out with the help of the Weka software (Azuaje, 2006). We performed a numeric to nominal modification on the classification columns to prepare it for highlight preprocessing. As a result of the imbalanced distribution of class examples in our database, we have recruited examples from the minority classes and deleted examples from the majority class to correct the mismatch in the class memberships distribution.

1. 'Principal Component Analysis (PCA)' and Linear Discriminate Analysis (LDA) are two types of statistical analysis.
2. PCA and 'Support Vector Machine with radial kernel (SVM)'
3. K-Nearest Neighbours (n = 1) is the closest neighbours (kNN)
4. Logistic Regression is the fourth kind of regression.
5. Forest of the Unknown or 'Random Forest (RF with 100 trees)'.
6. Perception with many layers of information.

2.4. Identifying Personnel Stress Using ML in Human Resource Systems

ML is a modern technological equipment of the next generation that can sense, understand, organise and accomplish activities that increase human efficiency without placing any constraints on the worker (Jain and Kumar Pandey, 2019). Motion Identification, BOTS and Algorithms are the three basic types of ML systems that are concerned with the human resource system.

BOTS	When Google and other search engines scan the web for relevant keywords, they use the BOTS program to help them do it. This device is excellent for extending conversations, asking questions, conversing, giving instructions, providing directions, tracking and doing other useful tasks. The current machine-learning system requires improvement via a number of improvements to address intricate and complex situations (Murgai, 2018).
ML Algorithm	ML algorithms are sets of coding and commands that must be followed stage per phase to lead ML operations. A smart algorithm may be used to automate a variety of HR processes, including intelligence gathering, transmission of data to users, monitoring key performance metrics and monitoring actions of current and prospective workers.
Motion Recognition	In most firms, ML assistants to human resource managers are responsible for these technologies. Gesture recognition, pulse rate, hypertension and other basic related activities (Murgai, 2018) are the key functionalities of these technologies, and they are used to initiate actions.

2.5. Healthy Office: Using Mobile Phones and Wearable Sensors, Employees May Recognise Their Moods at the Workplace

Employees who suffer from workplace stress, anxiety or depression are less productive and have worse morale, which has serious financial consequences for the company. In this chapter, we investigate the feasibility of employing such technologies for mood identification in the workplace, with a particular emphasis on the workplace. We also propose a Smartphone application (Healthy Office), which is intended to assist consciousness in an organised way and to contribute data to our models that are grounded in reality (Jain *et al.*, 2020).

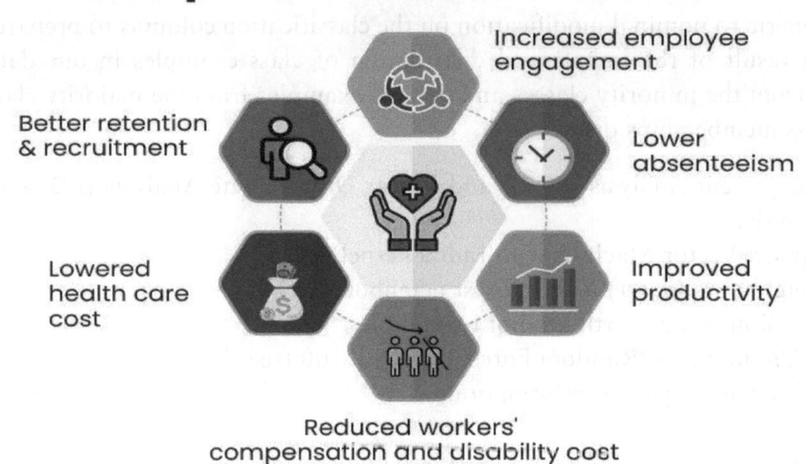

Figure 3. Healthy office application data.

2.6. Heart Rate Diversity Analysis, Both Linear and Non-Linear, is Used to Differentiate between Acute Stress and Chronic Stress

Chronic stress identification is a critical component in identifying and decreasing the incidence of cardiovascular disease in the general population. As a pilot project, the authors are concentrating on establishing a mechanism for distinguishing short-term psycho physiological changes from one another using the heart rate variability (HRV) properties (Jain *et al.*, 2020), this pilot research is being conducted. It was discovered that visuals and noises, as well as cognitive and physical activities and rest, generated four distinct forms of arousal, which were then categorised using linear or non-linear HRV characteristics from 'electrocardiograms (ECG)' obtained by the wireless wearable patch recorders. Sampling sensitivity, volume decreases fluctuations analysis and normalised high-frequency characteristics were used to achieve the greatest identification rates for the neutral stage (90%), the acute 'stress stage (80%)' and the 'background stage (80%)'.

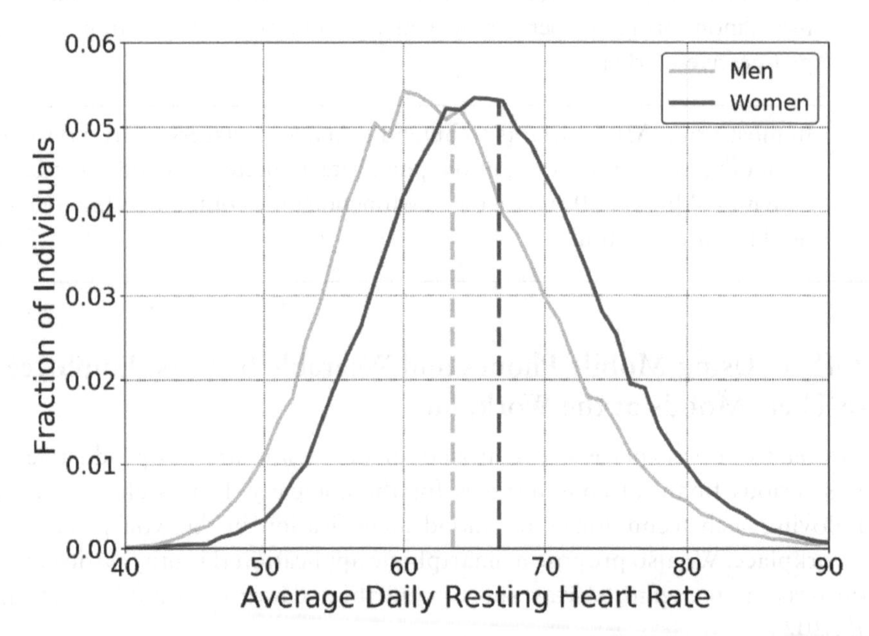

Figure 4. Heart rate diversity analysis.

3. DISCUSSION AND ANALYSIS

Based on the above discussion, it can be inferred that ML-based human resources management can play a crucial role in terms of detecting personal stress. A well-developed ML system can detect personal stress with a high accuracy rate of around 85%. Some additional monitoring systems can also be implemented such as wearable sensors, physiological indicators and analyzing sleep patterns. From the generated statistical data using ML-based systems valuable insights can be collected regarding the personal stress of employees. In the context of some general interpretation, by identifying the sleeping patterns employee behaviour and sentiments can be identified that are associated with employee stress levels. This can help organisations to take proactive measures to resolve common issues.

4. CONCLUSION

In this study, we examine what exactly stress is after that we are discussing about detecting anxiety and stress in video using face clues. In this research, we also study about stress tracking or monitoring in the workplace using ML. We also discuss ML equipment. To evaluate the effectiveness of ML-based stress reading systems, it is essential to consider primary data collection along with secondary data and the applications of appropriate research tools.

REFERENCES

Azuaje, F. (2006). "Witten IH, Frank E: Data Mining: Practical Machine Learning Tools and Techniques 2nd edition," *BioMedical Engineering OnLine*, 5(1). doi: 10.1186/1475-925x-5-51.

Baheti, R. B., and S. Kinariwala (2019). "Detection and analysis of stress using machine learning techniques," *International Journal of Engineering and Advanced Technology*, 9(1), 335–342. doi: 10.35940/ijeat.f8573.109119.

Baumer, E., V. Khovanskaya, P. Adams, J. Pollak, S. Voida, and G. Gay (2013). "Designing for engaging experiences in mobile social-health support systems," *IEEE Pervasive Computing*, 12(3), 32–39. doi: 10.1109/mprv.2013.47.

Colligan, T., and E. Higgins (2006). "Workplace stress," *Journal of Workplace Behavioral Health*, 21(2), 89–97. doi: 10.1300/j490v21n02_07.

Daugherty, L. (2011). "Stratification according to psychosocial risk factors: implications for future nutrition and coronary heart disease research," *The American Journal of Clinical Nutrition*, 93(6), 1386–1386. doi: 10.3945/ajcn.111.014746.

England, M., C. Liverman, A. Schultz, and L. Strawbridge (2012), "Epilepsy across the spectrum: Promoting health and understanding," *Epilepsy & Behavior*, 25(2), 266–276. doi: 10.1016/j.yebeh.2012.06.016.

Kiranashree, B., V. Ambika, and A. Radhika (2021). "Analysis on machine learning techniques for stress detection among employees," *Asian Journal of Computer Science and Technology*, 10(1), 35–37. doi: 10.51983/ajcst-2021.10.1.2698.

McEwen, B. S. (2006). "Protective and damaging effects of stress mediators: central role of the brain," *Dialogues in Clinical Neuroscience*, 8(4), 367–381. doi: 10.31887/dcns.2006.8.4/bmcewen.

Murgai, D. (2018). "Role of artificial intelligence in transforming human resource management," *International Journal of Trend in Scientific Research and Development*, 2(3), 877–881. doi: 10.31142/ijtsrd11127.

Nashef, L., S. Lhatoo, L. Bateman, J. Bird, and T. Tomson (2013). "Incidence and mechanisms of cardiorespiratory arrests in epilepsy monitoring units (MORTEMUS): a retrospective study," *The Lancet Neurology*, 12(10), 966–977. doi: 10.1016/s1474-4422(13)70214-x.

Picard, R. (2016). "Automating the recognition of stress and emotion: from lab to real-world impact," *IEEE MultiMedia*, 23(3), 3–7. doi: 10.1109/mmul.2016.38.

Rosch, P. (1997a). "Book Review: Measuring Stress: A Guide for Health and Social Scientists, Cohen, S., Kessler, R. C., and Gordon, L. U., eds., Oxford University Press, New York, 236 pgs. $39.95," *Stress Medicine*, 13(1), 67. doi: 10.1002/(sici)1099-1700(199701)13:1<67::aid-smi730>3.0.co;2-8.

Rosch, P. (1997b). "Book Review: Why We Eat What We Eat: The Psychology of Eating, Capaldi, S., ed., American Psychological Association, Washington, 1996, 339 pp., $49.95," *Stress Medicine*, 13(4), 268–268.

5. An Empirical Analysis of Machine Learning and Strategic Management of Economic and Financial Security and its Impact on Business Enterprises

Mahabub Basha S[1], Bhadrappa Haralayya[2], Dharini Raje Sisodia[3], Mohit Tiwari[4], Sandeep Raghuwanshi[5], K. G. S. Venkatesan[6], and Astha Bhanot[7]

[1]Assistant Professor, Department of Commerce, International Institute of Business Studies, Bengaluru

[2]Professor and HOD, Department of MBA, Lingaraj Appa Engineering College Bidar-585403, Karnataka, India

[3]Assistant Professor, Management, Army Institute of Management & Technology

[4]Assistant Professor, Department of Computer Science and Engineering, Bharati Vidyapeeth's College of Engineering, Delhi A-4, Rohtak Road, Paschim Vihar, Delhi

[5]Assistant professor, Amity University Gwalior Madhya Pradesh

[6]Professor, Department of C. S. E., MEGHA Institute of Engineering & Technology for Women Edulabad - 501 301 Hyderabad, Telangana, India

[7]PNU, Riyadh, KSA

ABSTRACT: Building a firm includes strategic management and corporate development. These elements can help a corporate organisation develop its financial and economic stability. With the help of this organisation, it is possible to plan out the business-related strategy that will be used to put ideas into practice and accomplish their organisational goals. An organisation can comprehend its accessibility, leads, revenue, sales, stability, etc., with the aid of this aspect. On the other hand, developing a corporate strategy is what is meant by strategic management. They carry out the strategic vision with the aid of this preparation and examine the outcomes by putting this planning into practice. The machine learning process helps in the strategic management of the financial sector of a company.

KEYWORDS: Corporate planning, business organisation, strategic management

1. INTRODUCTION

A business is designed to sell the product or services with a vision of expansion and earning a good profit. Corporate planning and strategic planning in the business's economic structure and financial security help to frame the basic structure of the business. It is considered an approach of the company and gives guidance to the company to achieve short-term or long-term goals as per requirement (Md, 2019). The financial structure is the most crucial division of the company, and strategic planning can act as a fruitful measure in developing the economic system of the company. Nevertheless, the machine learning system is quite significant in the financial sector. The information system is a crucial component of the machine learning system as a whole. Machine learning is essentially an artificial intelligence that makes it easier and more accurate for people to find solutions.

DOI: 10.1201/9781003532026-5

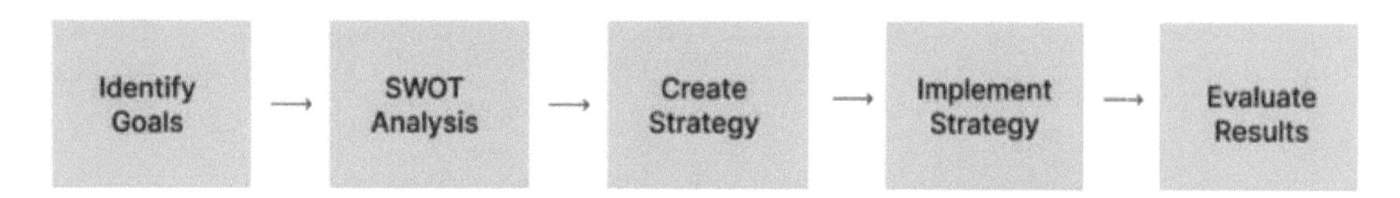

Figure 1. *Strategic management. Source: Md (2019).*

1.1. Objectives and Hypothesis

The objectives of this research study are,

- To identify the current strategic management and machine learning practice adopted by business enterprises for economic and financial security
- To examine the critical impacts of machine learning and strategic management on the economic and financial security of business organisations
- To evaluate the key factors associated with machine learning and strategic management that influence the economic and financial security of business organisations
- To analyse the effectiveness of machine learning and strategic management techniques to ensure financial and economic security

Hypothesis 1: *Implementation of machine learning and strategic management positively impacts the economic and financial security of business organisations*

Hypothesis 2: *Machine learning techniques have a significant impact on the risk management of business organisations*

Hypothesis 3: *Strategic management approaches can detect and prevent fraud as well as improve the economic and financial security of business organisations*

2. LITERATURE REVIEW

2.1. Strategic Management and Corporate Planning

The organisation uses different types of strategic management and corporate planning to execute the business procedure (Jelonek, 2023). They are tactical, operational, contingency and strategic planning. The primary motive for incorporating this planning in managing the economic and financial structure of the business is to get a clear objective and pathway of business execution.

According to the journal by Cherevko *et al.* (2019), the financial security of the business enterprise demands the concept of ensuring financial stability achieving high competitiveness in the different sectors of the industry. Tactical planning helps to break the strategy into short-term plan. It demands creating the flow chart from collecting resources to completing the slates to the end consumers in one business scenario. The second scenario helps to identify the problem and find a route for the solution to the preceding problem in the business structure. Operational planning helps a strategy break down into a detailed route map that helps outline the business's action. Millions of customer information are available to insurance and banking companies, which can be used to train machine learning algorithms and improve the approval processes. Organisations can save time and money by using machine learning algorithms to actually determine filtering and credit reporting instead of hiring human decision makers.

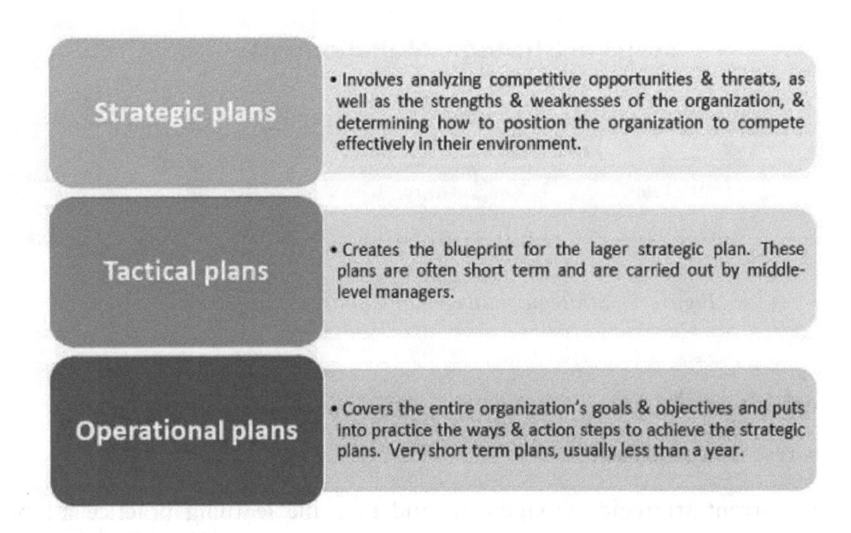

Figure 2. *Describes different types of corporate planning. Source: El Ammar (2020).*

2.2. Importance of Strategic Management and Corporate Planning in Businesses

According to the journal by Chia-Chi (2021), strategic management corporate planning helps the organisation be responsible for mapping out the strategy and identifying the measures to implement the planning to empower the administration's performance. It helps the organisation to strengthen the brand identity of its product or service. According to Najafi *et al.* (2018), strategic management studies the associated aspects before formulating the plan to execute. It is formulated with three components: aspects of the environment through scanning, how to formulate the strategy and its implementation and evaluation of the strategy. According to Putri *et al.* (2019), first, the organisation identifies the problem or the business goal, vision and objectives of the organisation and the future sustainable plan before formulating the strategic plan.

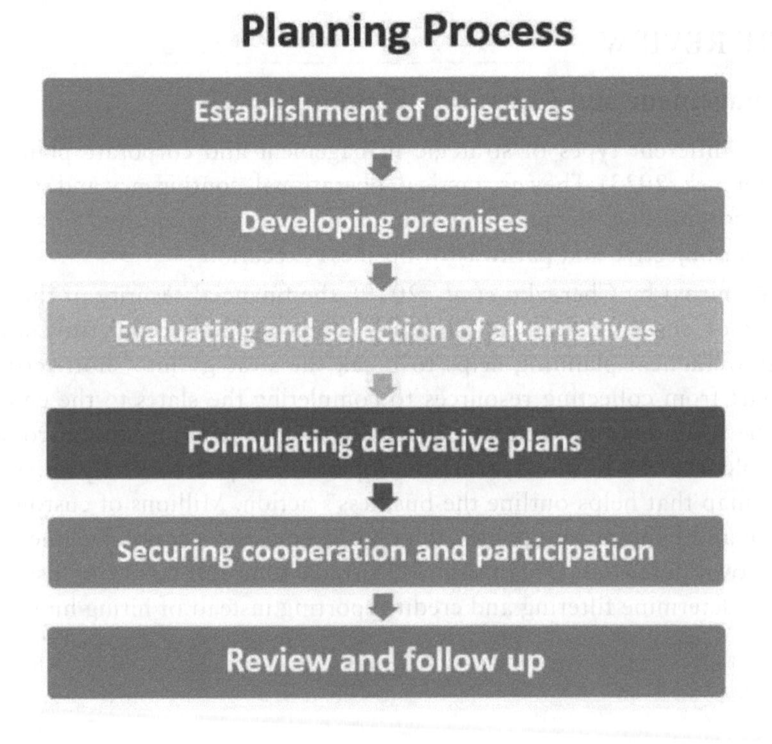

Figure 3. *Describe the different functions of corporate planning. Source: El Ammar (2020).*

2.3. Strategic management and corporate planning Focusing on Financial and Economic Security

According to Md (2019), strategic management and corporate planning help manage finance and maintain economic and financial security. Strategic planning is developed with different approaches, and financial tools and techniques are used to manage the company's finances. It has some specific features as follow:

- Its focuses on long-term management.
- It shows the organisation the route of profitability growth and strives to increase its wealth.
- It needs to be flexible and structured to execute systematic planning.
- The strategic plan is continuously evolving according to the precise situation and needs to adapt to the change.
- It has a multidimensional and innovative approach to solving the issue in finance.
- After the application constant supervision is required to meet the business objective Jelonek(2023).
- Before formulating the plan, the strategic approach analyses the factual information using financial tools and analytic financial methods.

2.4. Challenges of Corporate Planning and Strategic Management and Their Mitigation Strategy

According to Citraresmi and Haryati (2021), a few distinctive challenges are faced in executing corporate planning and strategic management in the business organisation. The strategy can be weak in its structure due to poor quality goal setting. The system can face the lack of alignment of different organs of the business structure. The unorganised strategy will not be able to track the plan's progress and determine the growth of the business structure Jelonek(2023). According to Gileva *et al.* (2021), the organisation's business structure may have a poor-structural policy framework, lack of skilled employees and managerial department, which can be a challenging factor in executing corporate planning.

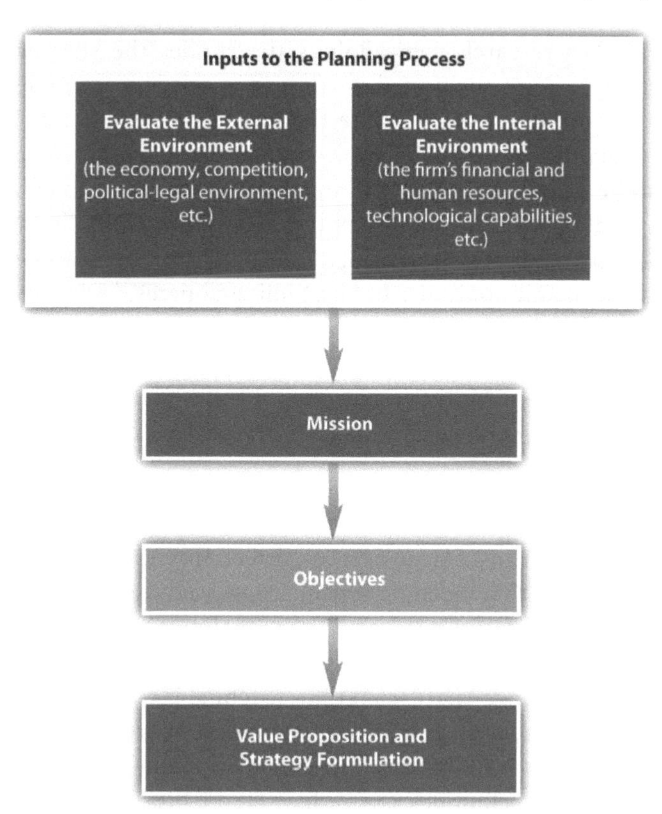

Figure 4. *Describes the work of corporate planning and strategic planning. Source: El Ammar (2020).*

2.5. Impacts on the Business Environment

Corporate planning and strategic planning have different forms of impact and help in the company's overall business. It helps to locate the problematic issue of the company. It helps to enhance the quality of the service of the enterprise. It brings significant change to meet the profitable level of the company. According to the journal by El Ammar (2020), it controls the intensity of the organisation's managerial, organisational and environmental factors.

Figure 5. Describes different components of the corporate strategy. Source: El Ammar (2020).

3. RESEARCH METHODOLOGY

According to Svatošová (2021), research methodology determines the specific procedure or technique used to conduct the research. It helps them identify the issue, select a research method, collect the data and analyse the information and data to formulate the research result. There are four general types of research methodology: observational, experimental, simulation and derived research after selecting a research topic. The researchers have followed a systematic research approach. This research study has adopted only a secondary qualitative research design by analysing existing literature, research papers and case studies relevant to the research topic. Moreover, all the data of this research study were collected from secondary sources such as academic journals, books, online sources, industry reports and government publications. These data have been analysed using the qualitative content analysis method by identifying and categorising themes and patterns. As the study critically focused on secondary data collection no sampling method has been employed. Moreover, throughout the research process, ethical considerations have been maintained by ensuring the integrity and accuracy of data.

4. ANALYSIS AND INTERPRETATION

Machine learning basically first identifies a large data set, detects any anomalies, predicts risks and enhances decision-making processes. By adopting machine learning potential frauds can be detected. Organisations that use various machine learning techniques and data analytics experienced a significant reduction in fraud losses by around 52%. Moreover, a 50% increase in fraud detection rate can be achieved by implementing AI and machine learning. Apart from that, machine learning and strategic management also help organisations in risk management and predictive analytics. More than 71% of business enterprises admitted that they are using predictive analytics for essential risk management purposes. Whereas machine learning-led strategic management techniques can reduce credit loss by 10–20%. Based on the analysis, it can be interpreted that by the end of 2024, more than 70% of large organisations will have to rely on machine learning algorithms and advanced analytics for financial planning which will subsequently lead to a 25% improvement in the financial planning cycle times.

Apart from that, to enhance the application of strategic management and machine learning for the economic and financial security of businesses various evolving tools and technologies can be used. Among them, one of the most commonly used tools in machine learning applications is Python. It is a common programming language that includes libraries PyTorch and TensorFlow to provide strong and robust frameworks to implement machine learning algorithms. By adopting this tool and leveraging the availability and versatility of these libraries, organisations can enhance the efficiency of data analysis and model development by 20%.

5. DISCUSSION AND FINDINGS

The data related to the subject helps the researchers understand the impacts of strategic management and corporate planning. With the help of an introductory study, analysis can identify that this business-related practice is helpful primarily for any business organisation (de Moura *et al.*, 2021). wherever a small parentage of people has no idea about this kind of factors. In this research, the paper researcher conducted this survey among 70 people connected with any corporate organisation. Along with that, researchers also can identify that most of the people who are working or relate with any corporate sector have such knowledge and also have some interest in the organisational systems and planning (Kitsios *et al.*, 2022). They have some ideas about the organisational-related factors as well. According to the survey, maximum people are told that corporate planning and strategy management positively impact the business organisation. People think that this kind of organisational practice can develop the business and increase the company's revenue rate.

6. CONCLUSION

This chapter has critically discussed the basic meaning of strategy management and corporate planning. Also, this research chapter has mentioned the impact on financial and economic security with the help of this organisational practice. With the help of proper financial and economic security, any business can quickly develop. This research chapter has also discussed some relevant information about this topic. However, the primary data collection process needs to be employed along with secondary data to gather more accurate and reliable information to draw an effective conclusion.

7. FUTURE SCOPE

Proper business planning and strategic management can assist an organisation in developing its economic and financial security to increase the brand value of the business organisation. With the help of a good brand image, the customers also can trust the organisations to consume their products or services. If an organisation follows some advanced corporate planning process and business strategy that can help the organisation develop all types of security, it helps make their business grow.

REFERENCES

Amiran, M., A. Asadi, and M. Oladi (2022). "Presentation of novel multiple regression model for accounting information quality, corporate investment, and moderating role of ownership structure in companies," *Discrete Dynamics in Nature and Society, 2022.*

Cherevko, O., S. Nazarenko, N. Zachosova, and N. Nosan (2019). *Financial and Economic Security System Strategic Management as an Independent Direction of Management.* Les Ulis: EDP Sciences.

Chia-Chi, L. (2021). "Analysis on the strategy of improving management consulting business performance: evidence on a management consulting company established by an accounting firm," *Asia Pacific Management Review*, 26(3), 137–148.

Citraresmi, A. D. P. and N. Haryati (2021). "The strategy of business model development in mushroom agroindustry," *IOP Conference Series. Earth and Environmental Science*, 924(1).

de Moura, D., and P. A. Tomei (2021). "Strategic management of organizational resilience (SMOR): a framework proposition," *Revista Brasileira De Gestão De Negócios*, 23(3), 536–556.

El Ammar, C. (2020). "iBalanced scorecard: an effective strategy implementation in Lebanese government authorities," *Revista De Management Comparat International*, 21(2), 146–164.

Gileva, T. A., M. P. Galimova, A. V. Babkin, and M. E. Gorshenina (2021). "Strategic management of industrial enterprise digital maturity in a global economic space of the ecosystem economy," *IOP Conference Series. Earth and Environmental Science*, 816(1).

Jelonek, D. (2023). Environmental uncertainty and changes in digital innovation strategy, *Procedia Computer Science*, 225, 1468–1477

Kitsios, F., E. Chatzidimitriou, and M. Kamariotou (2022). "Developing a risk analysis strategy framework for impact assessment in information security management systems: a case study in IT consulting industry," *Sustainability*, 14(3), 1269.

Md, S. H. (2019). "The impact of accounting information system on organizational performance: evidence from Bangladeshi small & medium enterprises," *Journal of Asian Business Strategy*, 9(2), 133–147.

Najafi, A. I., R. Mahmood, and S. B. Muhammad (2018). "Strategic improvisation and HEIs performance: the moderating role of organizational culture," *PSU Research Review*, 2(3), 212–230.

Putri, C. F., I. Nugroho, and D. Purnomo (2019). "Performance measurement of SMEs of Malang batik as a result of local wisdom with balanced scorecard," *IOP Conference Series. Materials Science and Engineering*, 505(1).

Svatošová, V. (2021). "Proposal and simulation of a business process model of strategic management in E-commerce 1," *Economic Casopis*, 69(7), 726–749.

6. The Implementation of the Internet of Things to Improve Industry 4.0 for Sustainable Growth in the Construction Sector

Avein Jabar AL-asadi[1], Sumeshwar Singh[2], Dilip Kumar Sharma[3], M. Kalyan Chakravarthi[4], Nilesh Singh[5], M. Sundar Rajan[6], and Rania Mohy ElDin Nafea[7]

[1]Assistant Lecturer, Sulaimani Polytechnic University, Technical College of Informatics, Department of Information Technology, Iraq – Sulaymaniyah

[2]Assistant Professor, Department of Computer Science and Engineering, Graphic Era Hill University, Dehradun, Uttarakhand

[3]Department of Mathematics, Jaypee University of Engineering and Technology, Guna (M.P.), India

[4]Associate Professor, School of Electronics Engineering, VIT - AP University, Amaravathi, Pin code: 522237, India

[5]Assistant Professor, Mechanical Engineering, Bharati Vidyapeeth's College of Engineering Lavale, Pune

[6]Associate Professor, Faculty of Electrical and Computer Engineering, Arbaminch Institute of Technology, Arbaminch University Ethiopia

[7]College of Administrative and Financial Sciences, University of Technology Bahrain, Bahrain

ABSTRACT: The Internet of Things (IoT) is a critical part of Industry 4.0. It has several applications for monitoring systems of production in both the services and manufacturing sectors. The key paradigms and variables that were cautiously evaluated were identified by a thorough examination of the appropriate literature. Utilising a quantitative research methodology, a close-ended questionnaire was created using the variables that had been obtained. According to the survey, industry 4.0 technology for smart building is the most popular in India's construction business. The data were evaluated using a Relative Importance Index (RII) approach. The primary snag to improvement is the absence of work and abilities in utilising Industry 4.0 and IoT advances. The key benefit of these innovations is that they will add sustainable administrative demands to bids. This study's main focus is based on an online analysis that defines a practical route for the structure company to operate locally and generates by involving major stakeholders and those affected by these developments.

KEYWORDS: IoT, sustainability, relative importance index, industry 4.

1. INTRODUCTION

The construction industry can use modern technologies to address problems under the banner of Industry 4.0. Young people from the technologically advanced Generation Z are drawn more and more to technological advancements that automate physical labour and increase either onsite or online productivity as they begin their careers (Ghosh *et al.*, 2020). Smart gadgets may now connect to the Internet and interact with one another due to the quick development of technology. Although they are smaller, they are more powerful and have more storage. Embedded systems are capable of real-time data access, gathering, storing and processing in today's smart devices. They can also interact, perceive and act. On the IoT, a variety of uses, features and services are accessible. It is a technology that is gradually gaining popularity. The IoT is used in many different industries and almost every facet of life.

DOI: 10.1201/9781003532026-6

Since its beginnings, Industry 4.0 has drawn a lot of interest, and researchers are trying to understand how it might be used so that it can be sustained, valid, reliable and secure. Thanks to contemporary technologies, employees are now under new demands. Therefore, the construction sector needs to improve the abilities of its current workforce. The major obstacles to the IoT in the construction industry right now are the cohabitation of numerous networks, massive data volumes, address constraints, automated address establishment, and security obligations like authentication and encryption. In the words of 'Made Smarter UK,' the construction industry might benefit from the industry 4.0 revolutions (Turner *et al.*, 2020) (Figure 1). The construction sector is scandalous for being delayed in taking on new technologies and advancements. Because of various elements, including the discernment that innovations are restrictively costly, the preparation required for carrying out mechanical change and improvements, and a longing to abstain from disturbing existing frameworks, cycles and methodology, the business has been delayed in embracing innovations.

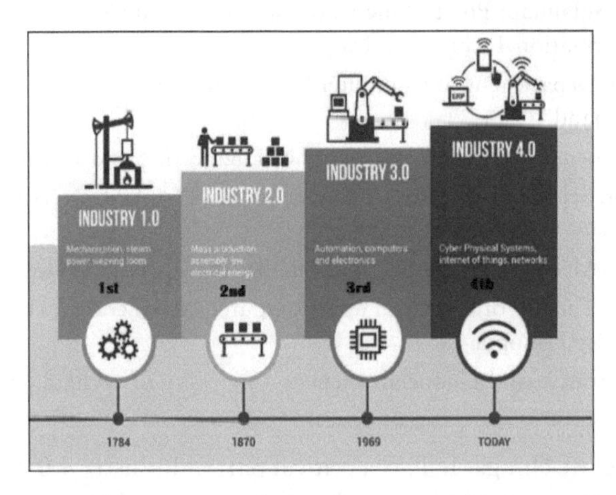

Figure 1. *Industry 4.0 revolution. Source: Author's compilation.*

Making an understanding of the whole app landscape is necessary. The Internet of Things (IoT) has focused on using several structural-related scenarios. Regarding these drawbacks, the construction sector must use IoT applications and advantages.

1.1. Objectives

RO 1 To obtain knowledge about some new industry 4.0 and IoT advances for the development area.

RO 2 To assess the advantages of applying Industry 4.0 and the IoT to the development area.

2. LITERATURE REVIEW

By utilising all of the accessible components of personalised approaches, Industry 4.0 helped to popularise the idea of the creative industry. In light of some of the most recent sustainability statistics, it is acceptable to be able to imagine, estimate and investigate innovative ideas and concepts without upending the status quo (Maqbool and Akubo, 2022). The nation's economy heavily depends on the building sector. Understanding the fundamental obstructions to utilising innovations that are intended for building projects is essential. Industry 4.0, a technique for improving the utilisation of digitalisation and computerisation fully intent on building an advanced worth chain all through the lifecycle of the item, from thought to improvement, manufacture, use, upkeep and removal, has gotten a ton of consideration as of late from the assembling area. This works with the creation of great items at lower costs and more limited opportunities to advertise, which helps general business execution.

2.1. Construction Industry and Industry 4.0

Researchers are still looking at Industry 4.0. The phrase is frequently used in the corporate sector outside of academia. The Industrial IoT is changing how organisations interact with and manage their manufacturing plants, machinery and warehouses. This shift has generated issues for the industry given the accessibility of digital data and the Internet, which may be used to autonomously collect and process digital data. The emergence of IR 4.0 will give rise to a whole new economic sector where all automated procedures will be able to run and communicate information without the aid of people, considerably enhancing efficiency.

2.2. Construction Industry Adoption of the IOT

A collection of 'Things' that connect to and interact with one another through a public or private network is referred to as the IoT. 'Things' are characterised as networked, intelligent devices that can communicate with one another without a lot of human intervention (Dilakshan *et al.*, 2021). The IoT may be divided into three groups based on how it is conceptualised: sensor layers, application layers and network layers.

2.3. IoT and Industry 4.0 in Sustainable Construction

Sustainability is a critical need in this digital age, as well as a technical challenge (Bajdor, Pawełoszek, Fidlerova, 2021) (Jelonek, Rzemieniak, 2024). The creation of intelligent technologies is crucial to ensuring the long-term viability of industrial systems in the future. Industry 4.0's IoT-enabled sustainable production has been thoroughly researched from the perspectives of technology, business, organisation and operations (Leng *et al.*, 2020). While ensuring the long-term viability of the current industrial structure, these 'industry 4.0' innovations have the potential to significantly boost innovation and competitiveness (Müller *et al.*, 2018).

3. METHODOLOGY

A broad approach to collecting information on a research topic is known as research methodology. It is a process for performing research that includes at least three parts, including data collecting, equipment development and sample techniques (Bhattacherjee, 2012).

3.1. Research Procedure

The first stage of this research was a thorough examination of the research on industry 4.0 practices in publicly financed programs. After that, a survey was done using a face-to-face and postal questionnaire approach. Statistical analysis is then performed on the collected data.

- Look for books, journals, periodicals, research theses and pertinent textbooks about industry 4.0 practices in libraries and online.
- Quantity surveying, construction and engineering companies that were registered with the Indian national institution were the participants in a questionnaire study that used sampling techniques.
- Statistical analysis of the survey data using SPSS 20.

3.2. Sample Selection

The study's sampling method was non-probability sampling, often known as purposeful sampling. Purposive sampling is essentially synonymous with qualitative research (Parker *et al.*, 2019). By observing a small portion of a population, sampling aimed to gather data about that population.

3.3. Data Analysis and Collection

Secondary as well as primary sources of data were employed to gather the study's data. Direct responses from survey participants served as the source of primary data. The study's primary data provided trustworthy, accurate data from first-hand sources. Research reports, the internet, logbooks, weekly newspapers and the library were all used to obtain secondary data. To guide the study's conduct and eventually validate or refute the original findings, secondary data are collected. Utilising closed-ended questionnaires, the information was gathered. We can accomplish the goal of the study with the aid of the information that we have collected. The recurrence reactions trademark clear statistics tool from SPSS Window version 20 is utilised to investigate Table 1. The study's second set of goals is discussed.

4. RESULT AND DISCUSSION

This part includes the analysis of data utilising the SPSS version 20 of data collected by distributed questionnaires, as well as a discussion of the findings. Closed-ended questions were included in the survey as a means of gathering data. The data information provides the data required to accomplish this research project's objective. Table 1 and Figure 2 were to look into the histories of the respondents to confirm that their responses were trustworthy in addressing the goals of this research.

Table 1: The highest degree of education and expertise.

Variables	Category	Frequency	Per cent
Highest educational qualification	PHD	26	16
	Masters	23	14
	First degree	87	54
	Diploma	21	13
	Others	4	3
Years of professional experience	1–5 years	78	48
	6–10 years	31	19
	11–16 years	27	17
	over 16 years	25	16

Source: Author's compilation.

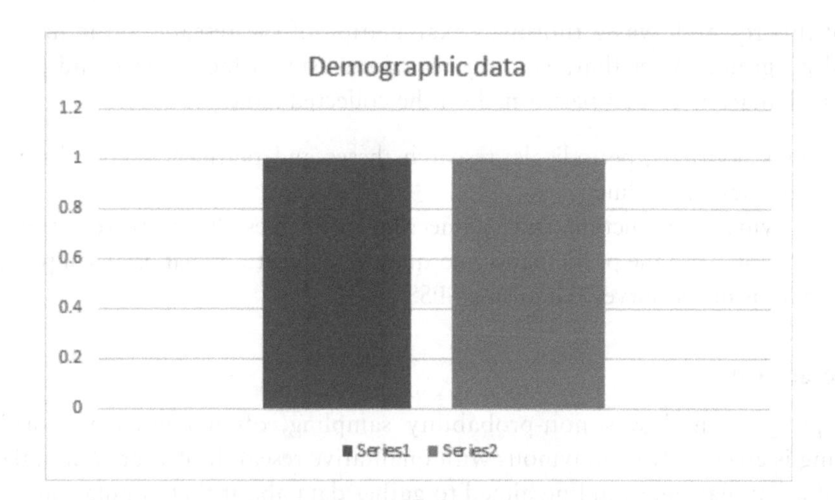

Figure 2

Drawing from Table 1, it can be seen that 78 out of the 161 respondents, or 48.0%, fell into the category of having no experience, followed by 31 respondents, or 19.0%, who had experience between 6 and 10 years, 27 respondents or 16.0%, who had experience between 11 and 15 years, and 25 respondents, or 15.0%, who had experience of more than 16 years. To assess each respondent's understanding of the survey and credibility, we asked them to state their highest degree of schooling. According to certain studies, having a degree can help you learn more information for your professional development and the development of your firm. The mean score situating and RII of impending industry 4.0 and IoT advances are displayed in Table 2.

Table 2: Ranking by average score and RII for developing industries 4.0 and IoT technology.

Innovation	1	2	3	4	5	RII
Big data	35	52	39	23	6	0.7232
Smart construction	32	51	37	25	4	0.7190
Cloud computing	31	50	34	39	5	0.6896
BIM	22	54	39	25	17	0.6547
Simulation	31	40	39	28	19	0.6520
Mobile computing	29	43	35	27	19	0.6459
Robotics construction	26	39	48	33	8	0.6445

Source: Author's compilation.

On a size of 1 to 5, where 1 addresses being very, extremely aware, 2 addresses being very aware, 3 represents being somewhat knowledgeable, 4 represents not being so aware and 5 represents not being at all informed, respondents were asked to rank their knowledge of the associated industry 4.0 and IoT advancements. The relative relevance index (RII) investigation and mean score positioning were utilised to check individuals' information on creating Industry 4.0 and IoT innovations. The dependability standards are met by the Cronbach's alpha coefficient, which, as indicated by Table 2, is 0.940. As displayed in Table 3, the relative significance index (RII) and mean score ranking (MSR) were utilised to evaluate the advantages of executing Industry 4.0 and Web of Things advancements in the development of the business.

Table 3: Benefits of using the rising industry, as measured by RII Internet of Things and 4.0 technologies in the building industry.

Merits	1	2	3	4	5	RII
Economic benefit	31	42	30	36	15	0.5457
Adding sustainable policies	31	34	39	31	20	0.5453
HRM	31	48	29	25	21	0.5406
Maintenance of machinery	29	43	34	29	18	0.5403
Improved project handling	27	46	35	24	21	0.5403
New opportunities and solutions	31	45	38	23	16	0.5352
Identification of flaws in construction	44	42	26	26	18	0.5138

Source: Author's compilation.

The consequences of the examination are displayed in Table 4. Like this, the reliability and validity of the information were checked before assessment to improve the nature of the data that was gathered.

Table 4: Reliability evaluation.

Reliability statistics	
No. of terms	Cronbach alpha coefficient
10	0.952

Source: Author's compilation.

The Cronbach's alpha coefficient, which can be found in Table 2, is 0.952, which is within the acceptable range for reliability. The scale is reliable as a result.

5. CONCLUSION

The study information was carefully analysed utilising a scope of methodologies that were generally suitable for the current review to feature inconsistencies or linkages between factors or gatherings. Frequency tables and charts were utilised as a component of the underlying assessment of the responder profile. The essential goal was to order some arising IoT and industry 4.0 advancements that are relevant to the development area. The following goal is to comprehend the advantages of integrating Industry 4.0 and IoT into the structure area. The reliability was evaluated using Cronbach's alpha and RII. This shows how the execution of these upgrades in the structure area will help the country's economy. The results show that professionals should be trained and educated on the IoT's general concept as well as the value of utilising IoT elements to increase adoption. Therefore, measures to reduce the costly nature of IoT devices and training are required.

REFERENCES

Bajdor P., Pawełoszek I., Fidlerova H. (2021), Analysis and Assessment of Sustainable Entrepreneurship Practices in Polish Small and Medium Enterprises, Sustainability Volume 13 Issue 7 DOI:10.3390/su13073595

Bhattacherjee, A. (2012). "Social science research: principles, methods, and practices," *Textbooks Collection*. 3. http:// scholarcom mons. usf. edu/ oa_ textbooks/3. Accessed 10 November 2022.

Dilakshan, S., A. P. Rathnasinghe, and L. I. P. Seneviratne (2021). "Potential of the Internet of Things (IoT), in the construction industry," *Proceedings The 9th World Construction Symposiuml* July (p. 446). University of Moratuwa, Sri Lanka. Available via: http://dl.lib.uom.lk/handle/123/16616. Accessed 23 December 2022.

Ghosh, A., D. J. Edwards, and M. R. Hosseini (2020). "Patterns and trends in Internet of Things (IoT) research: future applications in the construction industry," *Engineering Construction and Architectural Management*, 28(2), 457–481.

Jelonek D., Rzemieniak M. (2024). The Use of Artificial Intelligence in Activities Aimed at Sustainable Development - Good Practices, *Communications in Computer and Information Science*, 1948 CCIS, 277–284.

Jurczyk-Bunkowska M., and I. Pawełoszek (2015). "The concept of semantic system for supporting planning of innovation processes," *Polish Journal of Management Studies*, 11(1), 2015.

Leng, J., G. Ruan, P. Jiang, K. Xu, Q. Liu, X. Zhou, and C. Liu (2020). "Blockchain- empowered sustainable manufacturing and product lifecycle management in industry 4.0: a survey," *Renewable and Sustainable Energy Reviews*, 132,110112.

Maqbool, R., and S. A. Akubo (2022). "Solar energy for sustainability in Africa: the challenges of socio-economic factors and technical complexities," *International Journal of Energy Research*, 46(12), 16336–16354. https://doi.org/10.1002/er.8425.

Müller, J. M., D. Kiel, and K.I. Voigt (2018). "What drives the implementation of industry 4.0? The role of opportunities and challenges in the context of sustainability," *Sustainability*, 10(1), 247.

Parker, C., S. Scott, and A. Geddes (2019). "Snowball sampling. SAGE research methods foundations," Available via: https://eprints.Glos.ac.uk/id/print/6781. Accessed 23 December 2022.

Turner, C. J., J. Oyekan, L. Stergioulas, and D. Griffin (2020). "Utilizing industry 4.0 on the construction site: challenges and opportunities," *IEEE Transactions on Industrial Informatics*, 17(2), 746–756.

7. Artificial Intelligence as a Contributing Factor for Economic Growth and Sustainable Development among Emerging Economies

Shubhendu Shekher Shukla[1], Kapil Rajput[2], Sachin Tripathi[3], Vinima Gambhir[4], Geetha Manoharan[5], Syed Nisar Hussain Bukhari[6], and Dalia Younis[7]

[1]Assistant Professor, Department of Business Administration, SRM Business School, Lucknow

[2]Assistant professor, Department of Computer Science and Engineering, Graphic Era Hill University, Dehradun, Uttarakhand

[3]Symbiosis Law School, Nagpur, Symbiosis International (Deemed) University (SIU), WATHODA, Nagpur, Maharashtra, India

[4]Associate Professor, Department of Management, ATLAS SkillTech University, Mumbai, Maharashtra

[5]School of Business, SR University, Warangal, Telangana

[6]Scientist – C, National Institute of Electronics and Information Technology, NIELIT Srinagar, J & K

[7]College of International Transport and Logistics, AASTMT University, Egypt

ABSTRACT: Since the relationship between innovation and entrepreneurship is essential for organisational success in today's rapidly altering environment, it has been found that the two are complementary. To ensure the completion of the study and the establishment of a connection between the two forms of entrepreneurship and the organisations. Institutions are considered additional variables and utilise the methodologies used in. The purpose of this chapter is to implement a regulation strategy to lower joblessness, improve dependency on welfare and showcase income disparity. However, it is difficult to achieve these without economic growth influenced by the possibilities that are intelligent, competent, educated, inventive, innovative, all-encompassing, responsive human resources and constructive. Moreover, it was also essential to consider how AI can drive sustainable development and economic growth in emerging economies. The potential challenges and opportunities associated with leveraging technological advancements and the adoption of AI in policy formulation and decision making are also discussed in this research study. To gather pertinent and realistic information from many sources, researchers used a combination of methodologies, such as secondary qualitative methods. Knowledge-based on statistics and theory has been gathered to create a thorough examination of the subject. The research has found that AI plays a crucial role to conserve resources, preserve social well-being, foster innovation and offer sustainable economic opportunities. A significant link between economic growth and entrepreneurship has been found and highlighted along with explaining the importance of entrepreneurship and innovation to achieve ecological sustainability and economic growth.

KEYWORDS: Entrepreneurship, economic growth, innovation, emerging economy, sustainable development, environment

1. INTRODUCTION

Entrepreneurship and innovation administration have become critical for firms' long-term development and growth, regardless of industry or political location. Entrepreneurship is the sole worldwide phenomenon and order of the day in the 21st century since it has the capacity to accelerate countries' economic success and has genuinely become a social as well as a commercial phenomenon throughout the world (Pickernell *et al.*, 2013).

DOI: 10.1201/9781003532026-7

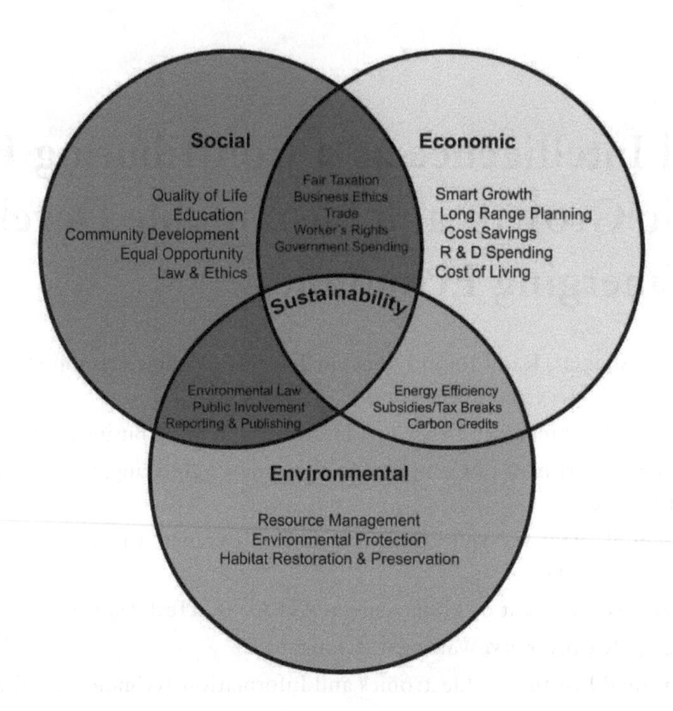

Figure 1. Concept of economic sustainability. Source: Reyad et al. (2018).

A fundamental aspect of Baumol's views was that entrepreneurship and institutions are crucial in understanding the variance in economic growth not explained by variations in factors of production.

The innovation of this chapter's objective is to highlight the critical significance of human capital in the creation of new economic systems, which in turn promote sustained entrepreneurial viability and quality economic expansion.

According to researchers, regulations that fail to discriminate between survivability and high-added-value enterprises may have long-term negative consequences. According to Blackburn and Ram, poorly handled entrepreneurship initiatives result in social alienation rather than inclusion, because new businesses fail to meet their objectives owing to a lack of markets and a supporting framework for social diversity, among other reasons.

1.1. Objectives and Hypotheses

The key objectives of this research study include,

- To identify the current state of emerging economies in terms of adopting AI for economic and sustainable development
- To analyse the significant impacts of AI adoption on economic growth, job creation, GDP growth and productivity in emerging economies
- To evaluate the impact of AI adoption and implementation to promote sustainable practices, social inclusion and environmental conservation
- To identify the key challenges and opportunities associated with AI adoption in emerging economies for economic growth and sustainable development

Hypothesis 1: *Economies that are growing can efficiently enhance economic growth by adopting AI*

Hypothesis 2: *The adoption and implementation of AI promote sustainable development practices in emerging economies*

Hypothesis 3: *There are many challenges and barriers faced by emerging economies to adopt and implement AI effectively for sustainable development and economic growth.*

2. LITERATURE REVIEW

There is a long history of research relating to entrepreneurship and growth, and considerable literature on the structure and economic development has arisen in the last 25 years. However, most research has concentrated on either entrepreneurship or organisations, with little attention paid to the combined impacts of entrepreneurship and institutions on economic growth. This prompts the world to wonder if, as an environment, entrepreneurship and institutions may be the missing link in understanding economic development disparities among countries. The concept is that the more robust the entrepreneurship environment, the more efficient technologies will be, and thus the greater the influence of technology on productivity expansion will be. As a result, entrepreneurs serve as agents who, by commercialising ideas, serve as a transmission mechanism for converting gains in information into economic growth. Even if the entrepreneurial initiative is evident, the institutional context can either hinder or help the transfer process.

North, Scott and Williamson contributed fundamental work to the area of institutional economics, which contends that official (constitutions, laws and regulations) and unofficial (norms, habits and social practise) rules play a crucial role in economic growth. Organisations, according to recent research on economic growth, are a basic driver of economic growth, influencing more proximal factors such as the build-up of physical and human capital (Naustdalslid, 2014).

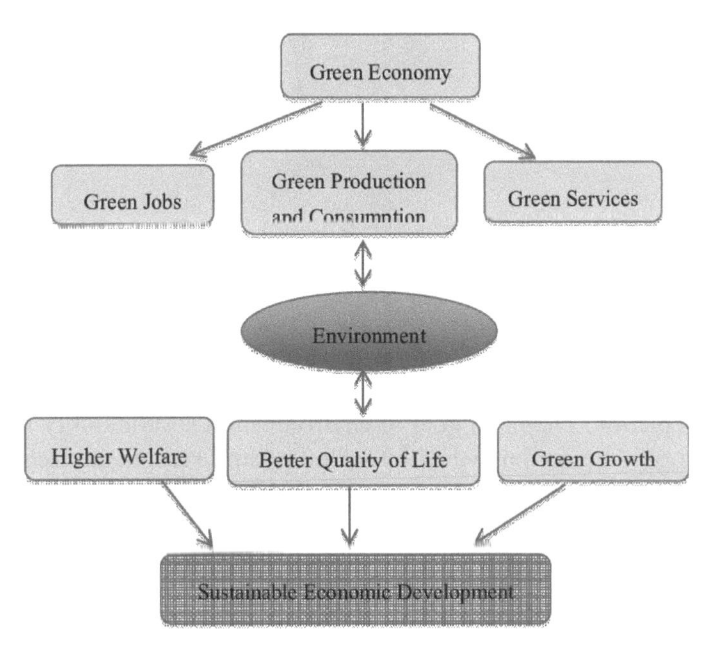

Figure 2. *Sustainable entrepreneurship as a contributing factor for economic growth.*
Source: Naustdalslid (2014).

Identifying the link between social capital utilisation and human resources is an essential subject of future investigation from a theoretical standpoint. Experts have suggested that improving current knowledge of the significance of social capital, human resources and social ties in the formation of new entrepreneurial networks, as well as discovering the best ways to promote them, is a key priority for developing entrepreneurial study. Given the emphasis on the standard of living, enterprises must balance sustainability features with social life components, especially social entrepreneurship and innovation have been highlighted as vital to defusing sustainability requirements.

The topic, which is of enormous interest to academics, the research community and policymakers, is obviously vast and complicated, and it must be handled from several angles (Apostolopoulos *et al.*, 2018). The chapter cannot and will not attempt to address all concerns; instead, it will serve as a starting point for debate, examining some of the numerous ways in which social entrepreneurship, creativity and structural environment are utilised in the framework of sustainable development. This topic is of significant relevance when viewed as a musical reality, because all civilisations, irrespective of their orientation, are still striving for the correct solutions that lead to the implementation (regional, state-wide, national and international). In this regard, scholars from all around the world are looking at how social entrepreneurship, advancements and structural environment might help accelerate the development of sustainable growth.

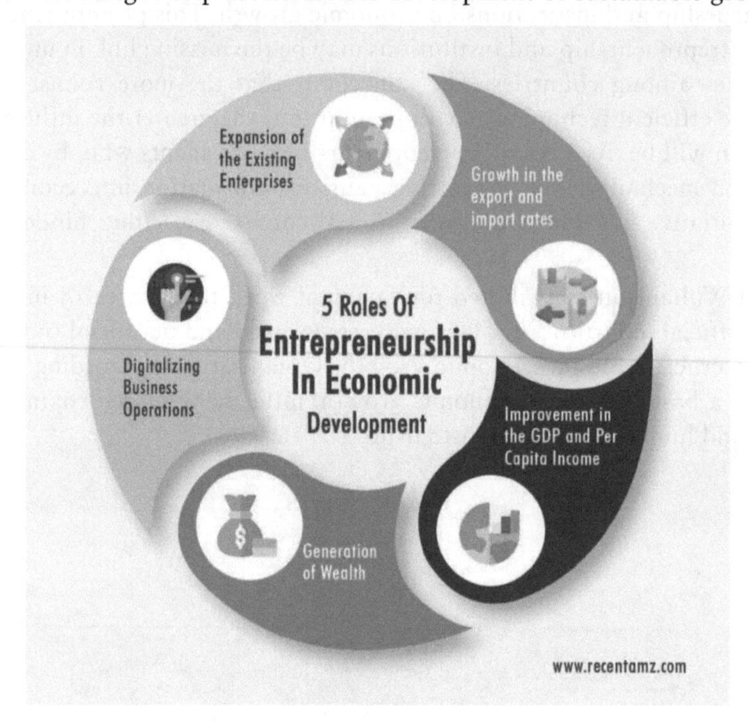

Figure 3. *Roles of entrepreneurship in economic development. Source: Alfaro et al. (2019).*

The idea of resource conservation for succeeding generations is one of the key components that sets sustainable development policies apart from traditional economic planning, which also seeks to consider the effects of resource depletion. The main goal of environmental sustainability is the long-term viability of the industry; this can only be accomplished by including and recognising economic and social issues throughout the decision-making process (Hysa *et al.*, 2020). Both social entrepreneurship as well as economic development are seen as viable options for ensuring the society's long-term growth. Still, the study issue of social entrepreneurship as well as sustainable development is comparatively new in the field of scientific fascinations, having emerged particularly in the previous two decades, with just a few articles in the field of sustainable development and entrepreneurial activities prior to 2010 (Cao and Shi, 2021). To tackle the significance of entrepreneurship development in addressing social challenges, the idea of sustainable growth has been introduced as an all-encompassing term. Moral conscience, economy, constant improvement, knowledge production and use, creativeness, dynamism and the pursuit of corporate gains that create social value are all characteristics of sustainable development (Khan *et al.*, 2019).

3. RESEARCH METHODOLOGY

This study explains how economics and entrepreneurship play a role in the sustainable growth of emerging economies. The goal of advanced economies is to increase growth while supporting producers as they switch from linear to circular economies through government assistance. As a result, waste products in manufacturing facilities are recycled or utilised again, which enables the no-waste approach to increase the effectiveness of using finite resources. To acquire pertinent and accurate data about the subject, researchers have taken into account a variety of ways. In this situation, statistical and theory-based data are managed according to the issue using qualitative methodologies. In addition, academics have looked at secondary qualitative processes in addition to primary quantitative methods to collect theory-based data from various journals and papers. To find journals and articles, search engines and databases like Google Scholar, ProQuest and PubMed are used. Concerning primary quantitative methods, data has been collected through online surveys and questionnaires. Using this method, primary data has been gathered directly from relevant stakeholders in emerging economies including industry experts, policymakers and business leaders. Moreover, it has also provided numerical information and statistical analysis to analyse the relationship between the variables of this research study.

4. ANALYSIS AND INTERPRETATION

In terms of analysis of the data collected from primary quantitative and secondary qualitative research methods concerning the research study about the role of artificial intelligence. All the theory-based data was identified and analysed critically to generate valuable insights relevant to the research topic. Based on the analysis, it has been found that artificial intelligence is one of the key drivers for economic growth and sustainable development more specifically in emerging economies. Along with revolutionising every industry, it also helps in enhancing productivity to address societal challenges. One of the key contributing factors in AI technologies is automation and machine learning which has significantly improved efficiency as well as productivity across industries. Based on the interpretation of various resources, it has been estimated that within the next few years by 2030, AI will contribute more than 15.7 trillion to the global economy. Apart from that it will also create lots of job opportunities in emerging economies. It has been estimated that more than 11 million AI-related jobs will be created by the end of 2025. Moreover, the potential opportunities and risks associated with artificial intelligence in economic and sustainable development in emerging economies are described in the below table.

Table 1: Opportunities and threats associated with AI technologies (Cao and Shi, 2021).

Opportunities	Threats
• It creates and introduces new products and business models	• Obsolescence of the traditional methods for economic growth
• It automated the core business process to improve productivity	• It increases the division between digital and technologies
• It helps human capital development	• It disrupted traditional job functions
• It fuels innovation	• It creates concern for public trust, security and privacy

5. DISCUSSION AND FINDINGS

Based on the findings above, researchers can conclude that sustaining economic opportunity, safeguarding the long-term importance of social life, conserving resources for future generations and entrepreneurial ventures (including social entrepreneurial, innovative thinking and organisational elements) all have significant economic benefits on ecological sustainability (Khan *et al.*, 2019). In the literature study on economic expansion, one of the characteristics that play an essential influence is entrepreneurship (Khan *et al.*, 2019). When businesses enhance their economic development, business model and innovations become more vital; as a result, innovation is one of the most important elements in their operations, having a substantial direct and indirect influence on total economic growth. Entrepreneurship is also seen as one of the factors that might contribute to economic progress (Saberi and Hamdan, 2019). Entrepreneurial activity, according to researchers, produces money, which expands the market, resulting in more revenue, new market possibilities and the establishment of market niches. As a result, entrepreneurship and economic expansion are inextricably linked (Saberi and Hamdan, 2019).

6. CONCLUSION

Since its inception, entrepreneurship research has progressed at a rapid pace. On the one hand, some researchers have examined the variables that stimulate business activities, as per the literature reviewed in this article. Entrepreneurship research, on the other hand, has concentrated on the impacts of new firm formation. The behavioural, organisational, institutional and economic aspects of the first issue have all been investigated. An institutional or economic framework might be used to investigate the second issue. All the major findings of this research study were also discussed which indicated AI plays a crucial role in developing sustainable opportunities, fostering entrepreneurship, driving economic growth and conserving resources. In emerging economies, entrepreneurship can be considered one of the important factors of economic growth. Hence, both entrepreneurship and innovation are essential to promote sustainability and drive economic growth.

REFERENCES

Alfaro, E., F. Yu, N. U. Rehman, E. Hysa, and P. K. Kabeya (2019). "Strategic management of innovation," *Proceedings of the Routledge Companion to Innovation Management*, Routledge, London, UK, 2019, pp. 107–168.

Apostolopoulos, N., H. Al-Dajani, D. Holt, P. Jones, and R. Newbery (2018). *Entrepreneurship and the Sustainable Development Goals*. Emerald Publishing Limited.

Cao, Z., and X. Shi (2021). "A systematic literature review of entrepreneurial ecosystems in advanced and emerging economies," *Small Business Economics*, 57(1), 75–110.

Hysa, E., A. Kruja, N. U. Rehman, and R. Laurenti (2020). "Circular economy innovation and environmental sustainability impact on economic growth: an integrated model for sustainable development," *Sustainability*, 12(12), 4831.

Khan, S. A. R., A. Sharif, H. Golpîra, and A. Kumar (2019). "A green ideology in Asian emerging economies: from environmental policy and sustainable development," *Sustainable Development*, 27(6), 1063–1075.

Naustdalslid, J. (2014). "Circular economy in China–the environmental dimension of the harmonious society," *International Journal of Sustainable Develepment and World Ecology*, 21, 303–313.

Pickernell, D., J. Senyard, P. Jones, G. Packham, and R. Ramsey (2013). "New and young firms: entrepreneurship policy and the role of government – evidence from the Federation of Small Businesses survey," *Journal of Small Business and Enterprise Development*, 20(2), 358–382.

Reyad, S., S. Badawi, and A. Hamdan (2018). "Entrepreneurship and accounting students' career in arab region: conceptual perspective," *The Journal of Developing Areas*, 52(4), 283–288.

Saberi, M., and A. Hamdan (2019). "The moderating role of governmental support in the relationship between entrepreneurship and economic growth: a study on the GCC countries," *Journal of Entrepreneurship in Emerging Economies*.

8. Detailed Investigation of Digital Transformational Strategies to Enhance Business Productivity

M. Sangeetha[1], Richa Gupta[2], Karnati Saketh Reddy[3], Ch. Sudipta Kishore Nanda[4], B. Raja Mannar[5], Geetha Manoharan[6], and Ivanenko Liudmyla[7]

[1]Assistant Professor in Commerce, Shri Shankarlal Sundarbai Shasun Jain College, University of Madras, Chennai, Tamil Nadu

[2]Assistant Professor, Department of Computer Science and Engineering, Graphic Era Hill University, Dehradun, Uttarakhand

[3]Assistant Professor, PG Department, IIBS Bangalore

[4]Assistant Professor – II, Commerce, School of Tribal Resource Management, Kiss Deemed To Be University, Higher Education Campus, Campus - 3, Bhubaneswar - 24, Odisha

[5]Lecturer - Business Administration, St. Theresa Intl College, Thailand

[6]School of Business, SR University, Warangal, Telangana

[7]Chernigiv Post Graduate Pedagogic Institute of K.D. Ushinsky, Ukraine

ABSTRACT: The need for greater research on digital transformation has become apparent, and scholars are actively interested in continuing to do so. People had to overcome many obstacles at first to grow and improve the digitisation process, but as time went on, the business and management completely depended on it. The report's primary goal is to examine and evaluate the key results and the effects of the digital transformational plan on the business's productivity rate. There are basically two main objectives of this research, first is to identify the current state of digital transformation across organisations and second is to analyse various crucial aspects of digital transformation to enhance business productivity. In terms of research methodology, a multi-method approach has been adopted. Primary data were collected from surveys and secondary data were collected from previous studies. By evaluating various aspects of digital transformation, the research study has provided a comprehensive understanding of effective strategies and practices for organisations to maximise productivity in this digital era. Whereas findings have provided practical guidance to improve digital transformation journey.

KEYWORDS: Business and management, AI, digital transformational strategy digitalisation

1. INTRODUCTION

The concept of digital transformation in this modern era has become the foremost trend, as not only the industrial business but also every field and industry were transforming into a digital platform. The reason behind the core changes in traditional business platforms to a digital platform is COVID-19, the global pandemic hit the world so hard, that there has been seeing certain changes in human life. People found the digital platform more reliable and significant as compared to the traditional platform. Talking about the business sector, the role of digitalisation in the corporate world has to be appreciated as it helps to enhance the structure of the business more significantly (Rêgo *et al.*, 2021).

Here, in this study, the discussion has been done based on the strategies related to digital transformation to improve business productivity. The digital transformational strategy has been enacted to strategize the action on how they should perform and function in the digital platform. In order to enhance the productivity level digital strategy, utilise collaborative performance, as it connects people from various regions and countries to operate in one platform (Wessel *et al.*, 2021).

DOI: 10.1201/9781003532026-8

1.1. Objectives and Hypotheses

- To evaluate the current state of digital transformation in organisations across different industries
- To examine how digital transformation can contribute to improving business productivity
- To identify the various critical aspects of digital transformation to drive organisational success
- To explore various challenges and opportunities associated with implementing digital transformation strategies and practices

Hypothesis 1: *Digital transformation can enhance business productivity*

Hypothesis 2: *Process Automation, data analytics and supply chain optimisation can drive business productivity*

2. LITERATURE REVIEW

2.1. Concept of Digital Transformational Strategies

The business management uses the 'digital transformational strategy' as a tool to assess the company's commanding advantage. If a lack of company productivity is noticed at certain stages, the manager might also take appropriate strategic initiatives. The 'digital transformative approach' can be used to successfully entice customers.

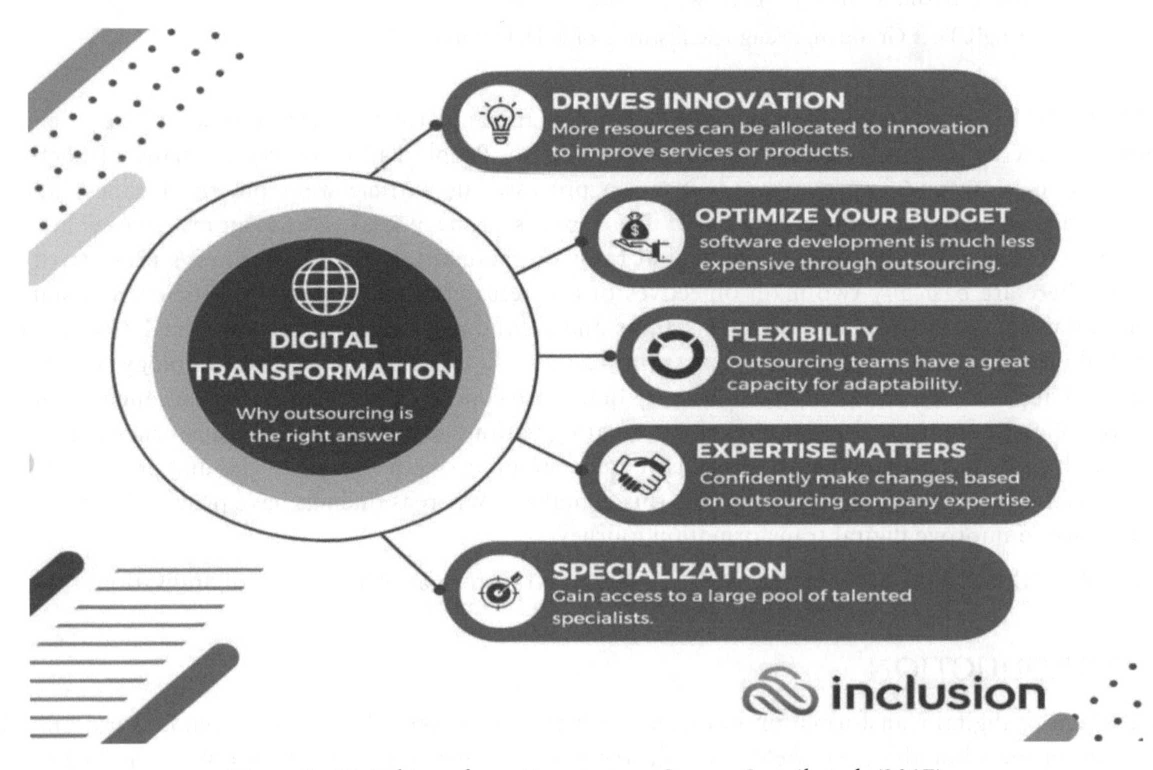

Figure 1. *Digital transformation strategy. Source: Ismail et al. (2017).*

It is necessary to initiate an innovative approach to ensure creativity in the business. The management can also improve the financial proceedings in this process. However, the transformational strategy is different from the traditional approach of the business. Implementation of proper technology is beneficial in this aspect (Ismail *et al.*, 2017).

To enhance productivity at desired level, it is necessary for the organisational management to ensure digitalisation in the business. They can also monitor the improvement in the business by implementing this strategy. The business analysis also signifies that the 'Digital transformational strategy' can also improve teamwork of an organisation.

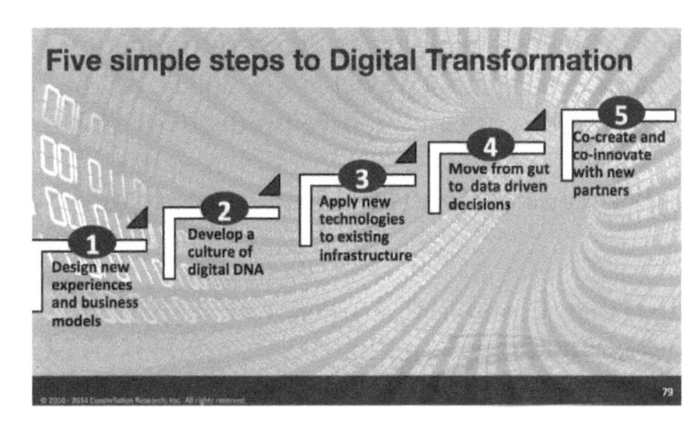

Figure 2. *Steps to initiate transformation. Source: Bouwman et al., (2019).*

2.2. Various Characteristics of Digital Transformational Strategy

There are some relevant characteristics of 'digital transformational strategy'. The major characteristics are discussed below:

Focus on customer experience: Most of the company's focus on customer experience to examine the digital transformation importance in the business. It is a crucial factor for detecting business sustainability. It is necessary to ensure customer satisfaction to enhance profitability (Bouwman *et al.*, 2019).

Evaluation of operational process: It can be beneficial to evaluate the operational process of the business critically before implementing 'digital transformation'. In the present scenario, the authority needs to determine the market fluctuation to maintain the operational process of the business. It plays a crucial role to meet the demands of the customers.

Integration between business data and process: It is essential to initiate integration between organisational data and business process to implement further strategy for the business (Ardi *et al.*, 2020). In this scenario, the authority has to make relevant decisions regarding the business tactics to improve productivity.

Improving value of the business: 'Digital transformation' can play a significant role in enhancing the business value. It helps the business to mitigate all the risks associated with the organisational activities. The authority can also develop a new and effective 'business model' by implementing this strategy.

2.3. Impact of Digital Transformation Strategy for Enhancing Business Productivity

The 'digital transformation strategy' helps the business to make effective decisions to reach the business goals. It results in enhancing the efficiency of the business. It has been detected that the business management focuses on 'higher success rate' by implementing developed technology. The financial management can also ensure monetary efficiency of the organisation through proper usage of 'digital technology'.

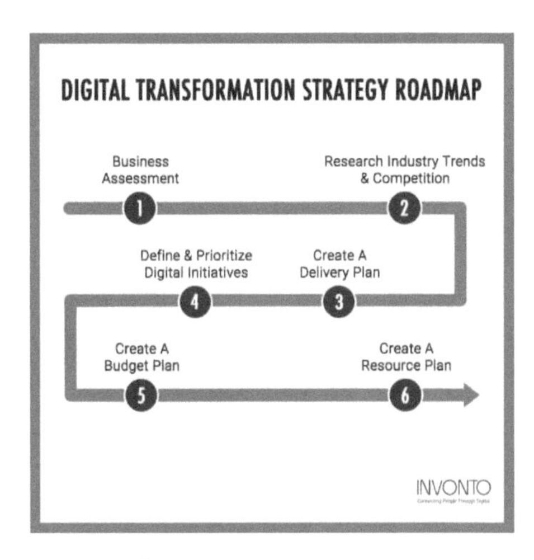

Figure 3. *Digital transformation strategy. Source: Rêgo et al. (2021).*

Based on the analysis, the financial management has to detect the monetary efficiency of the business after implementing 'digital transformation strategy' (Wessel *et al.*, 2021). As the management can access more resources from organisational operations, it can improve productivity of the business.

2.4. Advantage and Disadvantage of Digital Transformational Strategy

2.4.1. Advantage

The digital transformational strategy sustains the benefits of the workforce and increased the level of productivity (Pawełoszek, Wieczorkowski, Czarnacka-Chrobot, 2023). Nowadays, working remotely has become the most relevant thing to increase the rate of productivity as well as mental wellness. Enabling digital transformational strategy in the business evolves to adopt the latest technologies, more likely as online apps, new gadgets, artificial intelligence and various others (Chen *et al.*, 2021).

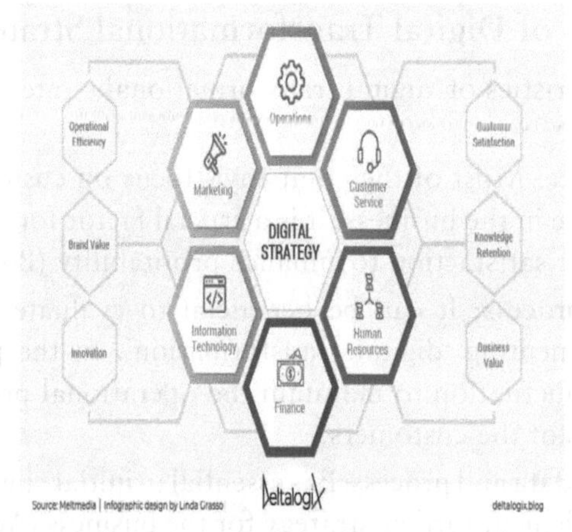

Figure 4. *Digital strategy. Source: Gomez-Trujillo et al. (2021).*

2.4.2. Disadvantage

Although digital transformation is a great platform to expand the business it also has negative sights as the report says that a maximum of the digital transformation projects has been failed. In the year 2018, most of the digital transformation programs has been failed (Gomez-Trujillo *et al.*, 2021). The ***resistance of the employees***, as the employees resist as they feel that their workplace is in danger and their skills also become irrelevant as it does not match with the mode of digital platform. Thus, it generates a negative impact on the workplace and also on other teammates and ***lowers the efficiency of productivity***.

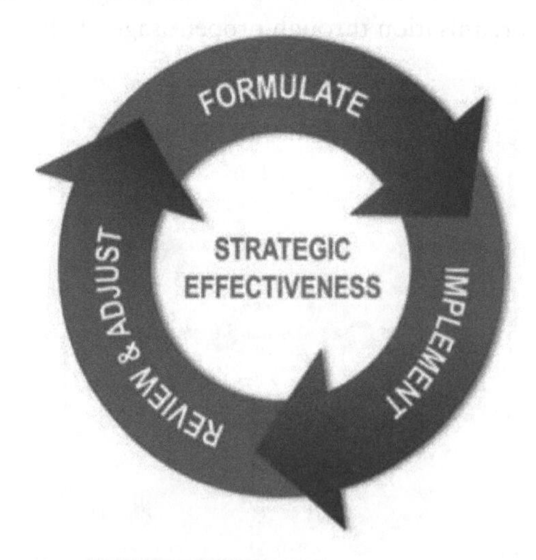

Figure 5. *Effective strategy. Source: Ziyadin et al. (2019).*

According to Ziyadin *et al.* (2019), digital transformation plays an important role for developing a sustainable business.

3. RESEARCH METHODOLOGY

The researchers can gather 'auxiliary information' regarding 'digital transformation strategy' implemented by the organisational management to initiate the research. In the present scenario, the researchers can initiate the 'examination approach' to gather relevant and trustworthy information related to the work. The initiative taken by the researchers can play a significant role in gathering auxiliary information and relevant information regarding 'digital technology'. It can provide some crucial information regarding changes in business processes in recent times. Usage of 'auxiliary information' is also necessary to develop relevant research strategies. In terms of data collection, both primary and secondary research methods have been used. Regarding primary data collection, online surveys have been conducted by researchers. The survey included relevant questions about digital transformation practices. Whereas, using a secondary data collection method, data were collected from various literatures, reports, journals and other online resources (Moreira *et al.*, 2018). This method is used to collect existing knowledge and theories about digital transformation. For quantitative data collected from online surveys, statistical software was used for data analysis.

4. ANALYSIS AND INTERPRETATION

In order to improve productivity to streamline organisational processes, digital transformational strategies play a vital role. Leveraging technology not only improves efficiency but also drives innovation within businesses. Using Cloud computing organisations can improve organisational computer resources to improve flexibility, collaboration and save costs. The market of cloud computing has been growing significantly and it will reach $567 billion by 2030 (Ardi *et al.*, 2020).

1
New Business Models
2
Enhance Customer Experiences
3
Modernizing IT Infrastructure
4
Operational Efficiency
5
Optimized Employee Skillset
6
External Partner Collaboration
7
Data-Driven decision-making

Figure 6. *Key drivers of digital transformation. Source: Ardi et al. (2020).*

It has been analysed that the global IoT market will reach the milestone of $1.5 trillion in 2024. On the other hand, e-commerce platforms provide solutions to businesses to reach a broader market and improve customer experiences. Within the last few years, global sales in e-commerce have increased from $4.38 trillion in 2020 to $5.4 trillion in 2022 (Wessel, 2022).

Based on the responses collected from surveys, it has been analysed that, currently organisations are at different levels of digital transformation across different industries. Some organisations have successfully embraced digital transformation by integrating the latest technological advancements into business operations.

Table 1: Communication or learning strategy archetype Chen *et al.* (2021).

Organisational Objectives	Communication/Learning	Unification/Optimisation	Automation
Governance	Bottoms-up	Supporting	Top-down
Management support	Observing	Hybrid	Involved
Interaction model	Guiding	Restrictive	Top-down
Education	Open	Selective	Individual
Tool support	Collaborative	Hybrid	Rigid

5. DISCUSSION AND FINDINGS

The business industry is wholeheartedly evolving the digitalisation in their business and making things more appropriate for the customers also by applying relevant digital strategies. The global pandemic has asserted the phenomenon and transferred the threat into an opportunity (António *et al.*, 2021). The development in digital transformation introduced a new process and display a new form of mechanism that generates an effect on the manufacturing process of the business. Digital transformation is nothing but a way that aims to develop significant changes in the business by contributing to the enhancing mode of digitalisation. The process of digital transformation renewal the use of advancement in technology and build the capabilities of the business organisation (Garcia-Garcia *et al.*, 2021).

6. CONCLUSION

The usefulness of 'digital technology' has been the subject of a thorough discussion in this regard. The main goal of this research is to ascertain whether a 'transformational strategy' has to be implemented to increase corporate productivity based on the 'business trend.' In the current situation, it is advantageous for organisational management to launch an innovative technique to more effectively gather business data. Thus, this research has successfully highlighted the role and impact of digital transformation on business productivity. This report has adopted a mixed-method approach to collect valuable insights regarding the research topic.

7. FUTURE SCOPE

It has been detected that 'digital transformation strategy' can play a crucial role from a business perspective. The researchers can use the information from the research to initiate future research. The research has not only fulfilled the monetary aspects of the business but has also developed the future scope of the study. The 'extensive plan' developed for the research can be beneficial for future researchers. Based on the analysis, the researchers can detect the study as highly prosperous for the future work.

REFERENCES

António, N., and P. Rita (2021). "COVID-19: the catalyst for digital transformation in the hospitality industry?" *Tourism & Management Studies*, 17(2), 41–46.

Ardi, A., S. P. Djati, I. Bernarto, N. Sudibjo, A. Yulianeu, H. A. Nanda, and K. A. Nanda (2020). "The relationship between digital transformational leadership styles and knowledge-based empowering interaction for increasing organisational innovativeness," *International Journal of Innovation, Creativity and Change*, 11(3), 259–277.

Bouwman, H., S. Nikou, and M. de Reuver (2019). "Digitalization, business models, and SMEs: How do business model innovation practices improve performance of digitalizing SMEs?" *Telecommunications Policy*, 43(9), 101828.

Chen, C. L., Y. C. Lin, W. H. Chen, C. F. Chao, and H. Pandia (2021). "Role of government to enhance digital transformation in small service business," *Sustainability*, 13(3), 1028.

Garcia-Garcia, J. A., C. A. Maldonado, A. Meidan, E. Morillo-Baro, and M. J. Escalona (2021). "gPROFIT: a tool to assist the automatic extraction of business knowledge from legacy information systems," *IEEE Access*, 9, 94934–94952.

Gomez-Trujillo, A. M., and M. A. Gonzalez-Perez (2021). "Digital transformation as a strategy to reach sustainability," *Smart and Sustainable Built Environment.*

Ismail, M. H., M. Khater, and M. Zaki (2017). "Digital business transformation and strategy: what do we know so far," *Cambridge Service Alliance*, 10, 1–35.

Moreira, F., M. J. Ferreira, and I. Seruca (2018). "Enterprise 4.0–the emerging digital transformed enterprise?" *Procedia Computer Science*, 138, 525–532.

Pawełoszek I., Wieczorkowski J., Czarnacka-Chrobot B. (2023), Digital Transformation of Polish Micro-enterprises: Lessons from the COVID-19 Era, 27th International Conference on Knowledge-Based and Intelligent Information & Engineering Systems (KES 2023), Procedia Computer Science, Vol. 225, pp 1572-1581, doi:10.1016/j.procs.2023.10.146

Rêgo, B. S., S. Jayantilal, J. J. Ferreira, and E. G. Carayannis (2021). "Digital transformation and strategic management: a systematic review of the literature," *Journal of the Knowledge Economy*, 1–28.

Wessel, L., A. Baiyere, R. Ologeanu-Taddei, J. Cha, and T. Blegind-Jensen (2021). "Unpacking the difference between digital transformation and IT-enabled organizational transformation," *Journal of the Association for Information Systems*, 22(1), 102–129.

Ziyadin, S., S. Suieubayeva, and A. Utegenova (2019). "Digital transformation in business," *International Scientific Conference "Digital Transformation of the Economy: Challenges, Trends, New Opportunities"* (pp. 408–415). Springer, Cham.

9. An Empirical Analysis of Machine Learning (Ml) for Developing Sustainable Agricultural Business and Supply Chain Performance

Nazneen Pendhari[1], Amit Gupta[2], Barinderjit Singh[3], Rajesh Deb Barman[4], Kousik Boro[5], Syed Nisar Hussain Bukhari[6], and Leszek Ziora[7]

[1]Assistant Professor, Computer Engineering Department, University of Mumbai, Mumbai, Maharashtra

[2]Assistant Professor, Department of Computer Science and Engineering, Graphic Era Hill University, Dehradun, Uttarakhand

[3]Department of Food Science and Technology, I.K. Gujral Punjab Technical University Kapurthala, Punjab, India 144601

[4]Assistant Professor, Department of Commerce, Bodoland University Kokrajhar BTR, Assam, India 783370

[5]Research Scholar, Department of Commerce, Bodoland University Kokrajhar, Assam, India

[6]Scientist – C, National Institute of Electronics and Information Technology, NIELIT Srinagar, J & K

[7]Faculty of Management, Czestochowa University of Technology, Poland

ABSTRACT: It is necessary to provide food security and protection in addition to generating a lot of crops. As a result, the engineers are concentrating on machine learning (ML) technologies that will offer an agricultural field that can be monitored in real-time. This research study has adopted both secondary qualitative and primary quantitative methods. Regarding secondary qualitative, various books, journals and research papers were reviewed. A survey was conducted to gather primary quantitative data. The research outcomes suggested that ML has immense potential to improve supply chain management and embrace sustainable agriculture. Various aspects of ML approaches were identified that directly contribute to agricultural business and supply chain management. Moreover, this research study will provide a brief insight into the applications of ML in the context of sustainable agriculture. By focusing on empirical analysis, this research has provided some evidence of the successful application of ML in agriculture. Moreover, the challenges and opportunities in the implementation of ML were also discussed regarding sustainable agriculture business and improving supply chain management.

KEYWORDS: ML, precision agriculture, sustainable agricultural business, demand and supply, food-security, agricultural supply chain

1. INTRODUCTION

The world population is getting larger day by day and thus, to ensure food security, a sustainable agricultural process is of utmost importance. According to the prediction, it is estimated that the total population of the world will rise to 9–10 billion. Therefore, it's essential to increase the food production rate by 70–110% in the future to eliminate the eradication of hunger and ensure *food security* (Liakos *et al.*, 2018). Scientists suggest that the conventional food production strategy is not as beneficial as the new technical approaches. The sustainable approach to producing foods not only focuses on the production; however, it will help to eliminate the '*environmental impact*' on the crops (Sharma *et al.*, 2020). For instance, many crops are destroyed by sudden rain, pest attacks and plant diseases.

DOI: 10.1201/9781003532026-9

1.1. Objectives and Hypotheses

- To evaluate the role of ML in promoting sustainable agriculture business and managing the supply chain.
- To analyse the critical impacts of machine learning (ML) on resource management, sustainability and agricultural productivity
- To explore the challenges as well as opportunities associated with the ML implementation in supply chain and agricultural business
- To provide appropriate recommendations for best practices and strategies to integrate ML into supply chain management and sustainable agricultural business

Hypothesis 1: *There is a significant impact of ML techniques to improve supply chain performance and agricultural business*

Hypothesis 2: *The integration and adoption of ML can contribute to supply chain performance and sustainable agriculture business*

2. LITERATURE REVIEW

Nowadays, the global supply chain practice is stepping towards digital inclusion and as a result, a huge amount of data related to the supply chain is getting generated. The data are also known as 'big data' which can be used to enhance the productivity of the supply chain. However, the structured, semi-structured and unstructured data needs to be analysed efficiently to extract meaningful information. Concerning this, '***Machine Learning*** (ML)', an 'Artificial Intelligence (AI)' technology has gained interest to execute '*big data analytics*', '*machine-to-machine (M2M) communication*', *IoT device operation* and *cloud computing* (Hosseinian-Far *et al.*, 2018). ML technology can be used to carry out 'decision-making' as well as intelligent decision-making; predicting the global market, monitoring supply chain performance (SCP); and to predict the risk and outcome. This research is going to analyse the roles and responsibilities of ML technology to enhance the sustainability of the agricultural business.

2.1. ML in Sustainable Agricultural Business (SAB)

SAB requires real-time monitoring technology to assess and predict external risk. The risk includes supply chain risk, climate change risk, crop disease risk and pest attack risk. Therefore, to execute real-time monitoring of different processes related to agriculture, developers suggest the use of sensors. The use of sensors and connected devices and networks is also known as '***precision agriculture***' (Figure 1) (Sharma *et al.*, 2020).

Figure 1. *Precision agricultural system. Source: Sharma et al. (2020).*

Previously, this system was developed by *John Deere* which allows the farmers to know where to sow seeds and where to apply fertilisers by using a 'Global Positioning System' (GPS) (Yang, 2020). Recently, this system has been developed by several authors by integrating IoT devices, AI technologies, big data analytics and blockchain technology, cloud computing and remote-sensing technology. It helped the farmers to monitor the condition of the soil, crop growth, automating some processes (fruit pick-up by Agro-bot) and many more (Sarri *et al.*, 2019).

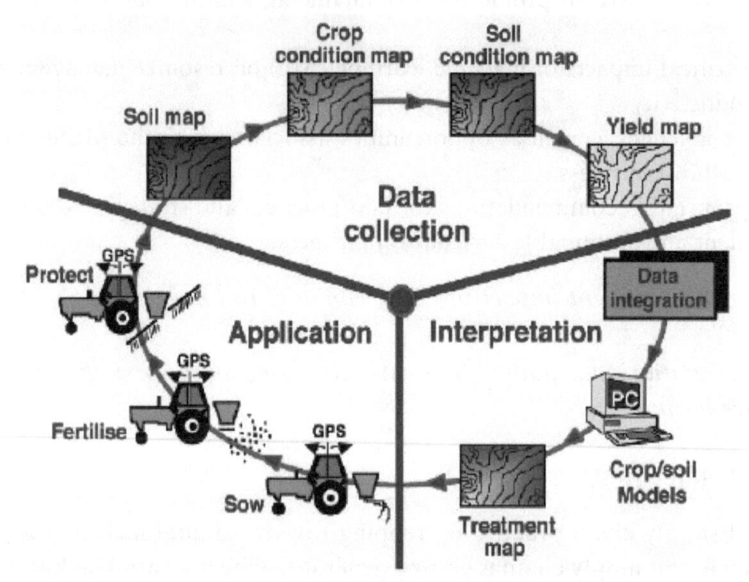

Figure 2. *Configuration of precision agriculture. Source: Sarri et al. (2019).*

2.2. ML in Sustainable SCP

Glória and colleagues used ML algorithms to predict the supply chain risk. The authors also demonstrated that ML algorithms are useful in decision-making, autonomous processes, dealing with uncertainty and environmental issues (Ryu *et al.*, 2020). The process is carried out by the operations shown in Figure 3.

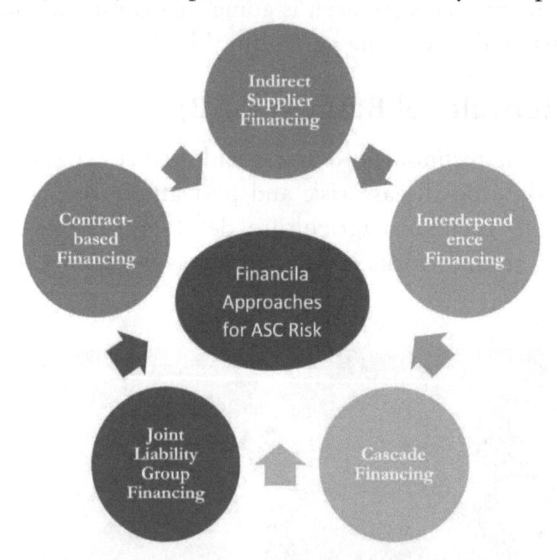

Figure 3. *Operations in 'agricultural supply chain' (ASC). Source: Sarri et al. (2019).*

The figure suggests that the ML is required in managing various stakeholders to carry out the decision-making. For example, ASC consists of consumers, distributors, agencies, farmers, retailers and traders (Oliver *et al.*, 2018).

Concerning this, it is essential to establish an ML-driven '*physical-agricultural*' monitoring system (wireless sensing technology) . The Internet of Things (IoT) devices will allow farmers and other nodes to collect real-time data on the agricultural environment. Below Figure 4 shows the ML system configuration for decision-making in *sustainable agricultural supply chain performance* (SASCP).

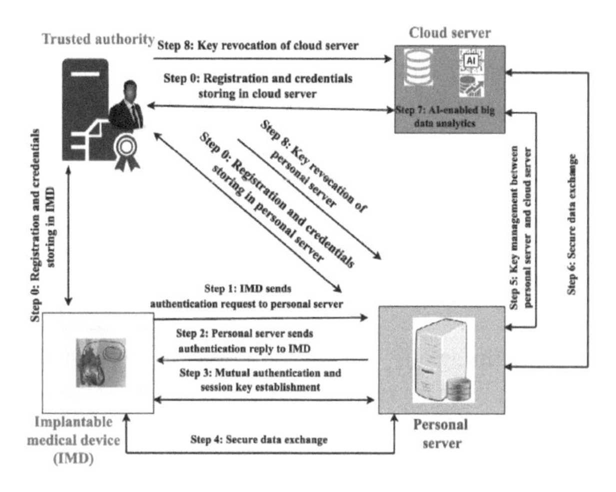

Figure 4. *Configuration of ML in decision-making for SASCP. Source: Ha et al. (2018).*

Several ML algorithms are used in SASCP; such as 'Bayesian networks', Random Forest, Regression analysis and decision trees which are learned by considering 'supervised learning' (Ha *et al.*, 2018). On the contrary, Artificial Neural Network (ANN) and 'Deep Learning' (DL) algorithms are used for unsupervised learning.

3. RESEARCH METHODOLOGY

Data has been collected through the secondary qualitative method as well as the primary quantitative method. A social media post was shared on social media and many respondents agreed to provide responses to the survey questions (Glória *et al.*, 2021). The data were then analysed, and the percentile values were calculated to understand the prevalent options among the provided options. The following section describes the interpretation of the analysed results. Besides, analysing the secondary findings, the research also analyses the various other sources to establish a meaningful and valid discussion (Ip *et al.*, 2018).

4. ANALYSIS AND INTERPRETATION

ML algorithms are effective in maintaining sustainability in agriculture. The survey responses have proved that ML algorithms are effective in reducing the risk of the supply chain, price prediction etcetera. Different ML algorithms such as SVM and ANN are the most effective ML algorithms as they help researchers in getting optimum sustainability in terms of crop production, consumption and transport. Numerous studies have proved that ML algorithms are necessary for 'price prediction', 'stock market prediction', 'inventory management', 'plant disease control', 'pest control', 'weed management', etcetera. Among various digital technologies, ML is extremely important in sustainable agriculture business and in improving supply chain management. In. ML-based yield prediction can achieve an accuracy of 92% .

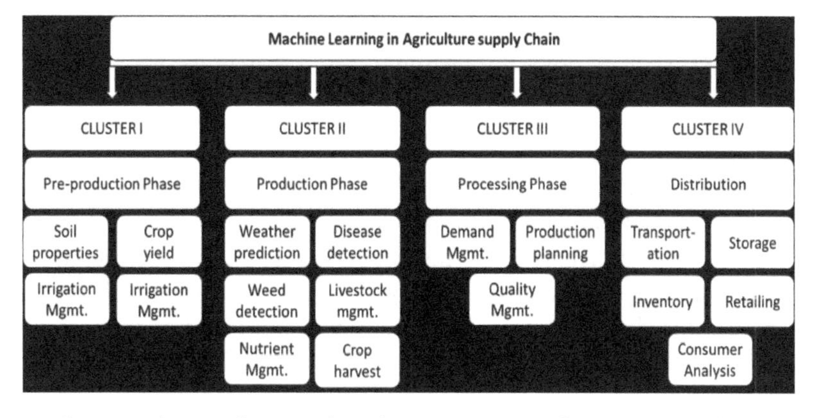

Figure 5. *SLR framework regarding machine learning in agriculture. Source: Sharma et al. (2020).*

Apart from that, ML algorithms are also used for precision agriculture and smart farming. It can be interpreted that the precision agriculture market will reach $12.8 billion by 2025 (Sharma *et al.*, 2020). Moreover, an overall view of ML algorithms in agricultural supply chain management is described using the following table.

Table 1: Agricultural supply–chain Operation (Oliver *et al.*, 2018).

Departments	Supplier	Processor	Distributor	Retailers or customers
Responsibilities	• Production planning • Predicting Supply chain risks • Demand forecasting • Supplier selection • Inventory management • Lead time forecasting • Performance prediction	• Process control • Scheduling • Quality control	• Distribution planning • Food delivery • Transportation	• Customers satisfaction • Consumption prediction • Buying behaviour
Processes	Production planning and control	Food processes	Distribution	Consumption

5. DISCUSSION AND FINDINGS

It has been stated that suitable agriculture can be achieved through ML algorithms used in production planning and control through big data analysis. Similar studies have been conducted by Sharma and other colleagues (Sharma *et al.*, 2020).

They have shown that in all four stages of ASC, are 'Production planning and control', 'food processing', 'product distribution' and 'product consumption' ANN and SVM are the most effective technologies. Also, it had stated that the majority of the surveyors have strongly agreed that ML can predict the risk of the supply chain. According to Sabu and Kumar, ML algorithms can predict the prices of agricultural crops. This helps entrepreneurs in analysing the risk of production and supply chain (Sabu *et al.*, 2020).

6. CONCLUSION

Thus, in the end, the research study can conclude that algorithms of ML have immense potential to improve sustainability and productivity in the agricultural sector. This research paper successfully analyzed the significance of ML in developing sustainable agricultural businesses and improving the performance of the supply chain. Throughout this study, the researchers have followed a systematic approach to meet the research objectives. Both primary quantitative and secondary qualitative research methods were used to collect data to provide valuable insights. Based on the data analysis and interpretation, it has been found that there are various aspects such as precision farming, resource management, supply chain management and demand forecasting which can be addressed by ML techniques of applications.

7. FUTURE SCOPE

Future AI technologies can be used to predict future population growth and according to that, the machines can predict the future need of foods. Apart from this, plant diseases can be predicted to mitigate large crop destruction. Reports suggest that pest attacks and plant diseases cause a major loss in the crop field. Thus, if AI technologies can be utilised to monitor pest attacks and diseases and subsequently can recommend intelligent solutions, food security will improve gradually.

REFERENCES

Glória A., J. Cardoso, and P. Sebastião (2021). "Sustainable irrigation system for farming supported by machine learning and real-time sensor data," *Sensors*, 21(9), 3079.

Ha, U., Y. Ma, Z. Zhong, T. M. Hsu, and F. Adib (2018). "Learning food quality and safety from wireless stickers," *Proceedings of the 17th ACM Workshop on Hot Topics in Networks*, 106–112.

Hosseinian-Far, A., M. Ramachandran, and C. L. Slack (2018). "Emerging trends in cloud computing, big data, fog computing, IoT and smart living," *Technology for Smart Futures* (pp. 29–40). Springer, Cham.

Ip, R. H., L. M. Ang, K. P. Seng, J. C. Broster, and J. E. Pratley (2018). "Big data and machine learning for crop protection," *Computers and Electronics in Agriculture*, 151, 376–383.

Liakos, K. G., P. Busato, D. Moshou, S. Pearson, and D. Bochtis (2018). "Machine learning in agriculture: a review," *Sensors*, 18(8), 2674.

Oliver, A., A. Odena, C. A. Raffel, E. D. Cubuk, and I. Goodfellow (2018). "Realistic evaluation of deep semi-supervised learning algorithms. Advances in neural information processing systems," 31.

Ryu, G. A., A. Nasridinov, H. Rah, and K. H. Yoo, "Forecasts of the amount purchase pork meat by using structured and unstructured big data," *Agriculture*, 10(1), 21.

Sabu, K. M., and T. M. Kumar (2020). "Predictive analytics in agriculture: forecasting prices of Arecanuts in Kerala," *Procedia Computer Science*, 171, 699–708.

Sarri, D., S. Lombardo, R. Lisci, V. D. Pascale, and M. Vieri (2019). "AgroBot Smash a robotic platform for the sustainable precision agriculture," *International Mid-Term Conference of the Italian Association of Agricultural Engineering* (pp. 793–801). Springer, Cham.

Sharma, A., A. Jain, P. Gupta, and V. Chowdary (2020a). "Machine learning applications for precision agriculture: a comprehensive review," *IEEE Access*, 9, 4843–4873.

Sharma, R., S. S. Kamble, A. Gunasekaran, V. Kumar, and A. Kumar (2020b). "A systematic literature review on machine learning applications for sustainable agriculture supply chain performance," *Computers & Operations Research*, 119, 104926.

Yang, C. (2020). "Remote sensing and precision agriculture technologies for crop disease detection and management with a practical application example," *Engineering*, 6(5), 528–532.

10. Development of Technological Advancement, Innovation and Machine Learning and their Influence on Entrepreneurship Business

K. Selvasundaram[1], Rishika Yadav[2], Aruna A[3], Bilal Ahmad Dar[4], Geetha Manoharan[5], Manyam Kethan[6], and Ivanenko Liudmyla[7]

[1]Professor and Head Department of Commerce, CSAF College of Science and Humanities SRM Institute of Science and Technology, Kattankulathur – 603203, Chengalpattu, Tamil Nadu, India

[2]Assistant Professor, Department of Computer Science and Engineering, Graphic Era Hill University, Dehradun, Uttarakhand

[3]Assistant Professor, Department of Information Technology, SNS College of Technology, Coimbatore

[4]Ph.D from Jaipur National University Jaipur, Srinagar, India

[5]School of Business, SR University, Warangal, Telangana

[6]Associate Professor, Department of Management, International Institute of Business Studies, Bengaluru

[7]Chernigiv Post Graduate Pedagogic Institute of K.D. Ushinsky, Ukraine

ABSTRACT: Information technology today significantly influences how new business ideas are thought about and developed. Blockchains, artificial intelligence, cloud computing and 'augmented and virtual reality' are just a handful of technological advancements that are changing how people connect and do business in our increasingly digital environment. It may be said that the evolution of technical innovation, advancement and digital technologies has grown to be a significant factor in the current situation. To collect data for providing essential and valuable insights to answer the research questions, the researchers have only considered a secondary qualitative research approach. A large data set has been collected from various reliable resources such as academic databases, research papers and website articles. This has helped the researchers to gather adequate information regarding the research topic. Moreover, by analysing the research data, it has been found that technology has significantly revolutionised entrepreneurship. Innovation has become an integral part of entrepreneurial success. Machine learning has emerged as the most effective and powerful tool to analyse data, optimise various business processes and automate processes for entrepreneurs.

KEYWORDS: Technological advancement, digital technologies, entrepreneurship business and innovation

1. INTRODUCTION

The primary focus of this particular research chapter will be on developing a critical debate while taking into account the advancement of technological innovation and digital technologies and their effects on entrepreneurship business (Beliaeva *et al.*, 2019). It was acknowledged that technological innovation increases economic output and enables the supply of new goods and services with the potential to improve efficiency and human lives. Significant use of a literature study will be made to comprehend this particular research issue and recognise it in a more critical approach. Even proper research methodologies will be used to enrich the study and make it more real and meaningful. Regarding the subject of the research, it was acknowledged that the pressure on business owners to continuously use new technology to remain competitive in the market over the long term has increased due to the intensifying market competition and ongoing development of new technology.

DOI: 10.1201/9781003532026-10

1.1. Objectives and Hypothesis

The key objectives of this research study include,

- To evaluate the impact of the latest technological advancements on entrepreneurship
- To explore the current role of innovation and technology to ensure that the success of entrepreneurship
- To analyse the significant effects of machine learning on various entrepreneurship practices
- To identify and discuss the potential opportunities as well as challenges in implementing technology, machine learning and innovation in entrepreneurship

Hypothesis 1: *There are positive influences of technological advancements on the growth and success of entrepreneurship business*
Hypothesis 2: *The integration of technology, machine learning, and innovation helps entrepreneurs in decision-making, broader market reach and optimise business processes.*
Hypothesis 3: *Collaboration and ecosystem, both play a crucial role to foster entrepreneurship growth, innovation and knowledge sharing.*

2. LITERATURE REVIEW

It may be said that technical innovation and advancement have made it possible for entrepreneurship to create new guidelines and ways of conducting business in a cutthroat atmosphere. The invention makes it possible to put fundamental findings and interventions into practice to create a useful product and approach. It has been acknowledged that a corporation can best utilise product and service development to attract new clients and retain existing ones with the help of technology improvement and innovation (Elia *et al.*, 2020).

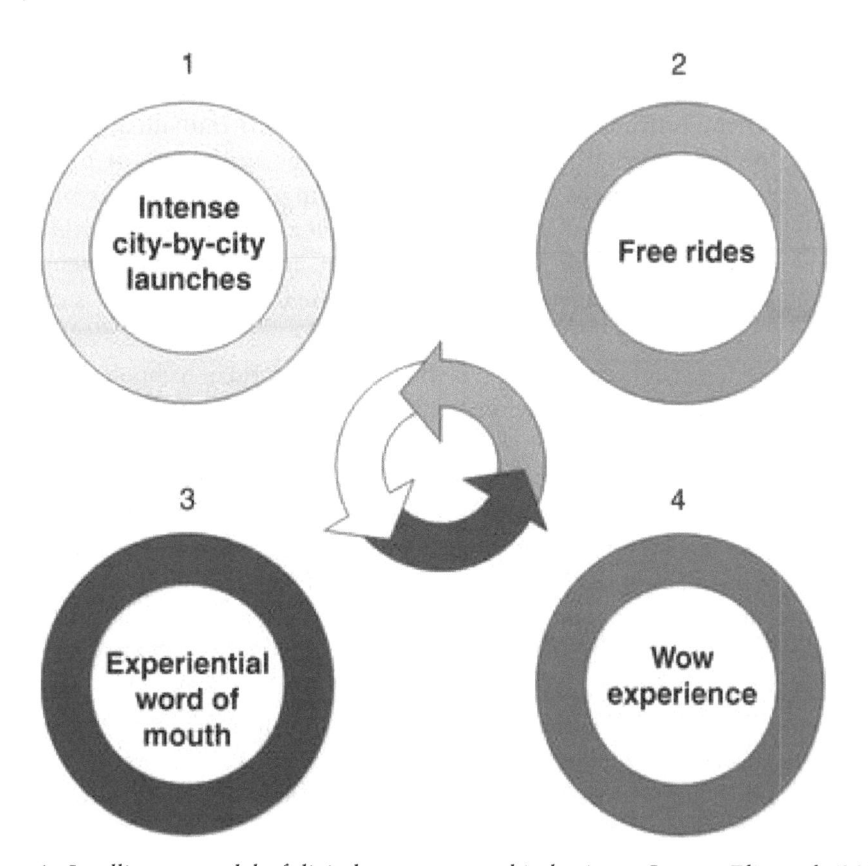

Figure 1. *Intelligence model of digital entrepreneurship business. Source: Elia et al. (2020).*

In the present situation, the continuous development of technology has transformed the overall process of conducting business and has created ease in undertaking many operations in the organisation.

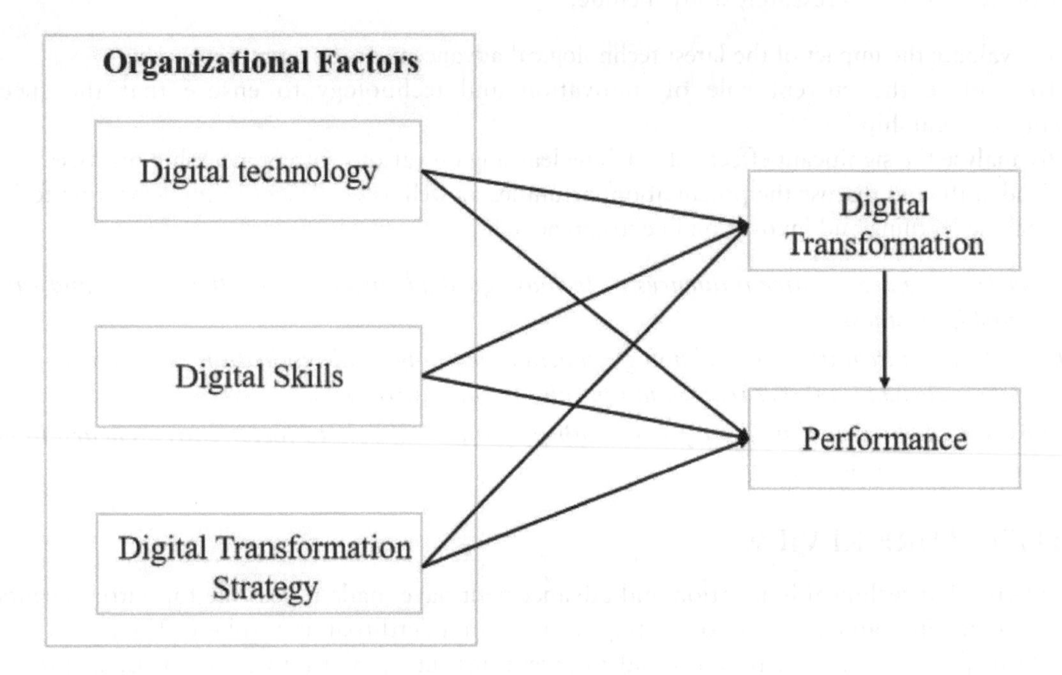

Figure 2. Implementation of digital transformation and impact on entrepreneurship business.
Source: Bican and Brem (2020).

It was recognised that the technology has been able to exceed the efficiency of the system and the services of the organisation. It enables monitoring and streamlining methods as well as managing data flow and contact (Bican and Brem, 2020). Even digital technology has enabled to keep records of staff which can be utilised for monitoring the performance of each staff individually and making required developments in their performance. It can be recorded that the utilisation of digital technology has upgraded the entrepreneurship business. It has been able to upgrade the process of communication as the technology provides a faster and highly efficient process of communicating within the organisation. this incorporates interaction among the organisation's staff, investors, and potential customers. the technology associated with digital incorporating video conferencing technology such as Skype as well as Zoom.

It can be regarded that there has been tremendous change in the world of technology due to the continuous development of technology and innovation (Jelonek 2023). It can be regarded that the technology-enabled organisation's online operation enables an organisation to create significant brand awareness and connect with customers living in different areas of the world. Even it was recognised that technology has facilitated global communication which was highly important in the world of business to enable fast growth. Not to be forgotten are companies having affiliates in other nations. It is possible to conduct corporate seminars and events irrespective of the place of the attendees thanks to software like Skype (News – Alert) (Jain *et al.*, 2019). As a result, the company is in a good situation to grow and prosper. Even productivity and work have increased (George *et al.*, 2018).

Technological reliance may lead to undesirable effects and issues in the most dire of circumstances. Hardware encryption is a state practice that might arise in the technology sector because of the reliance of many companies on specialised software as well as hardware. For some customers, switching from one provider to another is a difficult process. In many cases, companies just accept the supplier's technology's peculiarities and weirdness because it is too hard to alter.

To boost their own profits, a number of technology providers develop new or enhanced goods in a sequence of rapid succession. It is necessary to change operating technology with new versions on a regular basis to stay current with technology. Changes in technology might be advantageous, but the price and work necessary to implement them can put a strain on the firm's activities. In today's world, a hacker is an inevitable part of everyday life. Cyber hackers seem to be stealing the information of another large company almost every day, according to news outlets.

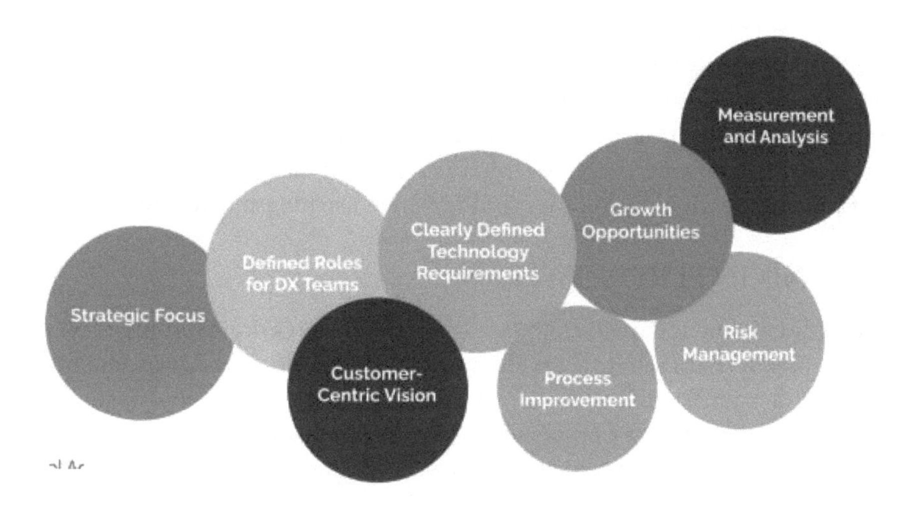

Figure 3. Framework development of digital transformation part 1. Source: Bican and Brem (2020).

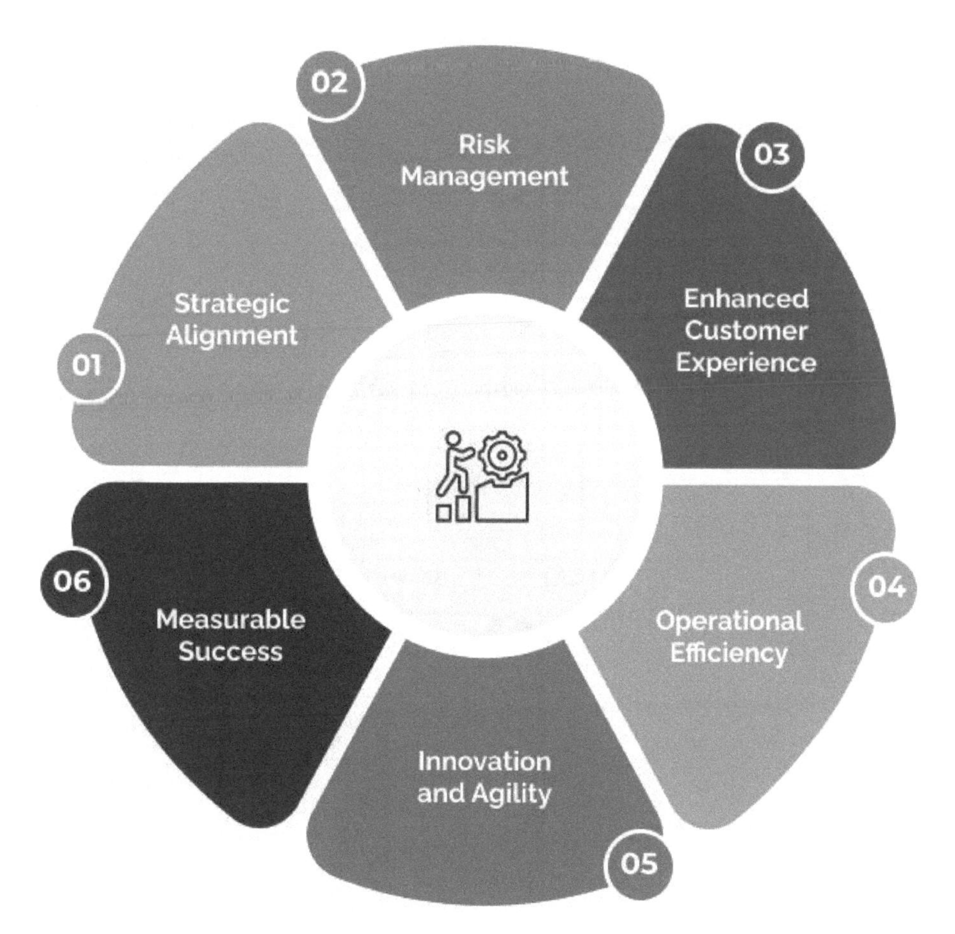

Figure 4. Framework development of digital transformation part 2. Source: Bican and Brem (2020).

3. RESEARCH METHODOLOGY

This particular research work will make significant utilisation of the secondary research process. These particular processes enable to make significant utilisation of information and will also ensure that the conducted research is authentic. Moreover, the researchers used a secondary qualitative research methodology for this study. It entails gathering and studying previously published articles, papers and case studies concerning the growth of innovation, machine learning, and technical improvement in entrepreneurship. To identify important themes and trends, the data are assessed, chosen, and analysed utilising qualitative analytic approaches. Moreover, the results are interpreted, compiled and presented. The methodology places a strong emphasis on the value of using current knowledge to acquire an understanding of how innovation, technology and machine learning impact entrepreneurship.

4. ANALYSIS AND INTERPRETATION

Based on the above discussion and analyzing data collected from various resources related to the research topic, it can be interpreted that the latest developments in technological advancements, machine learning and innovation have a significant influence on entrepreneurship and the business ecosystem. These advancements not only created new opportunities but also helped streamline operations and accelerate business growth. In the last few years, technological advances have led emergence of numerous start-ups. In comparison with 2010, the total number of startups across the world has increased rapidly from 300,000 to 570,000 in 2022.

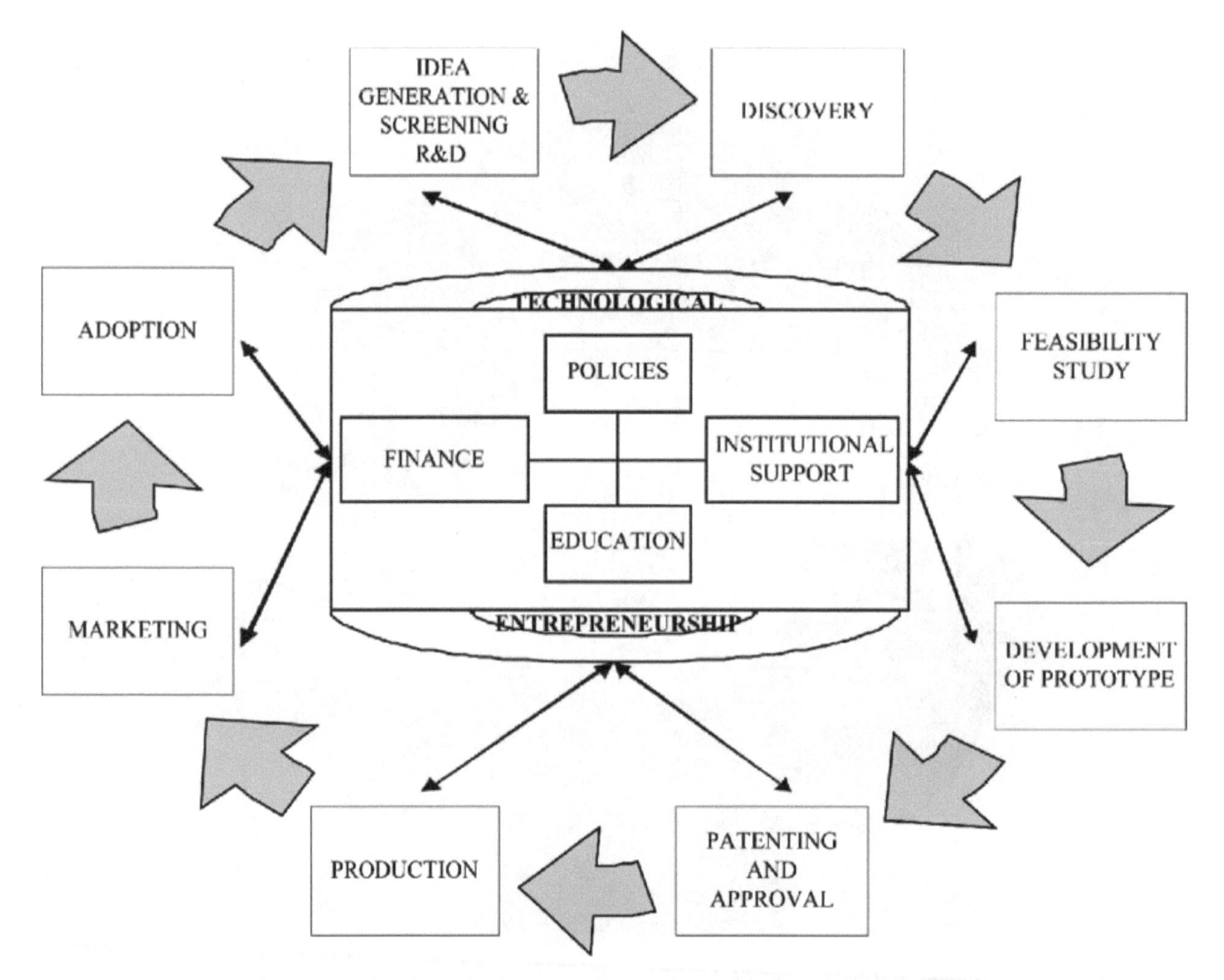

Figure 5. Link between technological advancements and entrepreneurship. Source: Panwar et al. (2021).

Along with this, the rise of investment in innovation is another crucial factor for strategy success. In 2021, venture capitalists have funded over $434 billion in innovative startups. Moreover, it can also be estimated that artificial intelligence and machine learning enhanced business productivity by 40% and will increase further. In addition to that, technological advancements also provide competitive advantages to businesses by leveraging data-driven decision-making. Whereas predictive analytics empower entrepreneurs to forecast current market trends, and business outcomes, and assess customer behaviour. Among various successful business leaders, 89% believe that predictive analytics can drive business growth. The below table shows the available intelligence machines and services in the current world and their user base.

Table 1: Commercially available intelligent machines and their users.

Task	Intelligent Machines or Software	Developer	Users
Virtual Assistance	Alex	Amazon	200 million
	Siri	Apple	600 million
	Google Assistant	Google	500 million
	Cortona	Microsoft	700 million
Mapping	Google Maps	Google	1.2 billion
Riding Apps	Uber	Uber.com	100 million
	Lyft	Lyft.com	50 million
Humanoid Robots	Pepper	Soft Bank	5000 companies
Collaborative Robots	Robot CR	FANUC	150,000

Source: Thakar et al. (2022).

5. DISCUSSION AND FINDINGS

The above data and different other discussions conducted associated with technological development, Innovation and digital technology utilisation in entrepreneurship and business have been elaborated. The utilisation of secondary data has passive five different kinds of advancement and disadvantages of the utilisation of technology in the entrepreneurship business. It was recognised that in today's competitive environment, it is important for the organization to adopt digital technology and different technological advancements (Beliaeva *et al.*, 2021).

6. CONCLUSION

From the above discussion, it can be concluded various factors such as innovation, technological advancements, digital technology utilisation and entrepreneurship help businesses develop innovative products and services based on customer preferences. The results of this research study indicate that technology has altered the business landscape by enabling greater market reach, lowering entry barriers and encouraging an entrepreneurial culture.

7. FUTURE SCOPE

Technological development, innovation and digital technology have become an important part of the organisation. Even it can be stated that not a single organisation can be considered operating business nowadays without making utilisation. it can be stated that many organisations or making utilisation of Technology to fulfil the requirement of worker by making robots do all the work. All these specifications take this particular research work in the light of the technological environment in which all the operations to be conducted in the organisation is completely reliant on technological adaptation. Even significant entrepreneurs make utilisation of new technology to upgrade their skills and organisation to grow and survive.

REFERENCES

Abu Amuna, Y. M., S. S. Abu-Naser, M. J. Al Shobaki, and Y. A. Abu Mostafa (2019). *Fintech: Creative innovation for Entrepreneurs.*

Akpan, I. J., E. A. P. Udoh, and B. Adebisi (2020). "Small business awareness and adoption of state-of-the-art technologies in emerging and developing markets, and lessons from the COVID-19 pandemic," *Journal of Small Business & Entrepreneurship*, 1–18.

Beliaeva, T., M. Ferasso, S. Kraus, and E. J. Damke (2019). "Dynamics of digital entrepreneurship and the innovation ecosystem: a multilevel perspective," *International Journal of Entrepreneurial Behavior & Research.*

Belik, E. B., E. S. Petrenko, G. A. Pisarev, and A. A. Karpova (2019). "Influence of technological revolution in the sphere of digital technologies on the modern entrepreneurship," *Institute of Scientific Communications Conference* (pp. 239–246). Springer, Cham.

Berger, E. S., F. von Briel, P. Davidsson, and A. Kuckertz (2021). "Digital or not–The future of entrepreneurship and innovation: introduction to the special issue," *Journal of Business Research*, 125, 436–442.

Bican, P. M., and A. Brem (2020). "Digital business model, digital transformation, digital entrepreneurship: Is there a sustainable 'digital'?" *Sustainability*, 12(13), 5239.

Elia, G., A. Margherita, and G. Passiante (2020). "Digital entrepreneurship ecosystem: How digital technologies and collective intelligence are reshaping the entrepreneurial process," *Technological Forecasting and Social Change*, 150, 119791.

George, G., R. K. Merrill, and S. J. Schillebeeckx (2021). "Digital sustainability and entrepreneurship: How digital innovations are helping tackle climate change and sustainable development," *Entrepreneurship Theory and Practice*, 45(5), 999–1027.

Jelonek, D. (2023). Environmental uncertainty and changes in digital innovation strategy, *Procedia Computer Science*, 225, 1468–1477.

Kraus, S., C. Palmer, N. Kailer, F. L. Kallinger, and J. Spitzer (2018). "Digital entrepreneurship: a research agenda on new business models for the twenty-first century," *International Journal of Entrepreneurial Behavior & Research.*

11. Implementation of Machine Learning On Entrepreneurship Business for Economic Development

Ch. Sudipta Kishore Nanda[1], Lisa Gopal[2], Sudha Rajesh[3], Dilip Kumar Sharma[4], Tejpal Jaysing Moharekar[5], Geetha Manoharan[6], and Astha Bhanot[7]

[1]Assistant Professor – II, Commerce, School of Tribal Resource Management, Kiss Deemed To Be University, Higher Education Campus, Campus - 3, Bhubaneswar - 24, Odisha

[2]Assistant Professor, Department of Computer Science and Engineering, Graphic Era Hill University, Dehradun, Uttarakhand

[3]Assistant Professor, Department of Computational Intelligence, College of Engineering and Technology, School of Computing, SRMIST, Kattankulathur, Chennai

[4]Department of Mathematics, Jaypee University of Engineering and Technology, Guna (M.P.), India

[5]Assistant Professor, Department of Commerce, Shri Shahaji Chhatrapati Mahavidyalaya, Kolhapur

[6]School of Business, SR University, Warangal, Telangana

[7]PNU, University, Riyadh, KSA

ABSTRACT: Entrepreneurship promotes efficiency as well as social change because it leads to an increase in demand for products and services while lowering their cost. Every organisationneeds a development process because it aids in guiding the various advancements anticipated during the business cycle and ways to effectively utilise the resources at hand. This helps to increase position and moreover works to meet people's expectations for basic conveniences. In this study, secondary research using qualitative techniques was the method of choice. The collection and analysis of previous research, scholarly publications, reports and case studies pertaining to the application of machine learning in entrepreneurship. The relevance, dependability and credibility of the data are assessed, and significant themes and discoveries are determined by applying qualitative analysis techniques, including content analysis and thematic analysis. The results of the study imply that using machine learning techniques in entrepreneurship has the potential to spur economic growth by fostering data-driven innovation and process optimisation. To fully reap the rewards of machine learning in entrepreneurship for economic development, however, problems must be solved and collaborative environments must be fostered.

KEYWORDS: Entrepreneurship, innovation, globalisation, SMEs

1. INTRODUCTION

Entrepreneurship is considered one of the important drivers for the growth and development of the economy. As entrepreneurship helps in creating new jobs and also includes innovative business ideas that increase market competition is therefore considered to be an important factor for emerging economic development. The introduction of innovative services, technologies as well as products helps in boosting the economy and also provides jobs on both a long-term and short-term basis. An increase in competition on the market challenges the existing competitors and therefore leads to greater research and innovation of the products and services. Entrepreneurship helps in bringing social change and also improves productivity as innovation leads to expands in demands and reduces the cost of the product or services.

DOI: 10.1201/9781003532026-11

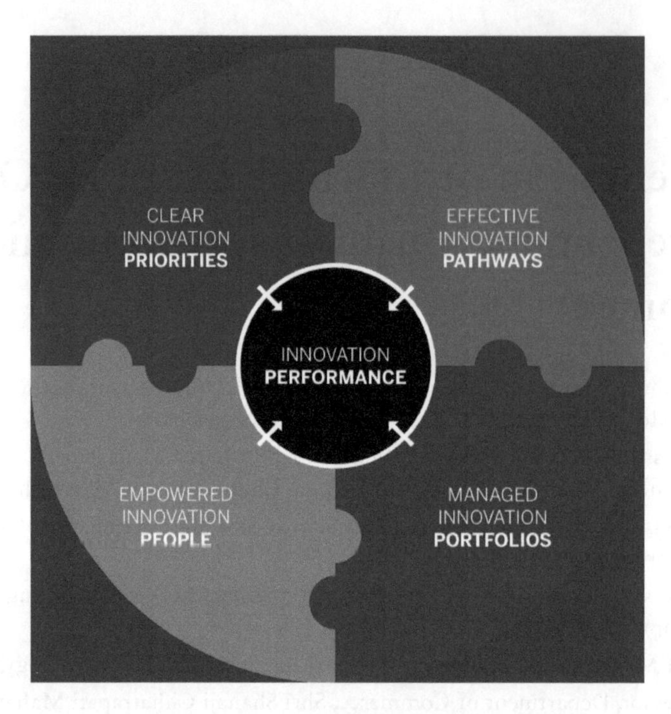

Figure 1. *Innovation performance strategy. Source: Johnson (2022).*

It is important to have an effective strategy for successful business management and to ensure the growth and development of the enterprise. Therefore, it is important to have flexibility and innovation while starting a business process. Monitoring the competitors and having a clear business vision and goals helps in assessing the risk in the market and one can strategise the market development according to it.

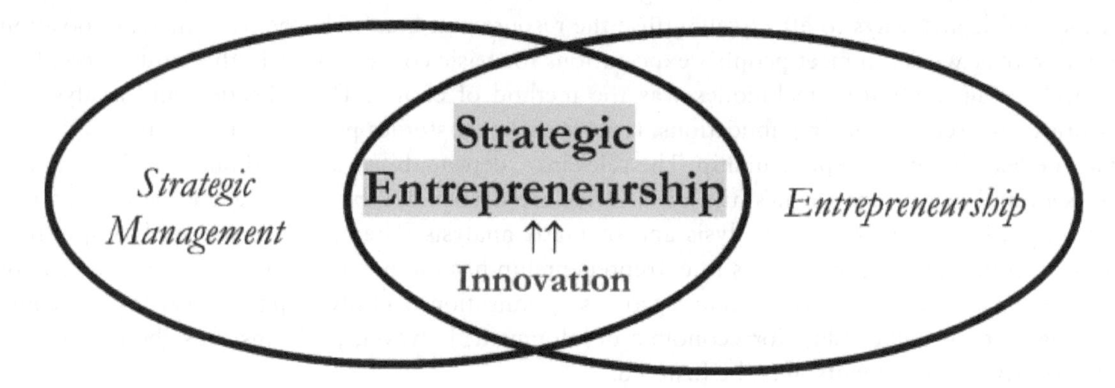

Figure 2. *Strategic entrepreneurship. Source: Self-developed.*

There are four types of innovation strategies, which can be classified as ***proactive, active, reactive*** and ***passive***. The proactive innovation strategies are mainly strong research-oriented and involve in constant improvement of technology. The active innovation strategies include protecting the available technology and using incremental innovation. Furthermore, the reactive innovation strategy uses a wait-and-see approach, which acts after examining the market position to make possible actions.

1.1. Objectives and hypothesis

- To evaluate the possibilities for the economic development of using machine learning techniques in entrepreneurship.
- To investigate how machine learning might improve entrepreneurial ventures' innovation, efficiency and decision-making processes.

- To investigate the difficulties and factors involved in applying machine learning in entrepreneurial environments.
- To determine the advantages and opportunities brought about by the effective application of machine learning in entrepreneurship.

Hypothesis 1: *By facilitating data-driven decision-making, creativity and process optimisation,, the use of machine learning techniques in entrepreneurship makes a favourable contribution to economic growth.*

Hypothesis 2: *The use of machine learning in business has a favourableeffect on key performance measures, resulting in higher revenue growth, happier customers and more efficient operations.*

Hypothesis 3: *Machine learning techniques used well in business can lead to competitive advantage, market differentiation and growth potential.*

2. LITERATURE REVIEW

2.1. Concept of Innovation Strategies on Entrepreneurship Business

Innovation is the process of developing an advanced and original concept regarding technology or service. For every business process, it is important to innovate new strategies in order to ensure an increase in profit as well as to meet customer satisfaction. Therefore, the implementation of innovation strategies on entrepreneurship business has been quite effective in developing the economy. The effective innovation strategies have helped small- and medium-scale enterprises to work efficiently and engage in profit maximisation. Many SMEs have experienced profit due to the innovative strategies and also put a positive impact on the economy of the country (Hakala *et al.*, 2020).

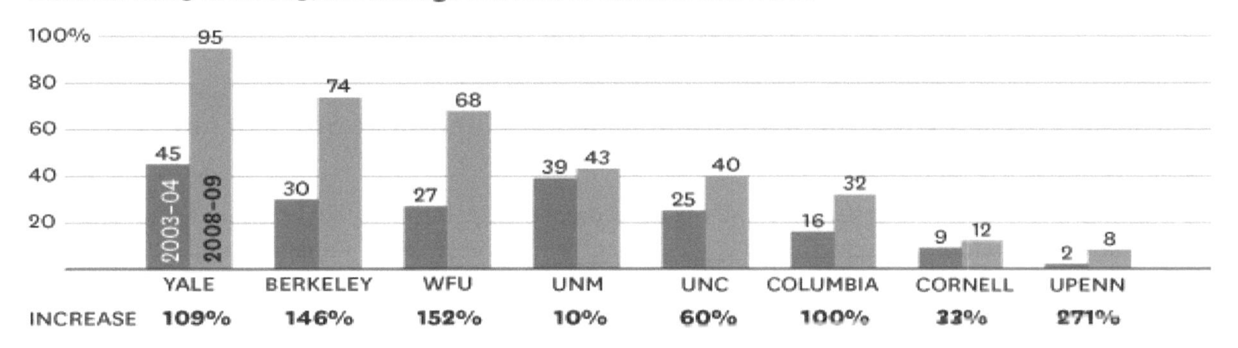

Figure 3. Graph representing a rate of increase in social entrepreneurship. Source: Cukier et al. (2011).

Innovation as per the demand of the customers is also done in service, product and technology to increase the customer's satisfaction rate. Innovation in marketing and advertisement is also an important process as it helps in increasing the market for the business, which leads to growth and development of the organisation.

2.2. Impact of Innovative Entrepreneurship on Economic Development

Entrepreneurs are considered to be an important factor for the growth and development of the economy. The entrepreneurs are responsible for boosting the rate of employment and introducing innovative products and services, as well as technologies, in the market. This creates competition between the new and existing firms, which leads to further development in the market economy. Therefore, it can be said that a positive impact of innovative entrepreneurship can be seen on the market economy of a country or economy as a whole. One of the major positive impacts of innovation is the increase in the rate of productivity due to the introduction of new skills and methods.

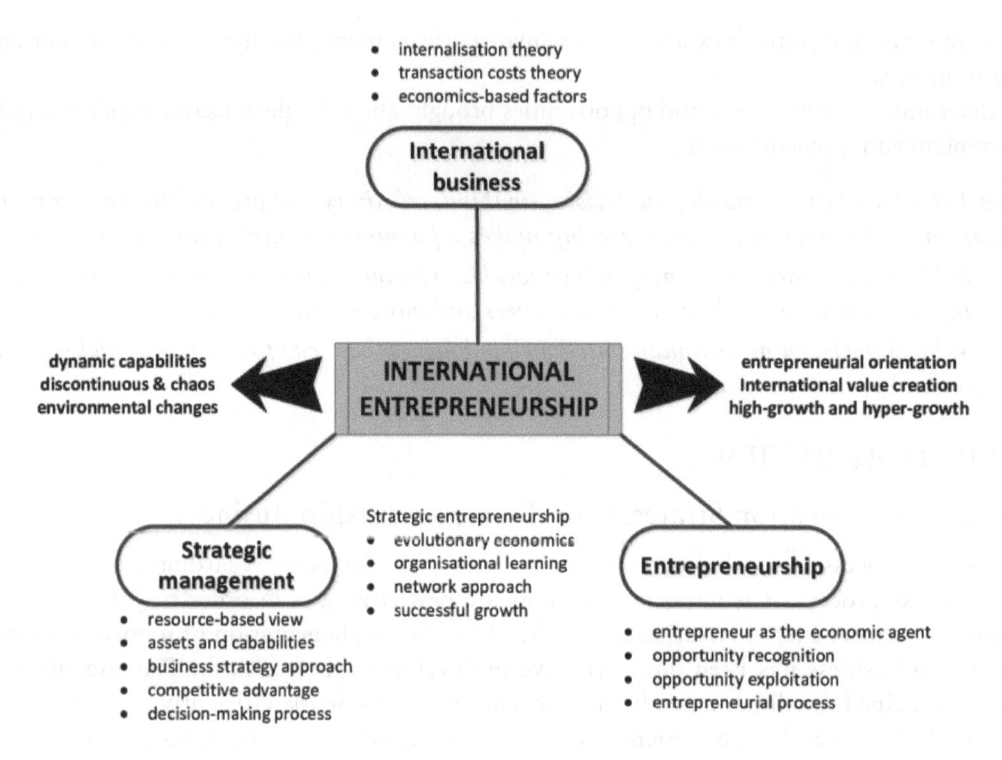

Figure 4. Impact of global entrepreneurship. Source: Matheson (2015).

Innovative entrepreneurship not only has an impact on the market but also helps in increasing the living standards of the workers. This helps in the development of the society and the country as well. This helps in making the enterprises increase the internal competition and enhance the research and development department (Distanont and Khongmalai, 2020).

2.3. Advantages and Disadvantages of Implementing Innovative Strategies in Entrepreneurship Business

Implementation of innovative strategies in the entrepreneurship business has various advantages and disadvantages. Therefore, at first, the advantages are being discussed of implementing the innovative strategies which are.

- It helps in making a good reputation of the firm as new and innovative products and services are provided by the firm, which also helps in increasing the demand over the goods and services.
- Implementing innovation in the existing products increases the value of the goods and also increases the demand for the products in the market (Bajdor and Pawełoszek, 2020).
- Implementing innovative strategies increases flexibility and opportunity to grow its business among the global market can lead to an increase in the rate of employment and profit of the firm (Jelonek 2023).

However, there are also some disadvantages of implementing innovative strategies in business, which are discussed below.

- Business risk can lead to a decrease in the reputation of the firm and may also cause a negative impact on the growth of the business as well as the quality of the product.
- Implementing an innovation strategy is very time-consuming and requires proper knowledge of the market structure, as many social aspects are related to the business process that is required to be looked after before implementation.
- Proper resource allocation is important as an ineffective strategy can lead to waste of resources as well as capital, which can cause a negative impact on the entrepreneurship business.

3. RESEARCH METHODOLOGY

In this research paper, a systematic way has been used to guide the assessment to drop by the right eventual outcomes of the impact and worth of creative philosophies on innovation strategies in entrepreneur business. Therefore, in this investigation work, only secondary data are being used to drop by the normal results considering innovative systems on undertaking business. It is indispensable to appreciate the usage of various frameworks in business communication as it helps in growing the value of as of late developed stock and products, as well as its effects on the overall economy of the country. Various resources have been reviewed to get the idea about the thoughts of the people engaged in the business activity about innovative strategies in entrepreneurship.

Along with their views over the impact in the market as well as the economy due to these strategies. Furthermore, the secondary data collected through various articles and journals as well as websites were also used to understand the impact of innovation strategies on the market economy. Through the research, it was prominent that the strategies are very effective and have helped the entrepreneurs to increase the demand as well as employment rate in the market economy. The subjective assessment has helped with separating data that were accumulated from the fundamental investigation; however, the data that are assembled in the discretionary system were examined by the quantitative procedure. In order to discover the main themes and conclusions, qualitative analysis is used to assess the data's relevance and veracity. The methodology enables a thorough investigation of the function, difficulties, advantages and effects of machine learning in entrepreneurial endeavours. The findings advance knowledge of how machine learning can improve entrepreneurial decision-making, efficiency and creativity, which promotes economic growth.

4. ANALYSIS AND INTERPRETATION

Based on data collection and interpretation, it can be analysed that machine learning has huge influences on entrepreneurship business, and it has been potential to drive economic development by enhancing decision-making processes, driving innovation and enhancing operational efficiency. Thus, it can be said that implementing machine learning in entrepreneurship business can improve operational efficiency along with savings costs. Potentially, it can automate more than 50% of current work activities, which leads to increased productivity. Along with this, machine learning also stimulates business growth and can create job opportunities. Based on analysis, it can be estimated that machine learning and AI could contribute over $17 trillion to the global economy.

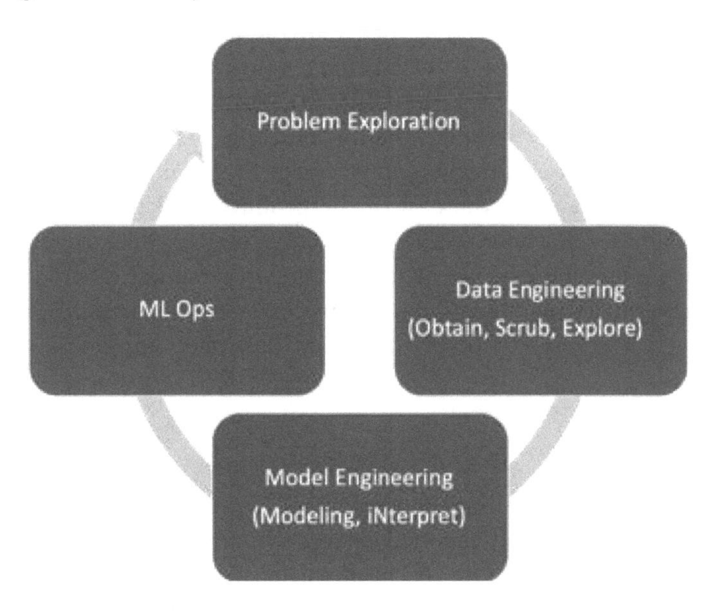

Figure 5. *Overview of machine learning process. Source: Author's compilation*

Moreover, it can significantly analyse customer data, behaviour and preferences to improve customer experiences and ensure customer satisfaction. More than 88% of marketers claimed that personalisation helped them improve. Whereas, 58% believed it had increased sales by more than 15%. However, the attributes of entrepreneurial competence are also discussed in the below table in the context of the implementation of machine learning.

Table 1: Attributes of entrepreneurial competence.

Knowledge	Skills	Attitudes
• Autonomous learning • Critical thinking • Experience • Possession of information	• Leadership • Persuasion • Creativity • Innovation • Problem-solving • Decision making • Exploiting Opportunities	• Self-efficacy • Motivation • Competitiveness • Risk-taking ability • Confidence • Internal focus • Tolerance of failure

Source: Pandey et al. (2019).

5. DISCUSSION AND FINDINGS

The above data provide the opinion of the people regarding the implementation of innovative strategies in entrepreneurship business as it is considered to be an important driver in the economic development of the emerging business. In this research paper, only secondary data are used in order to know and understand the effect and impact of these strategies on the economy. However, with the help of both these data, it can be said that the innovative strategy of entrepreneurship business has a positive impact on the economy and also helps in creating new jobs and expansion of the market. The use of new technology and product has increased the competition in the market, which enhances the research and development of the existing products and services.

6. CONCLUSION

Based on the above discussion, it can be concluded that the implementation of the innovative strategy in entrepreneurship business is important as well as have a positive impact on the market economy. This helps in increasing jobs and also improves the living standard of the people. The overall economy is largely effective through this strategy as more and more people are getting jobs, along with an increase in competition in the market that helps in the improvement of the existing products. The use of modern technology and innovation has lowered the cost of the products and therefore increases the demand as well as customer satisfaction level. Therefore, the innovative strategy has decreased the market risk and is effective at encouraging emerging entrepreneurs in the global market. Entrepreneurs can use machine learning to promote innovation, streamline company processes and make data-driven decisions, resulting in favourable outcomes like revenue growth, increased customer happiness and greater operational performance.

7. FUTURE SCOPE

The entrepreneurship business with implementing innovative strategy shows strong growth and development in the future. The new and diversified entrepreneurship will certainly help in the development of the economy as well as increase the rate of employment and diversified product in the market. The rapidly changing market and globalisation factor are what is considered to be the driving factor for the development of new technology and ideas, along with emerging enterprise business in the global market.

REFERENCES

Pawełoszek I., Bajdor P. (2020), A Statistical Approach to Assess Differences in Perception of Online Shopping, Procedia Computer Science, nr 176 (2020), pp. 3121-3132.

Cukier, W., S. Trenholm, D. Carl, and G. Gekas (2011). *Social Entrepreneurship: A Content Analysis*. Available at: http://www.na-businesspress.com/JSIS/CukierWeb.pdf (Accessed 13 March 2022).

Distanont, A., and O. Khongmalai (2020). "The role of innovation in creating a competitive advantage," *Kasetsart Journal of Social Sciences*, 41(1), 15–21.

Hakala, H., G. O'Shea, S. Farny, and S. Luoto (2020). "Re-storying the business, innovation and entrepreneurial ecosystem concepts: the model-narrative review method," *International Journal of Management Reviews*, 22(1), 10–32.

Jelonek, D. (2023). Environmental uncertainty and changes in digital innovation strategy, *Procedia Computer Science,* 225, 1468–1477.

Johnson, J. (2022). *High-Tech Innovation*. Sopheon. Available at: https://www.sopheon.com/high-tech-innovation/ (Accessed 14 March 2022).

Matheson, R. (2015). *New Report Outlines MIT's Global Entrepreneurial Impact*. MIT News | Massachusetts Institute of Technology. Available at: https://news.mit.edu/2015/report-entrepreneurial-impact-1209 (Accessed 14 March, 2022).

12. An Empirical Study of Artificial Intelligence on Financial Crisis for Human Resource Managers

Meeta Joshi[1], Shruti Bhatla[2], Bikram Paul Singh Lehri[3], Gurubasavarya Hiremath[4], Dhyana Sharon Ross[5], Sujay Mugaloremutt Jayadeva[6], and Rania Mohy ElDin Nafea[7]

[1]Associate Professor, Faculty of Management Studies, Marwadi University, Rajkot

[2]Assistant Professor, Department of Computer Science and Engineering, Graphic Era Hill University, Dehradun, Uttarakhand

[3]Assistant Professor, University School of Business, Chandigarh University, Gharuan, Mohali, Punjab, India

[4]Professor, School of Management Studies and Research, KLE Technological University, Hubballi

[5]Assistant Professor, Loyola Institute of Business Administration (LIBA)

[6]Assistant Professor, Department of Health System Management Studies, JSS Academy of Higher Education & Research, Mysuru 570009

[7]College of Administrative and Financial Sciences, University of Technology Bahrain, Bahrain

ABSTRACT: The COVID-19 pandemic has had an unanticipated impact on businesses all around the world, with a significant impact on human resource management. Both sides of HRM had to struggle with layoffs and reduced staff because to the pandemic shutdown. Today's businesses are coping with a unique problem that no other business has ever faced. This study's objective is to describe and evaluate how artificial intelligence is affecting the economic meltdown for human resource managers during the pandemic. This study has used statistical method. The essay also looks into the difficulties HRM has faced at work and the effects of human resources. The conclusions of this paper provide insights for human resource managers and the organisations that leverage AI. The results imply that AI technologies have the ability to help human resource managers handle financial crises in an efficient manner. Human resource managers may improve their strategic competencies, make data-driven decisions and help firms remain resilient and succeed in the long run during difficult economic times by utilising AI.

KEYWORDS: COVID-19, Artificial Intelligence, human resource, crisis, technology

1. INTRODUCTION

As the globe struggled with COVID-19, technology advancements and Artificial Intelligence (AI) bring businesses one step nearer to tackling this epidemic. AI is transforming how organisations run and perform. This pertains to all corporate divisions, particularly human resources. Human resource management is essential in enterprises. AI is assisting human resource managers in properly understanding and resolving the COVID-19 issue in the workplace Akour, et al. (2021). Furthermore, as even more firms shift away from conventional emails or towards public communications platforms, the possibilities for intelligent helpers to assume over functions like organising, task management or basic communication have grown significantly (Lopez and Alcaide, 2020).

1.1. Objectives and Hypothesis

- To investigate the function of AI and its prospective effects on financial crisis management from the viewpoint of human resource managers.
- To research the ways in which AI-based tools can help human resource managers prevent and address financial crises.

DOI: 10.1201/9781003532026-12

- To investigate how the use of AI in financial crisis management may affect many aspects of human resource management, including as recruiting, talent management, employee engagement and training and development.
- To determine the difficulties and factors to be taken into account when implementing AI in financial crisis management, as well as the ramifications for human resource managers.

Hypothesis 1: *The use of AI technology in financial crisis management has a positive influence on human resource managers' capacity to prevent and address financial crises.*

Hypothesis 2: *The application of AI technology in financial crisis management improves human resource management techniques like talent management, recruitment, employee engagement and training and development.*

Hypothesis 3: *Human resource managers have chances for better decision-making, proactive risk management and efficient workforce planning as a result of the deployment of AI technologies in financial crisis management.*

2. LITERATURE REVIEW

2.1. What is Artificial Intelligence?

AI is a concept that allows robots to interpret, develop or make judgements based on newly acquired knowledge. AI could be utilised to improve and ease processes performance in HR management in a variety of methods (Rada, 1986).

While companies are embracing AI, throughout their HR procedures at varying speeds, it is clear all of this as the technologies are becoming greater and widely adopted, will also have a longer effect on the market (Reese, 1985).

2.2. What is a Human Resource Manager

Human resource managers organise, lead or supervise an organisation's managerial work. He or she violates getting new staff hired, assessing them and firing them and also training existing employees (van Donk and Esser, 1992).

2.2.1. HR Manager's Everyday Duties

Coordination and supervision of the human resource crew's activity at a certain moment in Table 1.

1	Worker benefits plans must be planned and managed.
2	Oversee disciplinary processes or resolve worker disagreements.
3	Oversee all aspects of recruiting, including interviews, choices, recruitment and training.
4	Engage with additional departmental managers to learn approximately their personnel and training requirements.

HR managers are the frontline stages of assembling the workforce required to help a company prosper (Pawełoszek, 2017). Engaging with secretarial personnel to optimise worker worth and ensuring that everybody is operating as effectively and quickly as practicable is part of this (Guest *et al.*, 2017).

2.3. Leading AI Applications in Human Resources

AI can be used in a lot of different ways in the human resources field, but HR managers can expect to see advancements in recruiting and on-boarding, staff satisfaction, process optimisation and the automating of organisational activities.

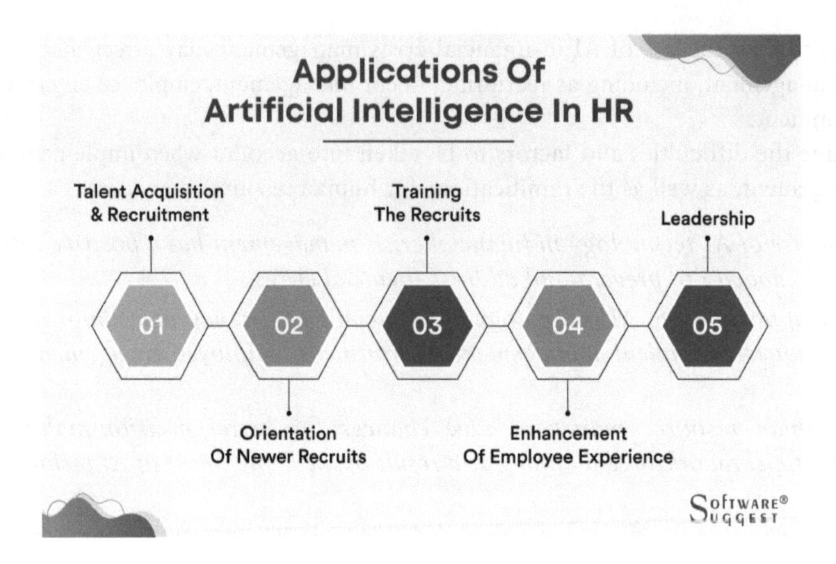

Figure 1. Applications of Artificial Intelligence' in Human Resources.Source: Gomathy 2022.

2.3.1. Worker Engagement and Internally Movement

In addition to enhancing the hiring process, HR professionals may employ AI to boost company rotation or worker engagement. Owing to personalised opinion surveys or worker acknowledgment programmes, HR businesses could now more efficiently analyse worker engagement and job satisfaction than ever before (Tapia and Turner, 2018). As per current HRPA research, certain AI software can analyse important markers of worker effectiveness in addition to identifying people who should be elevated. Consequently, underlying mobility is encouraged.

2.3.2. Administration Works Automation

One of the key benefits of introducing AI into various human resource activities would be the same as in some different professions or businesses: HR professionals have more opportunities to assist in corporate strategy planning by automating reduced, easily repeatable administrative duties (Turem, 1986). According to the Eightfold study, administrative activities were carried out 19% more efficiently by HR staff who used AI software than by departments that could not use the same technology (Choi *et al.*, 2021).

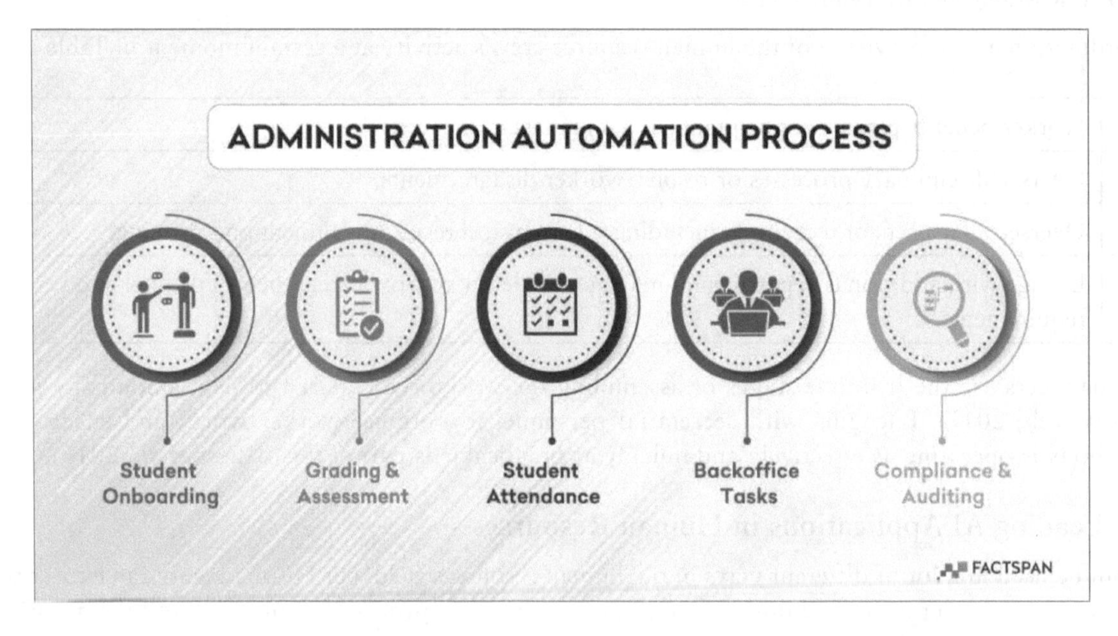

Figure 2. Administrative tasks of automation. Source: Self developed.

2.3.3. Onboarding and Recruiting

While several firms are currently using AI technologies in their recruitment attempts, the overwhelming bulk is not. In contrast, according to the 2019 'Global Human Capital' Insights report, just 6% of participants said their business had best-in-class recruiting procedures in technologies, while 81% thought their firm's procedures were ordinary or even below standards (Gomathy, 2022).

The induction process begins after hiring management selects the best candidates for their open roles. With the use of AI, this process need not be restricted to regular business hours, which is a big improvement over past onboarding practises (Furtmueller *et al.*, 2012).

2.4. Crisis Management during Pandemic

The COVID-19 pandemic was emerged and disseminated in December 2019 in Wuhan city of China, it has become a worldwide pandemic (WHO, 2020). As business organisations have been increasingly influenced by the international world, with its many, intricate or dynamic consequences from a technical standpoint, it is becoming vital to link crisis administration with organisational strategies in commercial organisations (Gospodinova, 2021).

3. METHODOLOGY

In a systematic literature review, scientists, experts and market researchers validate, appraise or synthesise their research in such a methodical, clear and reproducible way. This research used an experimental quantitative approach. The study uses secondary information. The information gathered from research, blogs, websites or books was reviewed and evaluated quantitatively. The COVID-19 outbreak has placed HR managers under tremendous stress. Organisations must react to the epidemic's global shifts and adjust accordingly (Akour *et al.*, 2021). Moreover, qualitative analysis is used to assess the data's relevance and veracity. The goal of the study is to comprehend how AI may help human resource managers respond to financial crises in an efficient manner. The research findings help human resource managers comprehend the role, difficulties, advantages and consequences of AI in managing financial crises.

4. ANALYSIS AND INTERPRETATION

Based on the analysis and interpretation of a large dataset, valuable information has been gathered. During the COVID-19 pandemic situation, the world has witnessed a significant economic downturn. Hence, both positive and negative impacts of AI concerning global financial crisis of HR managers have been identified. Using a table, the impact of the pandemic situation has been illustrated below.

Table 1: Industries affected during the pandemic.

Sectors	Percentage of Impact	Impact (Positive and Negative)
Food Industries	–22.8%	Negative
Chemical Industries	–18.8%	Negative
Furniture Industries	–28%	Negative
Leather Industries	–27.2%	Negative
Textile Industries	–26.1%	Negative
Engineering Industries	–21.3%	Negative
Other Industries	–17.3%	Negative
Overall	–25.4%	Negative

Source: Gospodinova (2021).

In terms of positive effects, AI has increased efficiency, improved workforce planning and analytics and enabled remote recruitment and onboarding. AI-powered tools helped HR managers to streamline various administrative tasks to ensure employee well-being. On the other hand, the pandemic has forced businesses to shift to remote work. Hence, by leveraging technology, AI recruitment platforms and virtual interviews are facilitated, which have improved the recruitment process by 40% (Loyola, 2021). Whereas, various AI algorithms are designed to help HR managers analyse a vast amount of data for better workforce planning, skill development and talent management. Most of successful businesses have admitted that AI helped them manage the financial crisis by improving efficiency by 50%. On the other side, in terms of negative effects, AI caused job displacement and around 33% of HR managers in multinational organisations have been struggling to leverage AI tools to provide strategic guidance.

5. FINDINGS OF THE STUDY

Table 2: Contains several solutions to Human Resource concerns.

S. No	HR Issues	Solution
1	Social distancing/ loss of employment	Worker adaptation, virtual employment, workplace flexibility
2	Payroll	The standardised procedure with genuine information aids in lowering run-time expenses.
3	Governments	Cognition call advice may help with problem resolutions and personnel satisfaction.
4	Insights for the employment	Models for determining the appropriate talent mix, recruiting goals and smart methods for simplifying complicated information handling
5	Job insecurity/stress/ layoff	Online meetings, internet conversations and chatbots are all possibilities.
6	Inadequate efficiency	Automated, digital service, digitised business and surveillance are all terms that may be used to describe what we do. Information in real-time

Source: Self-developed.

6. CONCLUSION

The effect of AI on human resources during the pandemic situation has been discussed by reviewing various studies. According to the research, AI can improve risk management, decision-making processes and workforce planning during financial crises. Human resource managers can use AI tools like machine learning, natural language processing and predictive analytics to extract meaningful insights from massive amounts of data and make defensible decisions to lessen the effects of financial crises.

7. FUTURE WORK

Furthermore, the article's primary results, along with their importance from both an academic or a commercial aspect, and also the inquiry's limitations may assist to motivate more study. The proposals provide several study topics for future research in this topic as well as suitable human resource methods to address developing human resource difficulties.

REFERENCES

Akour, I., M. Alshurideh, B. Al Kurdi, A. Al Ali, and S. Salloum (2021). "Using artificial intelligence algorithms to predict people's intention to use mobile learning platforms during the COVID-19 pandemic: artificial intelligence approach," *JMIR Medical Education*, 7(1), e24032. https://doi.org/10.2196/24032.

Choi, D., H. R'bigui, and C. Cho (2021). "Candidate digital tasks selection methodology for automation with robotic process automation," *Sustainability*, 13(16), 8980. https://doi.org/10.3390/su13168980.

Furtmueller, E., C. Wilderom, and M. Tate (2012). "Managing recruitment and selection in the digital age: e-HRM and resumes," *Human Systems Management*, 30(4), 243–259. https://doi.org/10.3233/hsm-2011-0753.

Gomathy, D. (2022). "Overview of recruitment and selection process in HRM," *International Journal of Scientific Research in Engineering and Management*, 06(03). https://doi.org/10.55041/ijsrem11714.

Gospodinova, S. (2021). "Influence of the COVID-19 crisis on labour productivity in the Bulgarian economy," *Socio-Economic Analyses*, 13(2). https://doi.org/10.54664/zowb3815.

Guest, D., R. Rodrigues, and K. Sanders (2017). "Manager HR attribution and employee outcomes: considering HR implementation," *Academy Of Management Proceedings*, 2017(1), 10100. https://doi.org/10.5465/ambpp.2017.10100symposium.

He, W., Z. Zhang, and W. Li (2021). "Information technology solutions, challenges, and suggestions for tackling the COVID-19 pandemic," *International Journal of Information Management*, 57, 102287. https://doi.org/10.1016/j.ijinfomgt.2020.102287.

Lopez, B., and A. Alcaide (2020). "Blockchain, artificial intelligence, internet of things to improve governance, financial management and control of crisis: case study COVID-19," *Socioeconomic Challenges*, 4(2), 78–89. https://doi.org/10.21272/sec.4(2).78-89.2020.

Loyola, B. (2021). "COVID-19 in Mexico: preparing for future pandemics," *Universal Journal of Pharmaceutical Research*. https://doi.org/10.22270/ujpr.v6i3.605.

Pawełoszek I. (2017), Process-oriented approach to competency management using ontologies, [in:] M. Ganzha, L. Maciaszek, M. Paprzycki (eds.), Proceedings of the 2017 Federated Conference on Computer Science and Information Systems, ACSIS, Vol. 11, pp. 1005–1013 DOI: http://dx.doi.org/10.15439/2017F441.

Rada, R. (1986). "Artificial intelligence," *Artificial Intelligence*, 28(1), 119–121. https://doi.org/10.1016/0004-3702(86)90034-2

Reese, D. (1985). "Artificial intelligence," *Artificial Intelligence*, 27(1), 127–128. https://doi.org/10.1016/0004-3702(85)90088-8.

Tapia, M., and L. Turner (2018). "Renewed activism for the labor movement: the urgency of young worker engagement," *Work and Occupations*, 45(4), 391–419. https://doi.org/10.1177/0730888418785657.

Turem, J. (1986). "Social work administration and modern management technology," *Administration in Social Work*, 10(3), 15–24. https://doi.org/10.1300/j147v10n03_02.

van Donk, D., and A. Esser (1992). "Strategic human resource management: a role of the human resource manager in the process of strategy formation," *Human Resource Management Review*, 2(4), 299 315. https://doi.org/10.1016/1053-4822(92)90003-9.

WHO Coronavirus Disease (COVID-19) Dashboard (2020). 10(1). https://doi.org/10.46945/bpj.10.1.03.01.

13. The Role of Machine Learning (ML) in Enterprise Architecture and its Impact on Business Performance

Joel Alanya-Beltran[1], Johan Silva-Cueva[2], Danes Niño-Cueva[2], Moises Niño-Cueva[2], Fanny Mancco-Rivas[2], and Adolfo Perez-Mendoza[3]

[1]Tecnologico de Monterrey

[2]Universidad Nacional de Educación Enrique Guzmán y Valle

[3]Universidad Tecnológica del Perú

ABSTRACT: One of the most significant and frequently applied concepts in the field of developing enterprise systems is system theory. In this context, the organisational architect's primary goal is to provide a map of the IT components that support the business activity. It should be emphasised that it is essential to personally record each business arrangement as else update errors frequently happen. The use of information systems is efficient in this situation. In this study, secondary research using qualitative techniques was the method of choice. The function of Machine Learning (ML) in organisational architecture and its effect on the performance of organisational are the subjects of a collection and analysis of existing literature, academic publications, reports and case studies. The relevance, credibility and dependability of the data are assessed, and significant themes and conclusions are determined by applying qualitative analysis techniques, including content analysis and thematic analysis. The results indicate that by improving productivity, decision-making and customer experiences, the integration of ML in enterprise design has a favourable impact on corporate performance. To fully realise the potential of ML in enterprise design, firms must overcome integration issues and make investments in teamwork and skill development.

KEYWORDS: Enterprise architecture management, strategic IS alignment, cluster analysis, organisational agility.

1. INTRODUCTION

Information system is a kind of sociotechnical, formal and organisational system that has been designed to execute business operations such as collection of information, processing of data, storing of data and distribution of the information. As per the perspective of sociotechnical, information systems are composed of four components, such as structure, task, people and technology (Pattij *et al.*, 2020). By following these, a firm can manage their internal structure and maintain transparency through the digitalised information system. The main advantage of the information system is that it allows the firm to operate the business over the online process (Anthony, 2021). For this reason, one can able to gather the business data like financial statement, stock management, competitors' data and many more through the digitalised media-based system (Nadeem *et al.*, 2018).

1.1. Objectives and Hypothesis

- To look into the application of machine learning (ML) in enterprise architecture.
- To examine how ML affects company performance in relation to enterprise design.
- To determine the difficulties and factors to be taken into account when incorporating ML technologies into enterprise architecture.
- To evaluate the advantages and opportunities brought about by ML usage in corporate architecture.

DOI: 10.1201/9781003532026-13

Hypothesis 1: *Enterprise architecture that incorporates ML technologies improves company performance.*
Hypothesis 2: *Data quality, integration and specialised knowledge are issues that arise with the introduction of ML technology in corporate design.*
Hypothesis 3: Organisations *have prospects for new income streams, resource allocation* optimisation *and the creation of cutting-edge products and services thanks to the integration of ML in enterprise architecture.*

2. LITERATURE REVIEW

2.1. The Mediating Role of Information System for Improving Agility through Enterprise Architecture

In order to maintain the existence of the business in the current volatile market, it is very important to adopting the changes of the market. On the other side, if a firm wants to float with the current of the market demands, then the organisation need to keep their dataset updated for understanding the current and upcoming market status (Kutzner *et al.*, 2018). However, the practical implementation of such concept is not that much easier as gathering data from whole market area through manual process with physically appearance is not only difficult but also can be impossible (Zimmermann *et al.*, 2018). By considering these things, the developer has introduced the AI-based information management system for business through which the system automatically detect the potential data and store them systematically for future use. Moreover, by utilising this, the enterprise architecture system could be able to analyse the overall status of the business along with market standard and based on that the system automatically suggest some desired changes for business to set the productive vision and outcomes (Pattij *et al.*, 2020).

2.2. Theory and Model Development

In order to get insight of the organisation's dynamic environment, appropriate theoretical lens is need to use that may help to underline the business development factors (Zimmermann *et al.*, 2018).

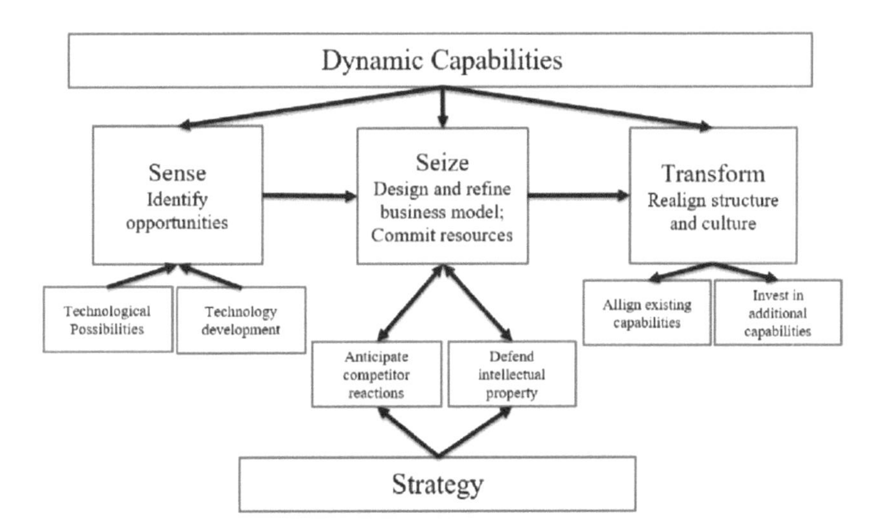

Figure 1. *Dynamic capabilities model. Source: Teece (2018).*

As per Teece (2018), a dynamic capability is a kind of firm's ability to configure and integrate with the internal and external competence that effectively helps to balance the business with the rapidly changing environments.

2.3. Enterprise Architecture (EA)

It can be defined as a conceptualised set of processes that effectively support the development of EA. The fundamental requirements of the enterprise architecture management are that it help to upkeeps the organisation's IT network and services by its technical intelligence. By utilising the enterprise management system a company can develop its business strategy with proper technology for achieving the business goals (Behúnová *et al.* 2018).

2.4. Organisational Agility

Agility is a term of business ability to upgrade itself by adapting changes of the market quickly and capable to withstand the unpredictable turbulence of the market (Zhang *et al.*, 2018). An organisational agility gets stronger by revolving around strengthening the relationship between direct reporters and managers for improving collaboration.

2.5. Strategic IS Alignment

Strategic IS aligner is a great tool for the business which shows how to develop an effective strategy for the business. At the initial level, the information system gather the critical data from the organisation's previous record and compare the same with the current market scenario.

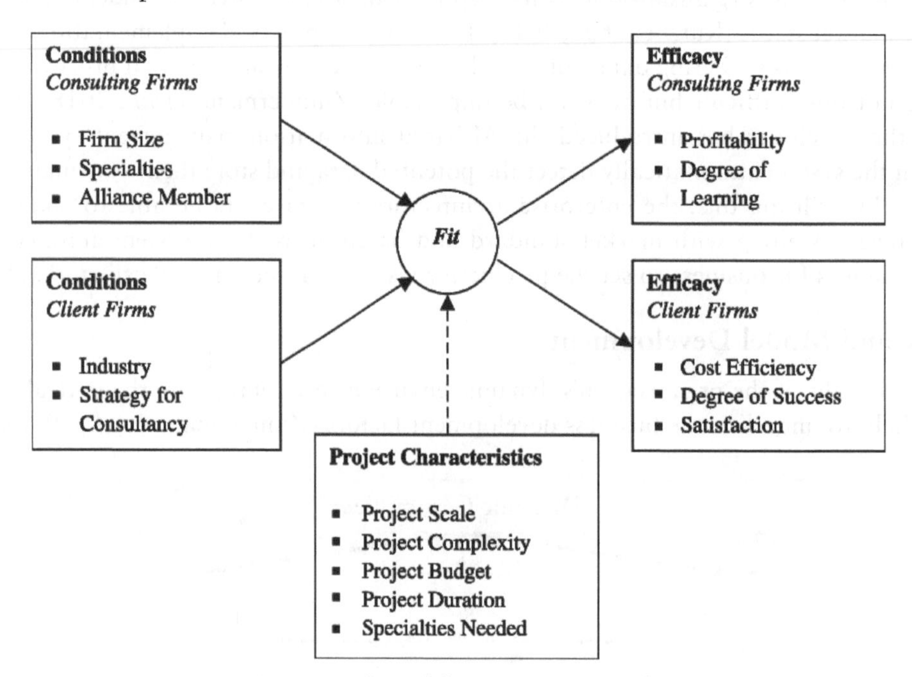

Figure 2. Research model. Source: Pattij et al. (2020).

According to Pattij *et al.* (2020), enterprise architecture management works like a bridge in between IT stakeholders, users and company employees.

2.6. Advantage and Disadvantage of the System Management

2.6.1. Advantage

- Fast in nature and shorter response time
- More accurate due to the integration of high-level algorithm
- More secure and easy to access from any where
- Observer of the data can be controlled by administration
- Helps to maintain transparency

2.6.2. Disadvantage

- Costly
- Required continuous maintenance to upgrade the configuration
- Inaccurate data can make the overall business process wrong
- Poor security may throw a threats to the organization

3. RESEARCH METHODOLOGY

Methodology of the study is a specific procedure or technique that used to identify the study analysis process based on which the data analysis and data collection process can be executed in a particular study (Zimmermann *et al.*, 2018). This study used secondary research with a qualitative approach as its research methodology. The function of ML in EA and the effect of EA on the performances of companies are the subjects of a collection and analysis of existing literature, academic publications, reports and case studies. The relevance and reliability of the data are assessed, and important themes and conclusions are identified using qualitative analysis techniques including content analysis and thematic analysis. Through the synthesis of already-existing knowledge and insights from reliable sources, the approach enables a thorough investigation of the study issue.

3.1. Research Design

This section is essential for understanding the current state of 'information system' technology around the globe to incorporate the operation of the business. Using a secondary data analysis approach, the technique section supported researchers in discovering relevant data on the chosen topic. For designing a specific study framework, this is one of the most well-known and widely used research methodologies. This part of the strategy entails the process of gathering information on a certain issue in order to complete the research. The primary purpose of this study is to develop a hypothesis based on the data acquired. The study design aids in the completion of thorough market analysis. The goal of this method is to include a piece of research that is both informative and entertaining.

3.2. Selection of Journals

Journal selection has been made based on their reliability and authenticity. Only journals and news articles published after 2018 are taken to gather data. Researchers have taken suggestions and some sort of basic data from the old secondary sources that were printed before 2018. However, most of them are from the 2018 to2022 timeframe. 'Google scholar' is used to search journals. Online websites are also analysed to gather current cases of ML handling car counting and other vehicle detection activities.

3.3. Research Philosophy

Appropriate integration of'secondary research' via a 'qualitative data collection' approach from appropriate sources improves examination into a certain research issue. The 'secondary qualitative research approach' simultaneously assists researchers in collecting the appropriate amount of relevant data in a short amount of time in order to gain a broad understanding of the full study findings. Obtaining all relevant data is vital for the researchers to provide the research findings with a smoother shape when performing secondary research using a '*positivism*' philosophy. This is useful in this study as this philosophy proposes a 'unified' methodology within the research (Zafary, 2020).

4. ANALYSIS AND INTERPRETATION

Upon analysing and interpreting data from various reliable and authentic sources, it has been found that ML has been considered the most effective technological advancement in recent years that made significant impacts on enterprise architecture and improved overall business performance by leveraging large data sets. Using customer analytics through ML algorithms, many organisations have secured a 120% profit increase (Pattij *et al.*, 2020).

Whereas, ML also helps to implement personalisation and improves customer experiences. On the other hand, ML algorithms also help organisations in fraud detection and risk management. Using ML, organisations have reduced fraud loss by 50% and increased detection by 58% . In addition to that, ML-based automation helps organisations achieve a 25% reduction in cost and a 65% increase in processing time. A table has been described below to facilitate the significance of ML in improving organisational performance.

Table 1: Impact of Machine Learning on different organizational aspects.

Categories	Overall Improvements
Cost analysis	10%
Corporate governance	6.9%
Strategic alignment	14.7%
Employee performance	19.5%
Risk assessment	18.8%
Strategy development	27.9%
Initiating change	12.9%
Strategy compliance	20.5%

5. DISCUSSION AND FINDINGS

The dependency on the digital technology is getting popular day by day due to the multivariate advantages. There are a series of advantages present in the information system that is useful for the business operation. By utilising the method, an organisation can operate their business intangible way. Starting from the data manipulation to storing of the data, everywhere information system is required. However, as per the result of this study, development and implementation of enterprise architecture (EA) are not a single time activity. It effectively requires continuous adaptation of the data to guide the constant reconfiguration of the organisational resources. This study has embraced a wider dynamic capability view in terms of the EA deployment (Ajer and Olsen, 2018).

The enterprise architecture provides a map of the business to support the reliable business operations. Besides this factor, enterprise architecture acts as a significant diagnosis tool for identifying the problem source occurred in the workplace (Nadeem *et al.*, 2018).

6. CONCLUSION

Conclusively, it is a set assumption that implementation of enterprise information system has some significant impact on the business performance indicator. In other words, the graph of the cost and revenue of the company is positively get changed after implementation of the enterprise information system. Based on this new indicator, it became easier to take essential decision quickly on the basis of the relevant data. This helps in establishment of the sustainable operation in the working environment. Moreover, the information system comes with high-level algorithm, through which the potential data can be manipulated and calculated efficiently as per the requirements. Thus, in the following way, the transparency of the business can also be maintained that further improve the relation between stakeholders and employees. Hence, with the establishment of the better communication, the productivity of the firm can be increased.

The enterprise architecture with implementation of the information system and innovative strategy shows strong growth and development of the business in the coming future. The new and diversified technology may bring a lot of changes in the business scenario through which diversified services can be provide at a same time without limitation. However, the rapid changes of the market globalisation of the business is the driving factor of this study that needs to be improved by developing the information system to some extent. Hence, by considering the above discussion and study scenario, it can be said that the demand of the information system will be increased for developing enterprise architecture.

REFERENCES

Ajer, A. K., and D. H. Olsen (2018). "Enterprise architecture challenges: a case study of three Norwegian public sectors," *ECIS*, p. 51.

Anthony Jnr, B. (2021). "Managing digital transformation of smart cities through enterprise architecture–a review and research agenda," *Enterprise Information Systems*, 15(3), 299–331.

Behúnová, A., T. Mandičák, K. Krajníková, P. Mesároš, and M. Behún (2018). "Impact of enterprise information systems on selected business performance indicators in construction industry," *MMS 2018: 3rd EAI International Conference on Management of Manufacturing Systems* (pp. 1–10). European Alliance for Innovation.

Kutzner, K., T. Schoormann, and R. Knackstedt (2018). "Digital transformation in information systems research: a taxonomy-based approach to structure the field," *ECIS*, 56.

Nadeem, S., A. K. Alvi, and J. Iqbal (2018). "Performance indicators of e-logistic system with mediating role of Information and Communication Technology (ICT)," *Journal of Applied Economics & Business Research*, 8(4), 217–228.

Pattij, M., R. Van de Wetering, and R. J. Kusters (2020). "Improving agility through enterprise architecture management: the mediating role of aligning business and IT," *AMCIS*.

Teece, D. J. (2018). "Business models and dynamic capabilities," *Long Range Planning*, 51(1), 40–49.

Zafary, F. (2020). "Implementation of business intelligence considering the role of information systems integration and enterprise resource planning," *Journal of Intelligence Studies in Business*, 1(1).

Zhang, M., H. Chen, and A. Luo (2018). "A systematic review of business-IT alignment research with enterprise architecture," *IEEE Access*, 6, 18933–18944.

Zimmermann, A., R. Schmidt, K. Sandkuhl, D. Jugel, J. Bogner, and M. Möhring (2018). "Evolution of enterprise architecture for digital transformation," *2018 IEEE 22nd International Enterprise Distributed Object Computing Workshop (EDOCW)* (pp. 87–96). IEEE.

14. Detailed Investigation of the Role of Artificial Intelligence and its Impact on Customer Relationship Management (CRM) to Enhance Customer Loyalty

Sarita Satpathy[1], S. Rajasulochana[2], Voruganti Naresh Kumar[3], V. Hima Bindhu[4], Basavaraj S. Mammani[5], Tripti Tiwari[6], and Dalia Younis[7]

[1]Professor, Department of Management Studies, Vignan Foundation of Science, Technology and Research, Vadlamudi, Guntur, Andhra Pradesh

[2]Assistant Professor, Information Technology, SNS College of Technology

[3]Associate Professor, Department of CSE, CMR Technical Campus, Hyderabad, Telangana, India

[4]Associate Professor, Department of Management Studies, Malla Reddy College of Engineering & Technology, Hyderabad, Telangana

[5]Assistant Professor, Faculty of Business Studies MBA, Sharnbasva University, Kalaburagi, Karnataka India

[6]Assistant Professor, Department of Management Studies, Bharati Vidyapeeth (Deemed to be University) Institute of Management and Research, Delhi, India

[7]College of International Transport and Logistics, AASTMT University, Egypt

ABSTRACT: This investigation's main focus is the agreement of 'customer relationship management', or CRM, as a group of technological advancements essential for effective production schedule. It is examined how various client information types, CRM systems and information-generating activities are related. In this study, secondary research using qualitative techniques was the method of choice. The function of AI in CRM and its effects on customer loyalty are discussed in the existing literature, academic publications, reports and case studies, which are gathered and examined. The relevance, credibility and dependability of the data are assessed, and significant themes and conclusions are determined by applying qualitative analysis techniques, including content analysis and thematic analysis. The results imply that by boosting customer experiences, personalisation and proactive customer care, the incorporation of AI technology in CRM significantly increases customer loyalty. To fully realise the potential of AI in CRM, firms must solve concerns about data privacy, ethics and competence.

KEYWORDS: Customer relationship management or CRM, productivity, profitability, customer loyalty, customer engagement.

1. INTRODUCTION

In today's world, there are an increasing number of firms or economic organisations that have popped up to take benefit of the benefits that every goods business gives to entice customers. Collaboration, functional and quantitative '*customer relationship management or CRM*' solutions are the three types of CRM solutions. Statistical tools greatly help the integration procedure, according to a review of CRM programmes in several firms. Outward manifestations benefit the most from cooperative solutions. Krizanova *et al.* (2018) have mentioned that client socialisation is facilitated by operative databases, whereas internal socialisation is facilitated by cooperative networks.

DOI: 10.1201/9781003532026-14

Figure 1. Block diagram of CRM model. Source: Yauwerissa and Putra (2021).

A company enhancement that will affect the company enhancement effort is among the elements that each company or corporate body must contemplate. The organisation is able to retain strong ties and concord with customers through successful interaction. Client support is among the business's most significant tasks since it affects the firm's prospective direction (Yauwerissa and Putra, 2021).

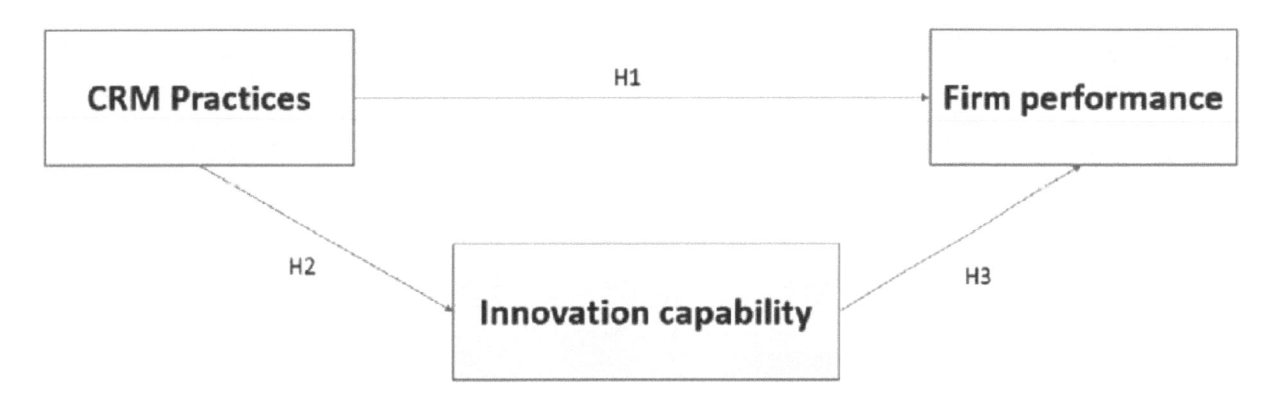

Figure 2. Model of CRM's impact on firm performance. Source: Guerola-Navarro et al. (2021).

Many of the company's problems could lead to a sour perception and reduced levels of client participation. To avoid ruining the company's image, the company must prioritise client service (Pawełoszek, Korczak, 2017). Because customers are indeed the company's greatest important asset and customer demand is vital in the expansion of the corporate image, customer assistance is essential in boosting customer satisfaction. In addition, according to McKenzie and Liersch (2011), clients must be pleased and committed as a result of enhancing the grade of these products and the corporation's profitability should improve as well. Aside from the nature of excellent operation, it has the ability to react to modify the picture or impression by channelling greater full data.

1.2. Objectives and Hypothesis

- To look into how customer relationship management (CRM) works with artificial intelligence (AI).
- To investigate how AI would affect CRM-related consumer loyalty.
- To determine the difficulties and factors involved in integrating AI technologies into CRM.
- To evaluate the advantages and opportunities brought about by the integration of AI into CRM for the purpose of boosting customer loyalty.

Hypothesis 1: *The adoption of AI in CRM enhances customer experiences,* personalisation *and proactive customer service, leading to increased customer loyalty.*

Hypothesis 2: *CRM integration with AI offers businesses the chance to run targeted marketing campaigns, streamline sales procedures and boost customer retention rates.*

Hypothesis 3: *The integration of AI technology into CRM raises issues with data privacy, ethics and the requirement for* specialised *knowledge.*

2. LITERATURE REVIEW

'Customer knowledge management' and 'innovation' are the two most important advanced factors for a variety of successful survival, expansion and strategic initiatives that improve business profitability, productivity and 'long-term competitive advantage'. 'Customer relationship management (CRM)' is among the most fundamental and important aspects of advertising strategy. As per the views of Arsic *et al.* (2018), CRM focuses on the client and their pleasure to the extent where all of the corporation's actions are directed towards them. The basic goal of CRM is to learn all there is to learn about a client so that a firm can provide faster, better suitable and greater value creation. Another type of engagement that may be utilised instead of subscriptions is customer solutions that comprises all attempts made by the company to retain existing people or customers who really are repeated consumers (Battor and Battor, 2010).

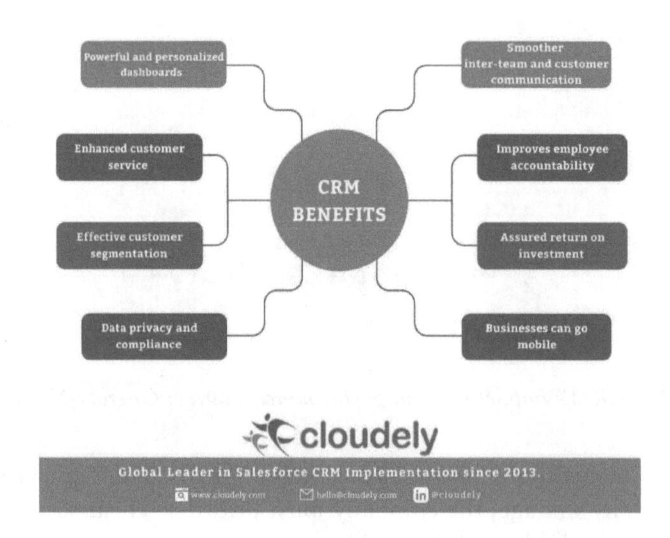

Figure 3. *Benefits of CRM map in business growth. Source: Gil-Gomez et al. (2020).*

'Customer relationship management (CRM)', a procedure of preserving and constructing client connections that benefits by offering significance, client contentment is substantial, so it will be prepared to boost client commitment by four components are recognised, obtain, preserve and grow is the responsibility of client support in constructing client allegiance by supplying top notch facility to its clients (Gil-Gomez *et al.*, 2020). The picture is itself a representation of one item's effect on some other item, which is created by combining data from a range of credible resources at any given moment effectiveness (Das and Hassan, 2021).

In terms of 'economic sustainable development', CRM may be seen as a key tool and an effective approach for more sustainable business models. Being able to achieve 'customer satisfaction' through making significant investments in 'customer relationship management' systems (CRM). Investments in 'research and development or R&D', 'disruptive technologies' and 'management information systems (MIS)' are used to achieve this (Gil-Gomez *et al.*, 2020). CRM allows businesses to recognise their customers, understand their particular needs and tailor their option or product supplies to match these needs in a sustainable manner, which provides significant additional shared value (Li and Xu, 2022).

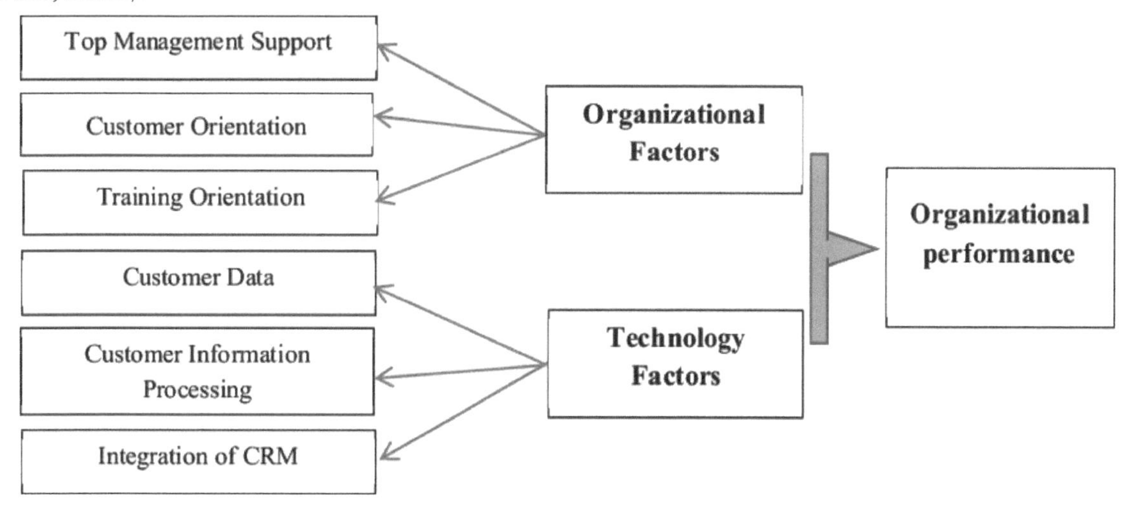

Figure 4. *Enhanced organizational productivity due to implementation of CRM. Source: Kumar et al. (2021).*

CRM is implemented in a business to save expenses, boost productivity and raise revenues through better customer satisfaction. To fully understand each customer's wants in a system model, this is essential. While interacting with clients in various areas and capacitance sensors, employees will be able to utilise this information to make rapid and accurate judgements (Kumar *et al.*, 2021).

- To develop protracted and lucrative connections with targeted clients
- To get nearer to all those consumers at each and every touchpoint; and
- To maximise the firm's part of the client's budget.

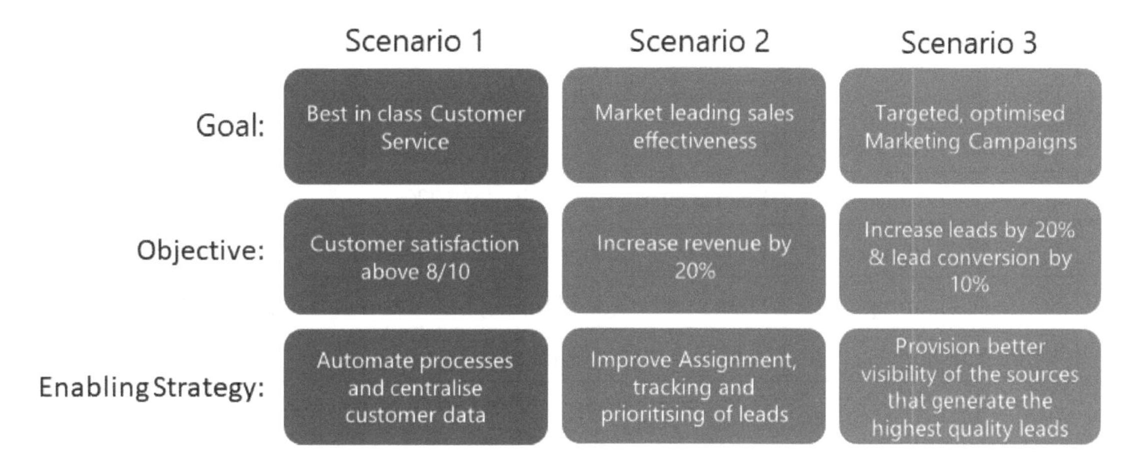

Figure 5. *CRM goals in business. Source: Kumar et al. (2021).*

Kumar *et al.* (2021) have stated that CRM is the process of locating, acquiring and client retention. As a result, CRM may be defined as a corporate approach that is more than a functional plan.

3. RESEARCH METHODOLOGY

Secondary approach is used to collect data on 'customer relationship management' and how it affects customer loyalty and business expansion. The secondary technique of data collecting was taken into consideration for this study since it aids in gathering comprehensive and current data from numerous journals, publications and newspapers. Secondary data helps add 'additional values and standards' and ensures the accuracy of primary data. Proper data collecting is essential for quality control, research credibility and making informed business decisions. There are considerations for methods, goals, time and money. To achieve their goals, researchers need to adopt a data-driven technique. Different databases such as Scholar, PubMed and ProQuest are used for collecting proper journals and articles to collect information. Keywords like 'CRM', Business productivity, customer loyalty are used to search proper articles related to work. The relevance and reliability of the data are assessed, and important themes and conclusions are identified using qualitative analysis techniques including content analysis and thematic analysis. Through the synthesis of already-existing knowledge and insights from reliable sources, the approach enables a thorough investigation of the study issue.

4. ANALYSIS AND INTERPRETATION

As a brief literature review of the research topic has been conducted, the researchers have analysed and interpreted a large dataset to gather efficient insights related to the role of machine learning in CRM. It has been found that artificial intelligence (AI) has transformed customer relationship management in the last few years by enhancing customer loyalty through personalised experiences.

It has been identified that AI-powered CRM management is so efficient that it can analyse a large set of customer data to provide adequate information about customer preferences, needs and behaviors. Based on that organisations can offer recommendations to meet those needs. Whereas, AI also plays a crucial part in customer retention and predictive analytics by predicting customer lifetime value, churn risks and others. Hence, it helped organisations to reduce risk and improve customer loyalty. It has been found that predictive analytics have improved customer retention by 27% . Structured AI algorithms are powerful enough to provide intelligent customer service by reducing costs by 30%, analysing customer feedback to ensure a 20% increase in customer retention and providing proactive recommendations to improve revenue by 37% (Kumar *et al.*, 2021). A table has been described below to showcase the impact of AI-powered CRM on organisations.

Table 1: The impact of AI on organizations.

AI-Powered CRM	Benefits Achieved
Developing Customers	Successfully enhance customer base by 20%
Customer Retention	Increase the rate of customer retention by 25%
Customer Loyalty	Ensure customer loyalty by 20%
Profit Maximization	Enhance overall profitability by 30%
Strategic Planning	Help in strategic planning to increase efficiency by 33%
Value Addition	Add significant value to the organisation as well to the brand

5. DISCUSSION AND FINDINGS

CRM has evolved as a useful technology for firms to manage and optimise marketing automation. Organisations, irrespective of its scale, are still encouraged to use CRM in order to better develop and maintain connections with their consumers. Client engagement and engagement, as well as earnings, may all benefit from improved relationships with clients (Kumar *et al.*, 2021).

Clients' requirements, ethnicities, lifestyles and purchasing habits will tend to evolve. Organisations who comprehend innovation and are in the forefront, frequently generating innovation, will be the organisations that endure and prosper. Many organisations work tirelessly to gain new clients, but the connection with those consumers typically ends there (Berezin *et al.*, 2015). The relevance and effect of software have been addressed on multiple occasions, as it has provided assistance that has led to the present degree of 'customer relationship management' capabilities and applications. It is fruitless and incomprehensible to fulfil the above listed requirements without the owner's cooperation as one of the most vital aspects (Fidel *et al.*, 2015).

6. CONCLUSION

A comprehensive 'customer relationship management' plan execution should include all phases that involve communication with the client in one firm. Organisations must first and foremost use a methodical tactical strategy in order for the supplied method to be effective. The results show how AI technology can improve CRM processes in terms of personalisation, proactive customer care and customer experiences. For businesses looking to use AI in CRM to increase customer loyalty and establish long-lasting client relationships in the dynamic business environment, the study offers insightful information.

7. FUTURE SCOPE

Due to the impact of CRM-related productivity on management procedures, it is indeed easy to see CRM as a critical tool in the quest of mutually beneficial relationships between customers and providers. The benefit of using the CRM Services component for sustainable construction might thus be quantified by establishing the relationship between improved customer service, increased 'customer–vendor relationship', and 'service-process management productivity'.

REFERENCES

Arsic, S., K. Banjevic, A. Nastasic, D. Rosulj, and M. Arsic (2018). "Family business owner as a central figure in customer relationship management," *Sustainability, MDPI, Open Access Journal*, 11(1), 1–19.

Battor, M., and M. Battor (2010). "The impact of customer relationship management capability on innovation and performance advantages: testing a mediated model," *Journal of Marketing Management*, 26(9–10), 842–857.

Berezin, A., N. Gorodnova, V. Matyushok, N. Plotnikova, and E. Shablova (2015). "Theoretical and applied aspects of transactional analysis of activities of enterprises [Paper Presentation]," *The 9th International Days of Statistics and Economics* (pp. 149–159).

Das, S., and H. K. Hassan (2021). "Impact of sustainable supply chain management and customer relationship management on organizational performance," *International Journal of Productivity and Performance Management.*

Fidel, P., W. Schlesinger, and A. Cervera (2015). "Collaborating to innovate: Effects on customer knowledge management and performance," *Journal of Business Research*, 68(7), 1426–1428.

Gil-Gomez, H., V. Guerola-Navarro, R. Oltra-Badenes, and J. A. Lozano-Quilis (2020). "Customer relationship management: digital transformation and sustainable business model innovation," *Economic research-Ekonomskaistraživanja*, 33(1), 2733–2750.

Guerola-Navarro, V., R. Oltra-Badenes, H. Gil-Gomez, and J. A. Gil-Gomez (2021). "Research model for measuring the impact of customer relationship management (CRM) on performance indicators," *Economic research-ekonomskaistraživanja*, 34(1), 2669–2691.

Pawełoszek I., Korczak J. (2017): From Data Exploration to Semantic Model of Customer, Proceedings of the 2017 Intelligent Systems Conference (INTELLISYS) London, pp. 382-388

Krizanova, A., L. Gajanova, and M. Nadanyiova (2018). "Design of a CRM level and performance measurement model," *Sustainability*, 10(7), 1–17.

Kumar, A., A. Soni, and K. Sahastrabuddhe (2021). "Implications on Customer Satisfaction and Loyalty in CRM," *Dr. Dy Patil B-School, Pune, India*, 950.

Li, F., and G. Xu (2022). "AI-driven customer relationship management for sustainable enterprise performance," *Sustainable Energy Technologies and Assessments*, 52, 102103.

McKenzie, C. R., and M. J. Liersch (2011). "Misunderstanding savings growth: implications for retirement savings behavior," *Journal of Marketing Research*, 48(SPL), S1–S13.

Yauwerissa, L., and J. S. Putra (2021). "The effect of service quality and customer relationship management towards customer loyalty," *International Journal of Review Management Business and Entrepreneurship (RMBE)*, 1(2), 339–345.

15. A Detailed Investigation of Implementing Artificial Intelligence and Business-to-Employee (B2E) e-business Model for Improving Customer Services

Jeidy Panduro-Ramirez[1], Jenny Ruiz-Salazar[2], Jady Vargas-Tumaya[3], Freddy Ochoa-Tataje[4], and Dante De-la-Cruz-Cámaco[5]

[1]Universidad Cesar Vallejo

[2]Campus Virtual Department, Universidad Privada del Norte

[3]Universidad Nacional de Educación Enrique Guzmán y Valle

[4]Universidad César Vallejo

[5]Universidad Tecnológica del Perú

ABSTRACT: One of the most significant changes to the corporate environment over the past ten years has been brought about by the development of digital technology and the internet. Internal organisational structures have changed as businesses migrate to a digital environment dominated by online business models and digital marketing (DM) strategies. In order to gather both statistical and factual data, this research article takes into account a hybrid approach of data collection (primary quantitative or survey and secondary qualitative). In order to determine the perspectives and experiences of both employees and consumers regarding the application of artificial intelligence (AI) in the Business to Employee (B2E) e-business model, a survey is undertaken. The integration of AI in the B2E e-business model and its impact on customer services are other topics that are the subject of existing research, academic publications, reports and case studies that are gathered and examined utilising qualitative analytical approaches. The results of this study's research project show how incorporating AI into the B2E e-business paradigm can enhance customer services.

KEYWORDS: E-business model, technology, resource orchestration theor, innovation, supply chain management, digital innovation

1. INTRODUCTION

E-business strategies are now boosting cooperative efficiency inside the supply network and creating substantial economic pay out by helping to improve digital interconnection all over bounds and combining different organisational resources and expertise, according to a massive amount of realistic proof from companies like Amazon, Dell and Lenovo (Harsono, 2014).

Instead of stepping into a business, an e-business may be a technique to perform modest and big commercial exchanges digitally. Furthermore, it advises that present trade forms be transformed into more profitable forms (Al-Gahtani, 2003).

The primary research question of this paper is,

1.1. How Does the Business's Resources Control System Change Over Time as a Result of the Time Spent Investing in e-business Technology?

Although a business plan delivers a wealth of business-related data about a certain firm, e-commerce, or electronic business culture, client expectations might cause this data to shift over time. The internet of

DOI: 10.1201/9781003532026-15

things (IoT) is used to record e-commerce company specifics as well as sales and processing information in order to overcome the aforementioned challenges. In 2020, the established IoT-based e-business method will boost the e-business sector to $2.5 billion. This article has been carried out the role of Business to Employee (B2E) E-business model for enhancing customer service and its impact on business productivity.

1.2. Objectives and Hypothesis

- To research how incorporating artificial intelligence (AI) into the B2E e-business model will affect how well customers are served.
- To investigate how AI technologies are incorporated into the B2E e-business paradigm and how that impacts the provision of customer support.
- To evaluate how customers and workers feel about the integration of AI into the B2E e-business paradigm for customer support.
- To identify the difficulties and factors to be taken into account when incorporating AI into the B2E e-business paradigm to enhance customer services.

Hypothesis 1: *The delivery of customer service is favourably impacted by the incorporation of AI technology into the B2E e-business paradigm.*

Hypothesis 2: *The integration of AI in the B2E e-business paradigm for enhancing customer services is well received by both employees and customers.*

Hypothesis 3: *Data privacy issues, training and skill development and striking a balance between AI automation and human contact are all difficulties involved with implementing AI in the B2E e-business paradigm.*

2. LITERATURE REVIEW

By supporting the growth of net profit, job performance, functional talent acquisition and high performance, e-business technology may aid in the development of a better chance of succeeding. e-business technologies can aid the company's ability to manage profitable product margins. Internet technology allows the company to exchange accurate and fast production costs and need data in real-time with external providers and end consumers, allowing it to effectively manage product margins. In the same way, e-business technologies may be used to boost staff productivity. Workers may acquire and exchange more heterogeneous or diverse material (e.g., data about the manufacturing method staff) and learn to execute numerous activities thanks to the company's web-based communications infrastructure (e.g., emails, Intranet). E-business tools can help with operations talent acquisition. The firm achieves relevant and complete information from the market using e-business technologies in order to attract and enrol excellent processability is required to formulate and combine its level of talent (Al-Ghaith *et al.*, 2010). Cortefiel, for example, employs web-based popular media websites like LinkedIn, Facebook and Twitter to acquire operationally qualified personnel who suit the profile required for its talent design procedure.

E-business methods, described as 'a type of business activity that depicts web data flows over organisational boundaries and connects suppliers and customers to promote digital transactions', are discussed in this article. IBM Ogilvy and Mather created this essential e-business operation in 1994, and it became a success by the year 2000 when IBM unveiled a $300 million ad for their e-business network. Furthermore, facilities and data may be specified without reporting revenues, which has a significant impact on the total business deal and raises the expenditure of the business strategy (Bharadwaj and Soni, 2007). As a result, the business strategy must have particular profit features that are utilised to increase the overall company strategy, item specifics, services, finance plan and end-user data. Companies spend millions of euros on information technology (IT) to improve their performance levels and profitability. However, not all IT initiatives provide the desired outcomes. This circumstance necessitates a thorough (re)evaluation of all IT initiatives by management.

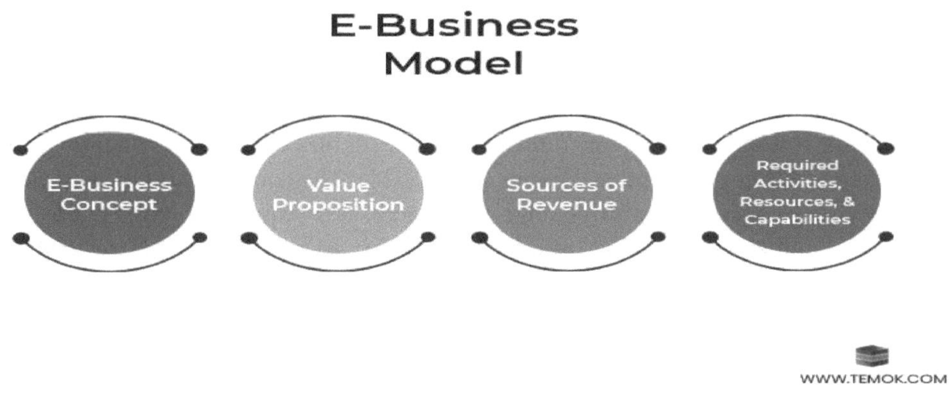

Figure 1. Structure of e-business model. Source: Saura et al. (2019).

A focal firm and its collaborators may accomplish flexible connection and real-time sharing across distant applications, improve coordinate development and distribution and develop business processes and operational procedures thanks to platform architectural adaptability. A focused firm may identify and source high-quality resources, enhance relationships with markets and consumers and swiftly understand and respond to consumer demands by boosting partner participation in e-business operations activities (Bajdor and Pawełoszek, 2020). The degree to which a focused firm has processes and practices in place to promote supply chain associates' participation in e-business activities is defined as partner involvement (Saura *et al.*, 2019).

Depending on the supply chain members with whom a focused firm engages, researchers classify diverse e-business operation capabilities into 3 groups:

- e-procurement competence,
- web-based channel management effectiveness and
- online service potential.

The process component lens proposes that the elements of business operations maintain structural linkage through integrating and utilising organisational assets to attain value creation (Patulak *et al.*, 2018). Identifying the links between components of e-business technologies for producing business performance is aided by resource orchestration theory.

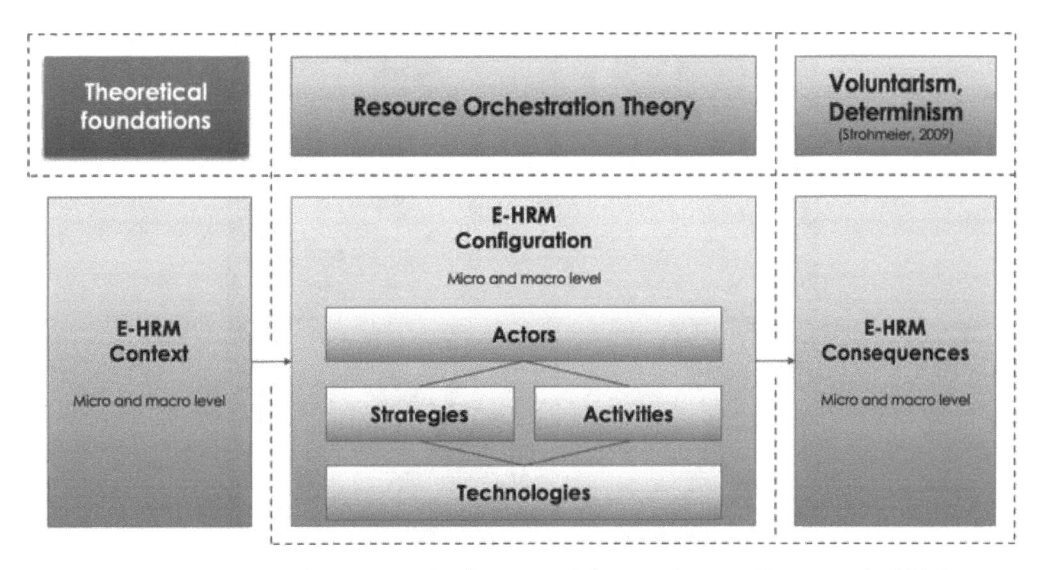

Figure 2. Concept of 'Resource Orchestration Theory'. Source: Huang et al. (2021).

The core logic of the resource orchestration concept, which extends from the resource-based approach, indicates that the success of organisational resource utilisation is established by utilising multiple resources via a sequence of managerial activities. A business can purchase asset portfolios and integrate the organised resources and generate unique innovations through successful resource restructuring activity. Capability leveraging is a set of management measures that enables companies to deploy competencies and capitalise on specific market possibilities (Huang *et al.*, 2021). The coordination of these two management acts is necessary for a firm's comparative edge to emerge. For supporting digital business operations, standardisation for online platforms allows business associates to quickly integrate, link and create automatic connectivity (Huang *et al.*, 2021).

3. RESEARCH METHODOLOGY

This research paper has considered mixed method of data collection, such as secondary qualitative to gather data from different sources. Implementing e-business model helps to create maximum advantageous environment within organisations. Secondary qualitative method was used to collect factual and theory-based information. For that, different journals and articles are considered by using keyword search. The study combines both primary quantitative research with secondary qualitative research techniques. To acquire information on employees' and customers' perspectives and experiences about the integration of AI in the B2E e-business model, the primary research comprises conducting a survey of both groups. The secondary research entails a thorough evaluation of the literature to assess the information already available, academic papers, reports and case studies pertaining to the application of AI in the B2E e-business model for customer services. A thorough investigation of the subject is possible thanks to the research technique, which takes into account both quantitative data and qualitative insights from the literature.

4. ANALYSIS AND INTERPRETATION

Based on analysing the collected data, it can be interpreted that AI is one of the most powerful technologies that has emerged. It can be implemented in more or less every industry to enhance performance. However, concerning this research topic about analysing the implementation of AI in business to employee e-business, it has been found that it has greatly influenced customer services by streamlining processes, improving efficiency and providing personalised experiences (Huang *et al.*, 2021).

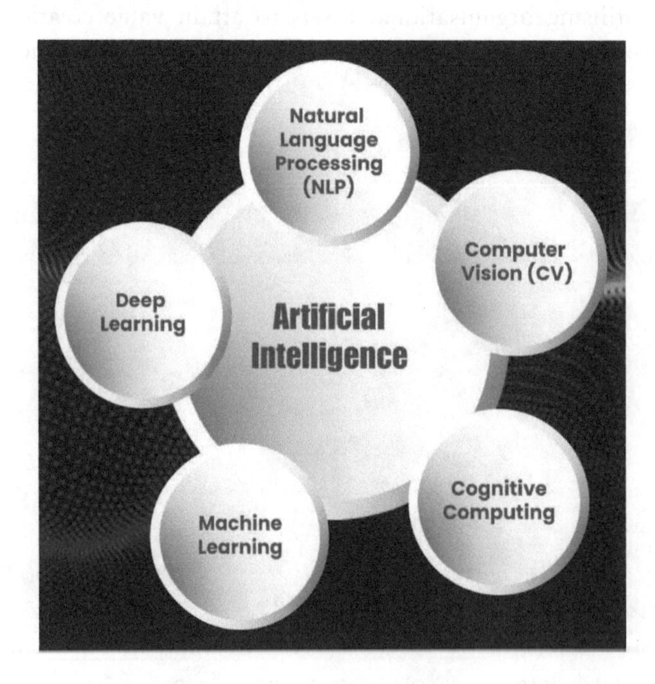

Figure 3. *Domains for implementing AI in e-business. Source: Zhu et al. (2020).*

Nowadays, every e-business has AI-powered chatbots and personal assistants that are able to provide personalised recommendations and solve customer queries. Hence, it has been predicted that by the end of 2030, 90% of customers will interact with emerging technologies. Whereas, predictive analytics for customer insights can identify patterns and trends to anticipate customer needs and preferences. In addition to that, customer sentiment analysis and voice recognition can also be evaluated using AI. As of now, more than 60% of businesses are ready to adopt AI to reduce customer handling time by 50% . However, a detailed discussion about the steps for implementing AI has been described using the table below.

Table 1: Steps for implementing AI in e-business.

Steps	Process
Stage 1	Defining anxiety gap
Stage 3	Understanding Organisational Capabilities
Stage 4	Improvising Value Transformation
Stage 5	Initiating change

5. DISCUSSION AND FINDINGS

A good resource structuring approach establishes an essential connection that brings together the resources of partners to increase digital operations competencies. To create e-business management skills in processes information, both framework design flexibility and partner interaction are required. These digital operations, on the other hand, cannot be carried out smoothly without the involvement of suppliers. Developing policies and processes to promote supplier participation can help to minimise partnership uncertainty while also increasing collaborative investments in vital tangible and intangible resources to enable digital transactions and cooperation. Open and trusted relationships will ensure ongoing supplier participation and minimise the risk of a long-term investment in e-procurement. A dynamic platform design, however, increases channel management performance by allowing the focal business to combine channel resources from many partners as well as coordinate activities in its promotional delivery mechanism. The digital platform allows the focal firm to establish an IT-enabled brand management strategy and develop potentials for continuous improvement by continuing to support collaborative efforts of advancement, concurrency control and order fulfilment across multiple operational networks. That is because, in a self-guided e-commerce journey, customer support is frequently, if not always, the consumer's primary point of contact. Customers interact with salespeople, see and touch things and ask questions in real-time at a brick-and-mortar store, while e-commerce is mostly a solitary experience. That is part of the attraction. Consumers may shop whenever they choose, wherever they want and in any manner they would like (Zhu *et al.*, 2020).

6. CONCLUSION

E-business innovation has a favourable influence on a wider range of experience that diminishes over time, eventually being non-significant and the firm's capacity to utilise a range of operational skills has a significant effect on financial performance that grows over time. The findings suggest that early IT expenditure is crucial for the operational success and long-term profits of the company. Because enterprises have more time and experience creating IT-enabled operational skills earlier in their growth, their income is maximised. In this work, researchers break down the e-business system into technological, operational and business elements and propose two methods to explain how to process components that lead to business value generation. According to the survey, both employees and consumers view and experience the integration of AI in the B2E e-business model for enhancing customer services favourably.

REFERENCES

Abid, A., and M. Rahim (2010). "Understanding factors affecting e-business technology introduction by Saudi small and medium enterprises (SMEs): toward developing a conceptual framework," *Australia Conference Paper. Caulfield School of IT, Monash University.*

Al-Gahtani, S. (2003). "Computer technology adoption in Saudi Arabia: Correlates of perceived innovation attributes," *Information Technology for Development*, 10, 57–69.

Al-Ghaith, W., L. Sanzogni, and K. Sandhu (2010). "Factors influencing the adoption and usage of online services in Saudi Arabia," *The Electronic Journal of Information Systems in Developing Countries*, 40(1), 1–32.

Bajdor P., and I. Pawełoszek (2020). "Data mining approach in evaluation of sustainable entrepreneurship," *Procedia Computer Science*, 176, 2725–2735. https://doi.org/10.1016/j.procs.2020.09.284.

Bharadwaj, P., and R. Soni (2007). "E-commerce usage and perception of e-commerce issues among small firms: results and implications from an empirical study," *Journal of Small Business Management*, 45, 501–521.

Enterprise Resources Planning and New Leverages: Knowledge Management and E-business.

Harsono, A. (2014). "The role of E-business in supply chain management," *Journal of Academia.Edu*, 1(4).

Huang, H., Y. I. Kaigang, R. L. Kumar, and V. Praveena (2021). "Category theory-based emotional intelligence mapping model for consumer-E-business to improve E-commerce," *Aggression and Violent Behavior*, 101631.

Patulak, I. M., M. B. Firdaus, and N. Dengen (2018). "Design of e-business furniture SMEs from commodity and waste utilization perspective," *2018 2nd East Indonesia Conference on Computer and Information Technology (EIConCIT)* (pp. 29–34). IEEE.

Saura, J. R., P. R. Palos-Sanchez, and M. B. Correia (2019). "Digital marketing strategies based on the e-business model: literature review and future directions," *Organizational Transformation and Managing Innovation in the Fourth Industrial Revolution*, 86–103.

Zhu, Z., J. Zhao, and A. A. Bush (2020). "The effects of e-business processes in supply chain operations: process component and value creation mechanisms," *International Journal of Information Management*, 50, 273–285.

16. The Contribution of Green Technology and Machine Learning Toward Developing Environmental Sustainability

Jesús Padilla-Caballero[1], Miguel Zubiaur-Alejos[2], Jorge Poma-Garcia[3], Claudia Poma-García[1], and Luis Rojas-Zuñiga[4]

[1]Universidad César Vallejo

[2]Universidad Nacional de Ingeniería

[3]Universidad Nacional Del Centro Del Perú

[4]Universidad Nacional de Educación Enrique Guzmán y Valle

ABSTRACT: At the national level, the creation of environmentally friendly and environmental plans depend heavily on green technologies. Due to the combined constraints of resources and the environment, green technology advancement has been seen to be becoming more and more significant. This study's main goal is to investigate how machine learning and green technology integration might support environmental sustainability efforts. In this study, secondary research using qualitative techniques was the method of choice. The collected and analysed existing research includes academic publications, reports, case studies and reports on how green technology and machine learning contribute to environmental sustainability. The relevance, credibility and dependability of the data are assessed, and significant themes and conclusions are determined by applying qualitative analysis techniques including content analysis and thematic analysis. This study has demonstrated the value of combining machine learning and green technologies to promote environmental sustainability. They emphasise the contribution of green technology to lessening the negative effects on the environment and the benefits of machine learning in decision-making and effective resource management.

KEYWORDS: Green technology, innovation, sustainability, environmental sustainability, development, eco role

1. INTRODUCTION

The establishment of a green lifestyle necessitates proper planning, analysis and elimination of environmental contaminants from polluted locations using environmentally friendly processes. High-strength management of water from home, commercial, as well as a multinational corporation that provides is a big concern across the worldwide. As a result, this work, Environmental Conservation as well as Green Technologies, focuses on the important role of low-carbon technologies in environmental protection, as well as the expansion of processes, practices and implementations that may upgrade or replace current solutions (Zhu *et al.*, 2021). On the one side, 'green technology innovation' generates economic benefits through conventional technical innovation, while on the other edge, it accepts outside environmental destruction.

It has been noted that ecological technological advancement is growing, which could have a significant impact on future atmospheric growth, based on the image above.

DOI: 10.1201/9781003532026-16

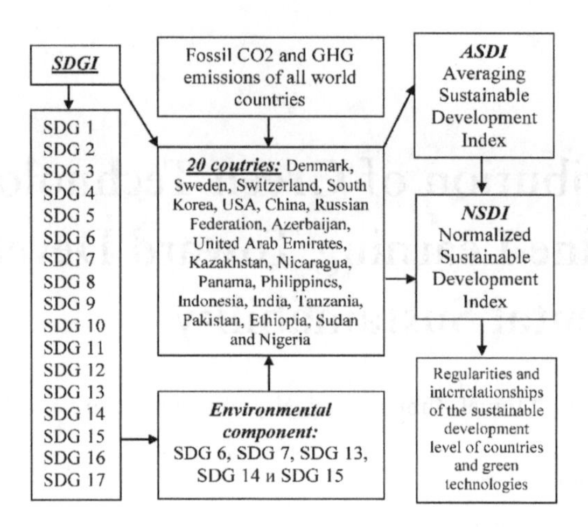

Figure 1. *Green technology and sustainable development. Source: Geddes et al. (2018).*

Despite the fact that a number of its forms have received widespread recognition, 'green technology' remains a broad and challenging topic to define. Many governments have stated their intentions to phase out single-use plastics, a goal which would necessitate significant expenditures in alternatives including 'paper replacements', 'bioplastics' and 'recycling technologies.' For instance, Singapore has promised to recycle 70% of its trash by 2030 (Geddes *et al.*, 2018).

Environmental sustainability has become a strongly shared goal as a result of the '***Brundtland Commission*** (WCED, 1987)', the '***Earth Summit*** (UNCED, 1992)' and '***Canada's attitude*** (1992)'. The phrase 'environmental sustainability (ES)' refers to the maintenance of the world's life-support systems, or more specifically, the maintenance of ecological sink capacity to absorb wastes and environmental source capacities to replenish fundamental resources like clean air (Wang *et al.*, 2021).

1.1. Objectives and Hypothesis

- To study how machine learning and green technologies can be used to create an environmentally sustainable future.
- To investigate how green technology may promote sustainable behaviours and lessen its influence on the environment.
- To find out how machine learning techniques are being incorporated into green technology and how this affects decision-making and resource management for environmental sustainability.
- To identify the difficulties and factors to be taken into account when implementing green technology and machine learning for environmental *sustainability.*

Hypothesis 1: *The application of machine learning and green technology helps in maintaining environmental sustainability.*
Hypothesis 2: *In order to lessen environmental effects and encourage sustainable habits, green technology is crucial.*
Hypothesis 3: *The fusion of green technology and machine learning opens up possibilities for creativity and the creation of long-lasting responses to environmental problems.*

2. LITERATURE REVIEW

Environmental stability is the commitment to protect global ecosystems and environmental assets to advance human health and well-being for today as well as for the future. Human well-being is intrinsically linked to the health of the ecosystem. 'World Health Organisation or WHO' stated that environmental impacts cause 24% of all fatalities worldwide.

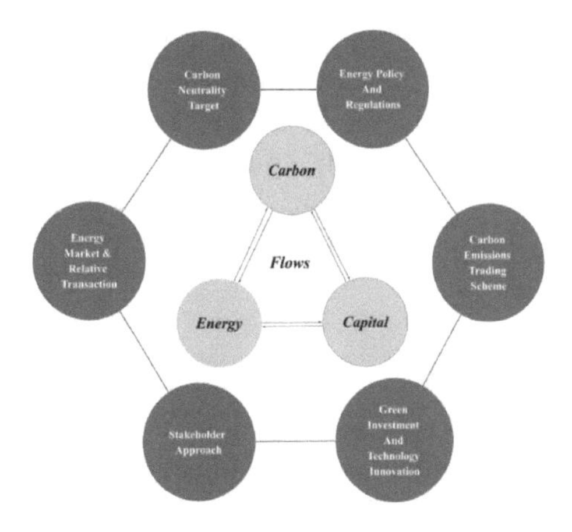

Figure 2. *Framework of 'Integrated Green Technology' flows. Source: Yin et al. (2021).*

To promote economic growth, modern manufacturing's development model relies on resource consumption, particularly non-renewable resource use. The unsustainable pursuit of GDP at the expense of resource waste and environmental devastation is unavoidable (Yin *et al.*, 2021).

Green technology innovation's emergence, growth and decline are all influenced by the surrounding ecosystem. From the standpoint of the entire process, production business green technology innovation is in step with the environmental niche shift. Niche is an inherent characteristic and it was initially presented by Grinnell (1917) in the context of bird population ecology, who described it as a creature's location and purpose in the ecosystem (Lisi *et al.*, 2020). The term 'Eco role' relates to an organism's unit's real domination or impact over its surroundings, like the speed of change of matter and energy, biological development rate and capacity to inhabit new habitats.

It has been identified from the above figure that green technology is used in different segments that help to achieve success in a successful manner. Science and research institutes, schools and universities and inventive skills make up the majority of research and design elements (Awan *et al.*, 2020). The 'Eco role' level is the ecological niche of green technical innovation in the manufacturing enterprise. This kind of analysis mostly represents the production industry's changing and developing green technology development tendency (Hottenrott *et al.*, 2016).

Small and medium businesses (SMEs) are often considered as the foundation of any business, contributing significantly to economic growth and job creation. 'Small- and medium-sized enterprises (SMEs)' make up the majority of established organisations in most emerging countries (Yacob *et al.*, 2019). Being forward-thinking is one of sustainable development's most important characteristics because many environmental acts have long-term effects. Likewise, the 'Environmental Protection Agency' of the United States describes it as 'meeting today's needs without jeopardising future generations' capacity to meet their own' (Williams and Schaefer, 2013).

Of course, a business must evaluate the potential advantages of a technological innovation against the expenditures associated with it (Wang *et al.*, 2021).

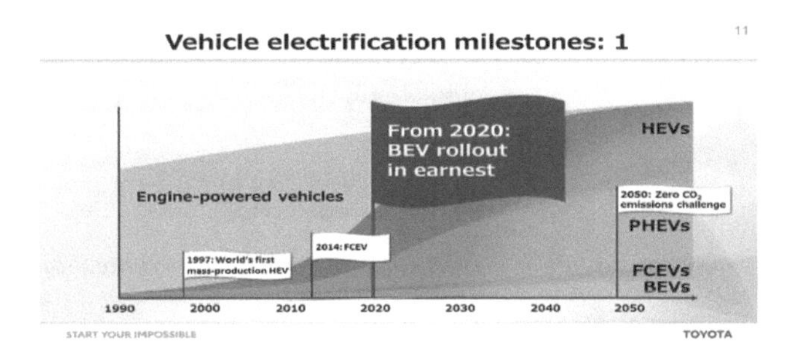

Figure 3. *Vehicle electrification milestones of toyota. Source: Toyota Motor Corporation (2020).*

When Toyota was choosing whether or not to build the Prius, for instance, it weighed the likelihood of future legislation as well as the chance of other companies catching up. On the one side, when more businesses implement emerging innovations, the benefit to a company's image and demand diminishes. When Toyota's Prius was first debuted, it nearly controlled the industry, but when other companies produced their hybrid vehicle models, the Prius' market share fell (Toyota Motor Corporation, 2020). Competing businesses' acceptance choices can deter a company from incorporating advanced technology in this situation (Cohen *et al.*, 2016).

3. RESEARCH METHODOLOGY

In order to gather information related to green technology implementation and develop sustainable environment, secondary qualitative method of data collection to develop quality analysis has been considered. Several databases such as Scholar, ProQuest and PubMed are considered to gather journals and articles regarding the topic. For this, researchers have used keywords such as green technology innovation, sustainable development, business growth and environmental sustainability to choose full pdf journals and quality articles from databases. Key themes and conclusions are identified using qualitative analytical approaches like content analysis and thematic analysis. The research methodology enables a thorough examination of how green technology and machine learning contribute to environmental sustainability by synthesising data from multiple sources. It offers information on the function of green technology, the incorporation of machine learning, difficulties and factors to take into account, as well as chances for innovation in sustainable solutions.

4. ANALYSIS AND INTERPRETATION

Upon reviewing the collected data, it has been found that both green technology and machine learning contribute significantly towards developing and achieving environmental sustainability by optimising energy efficiency, improving resource consumption and making smart decisions Based on analysis and interpretation, it can be identified that implementing machine learning algorithms can enhance the efficiency of wind turbines by more than 20% and solar panels by 10%. Moreover, it can also help analyse energy consumption patterns to identify opportunities for optimisation by leveraging a large set of data. Machine learning-based techniques can significantly reduce energy consumption by 35% .

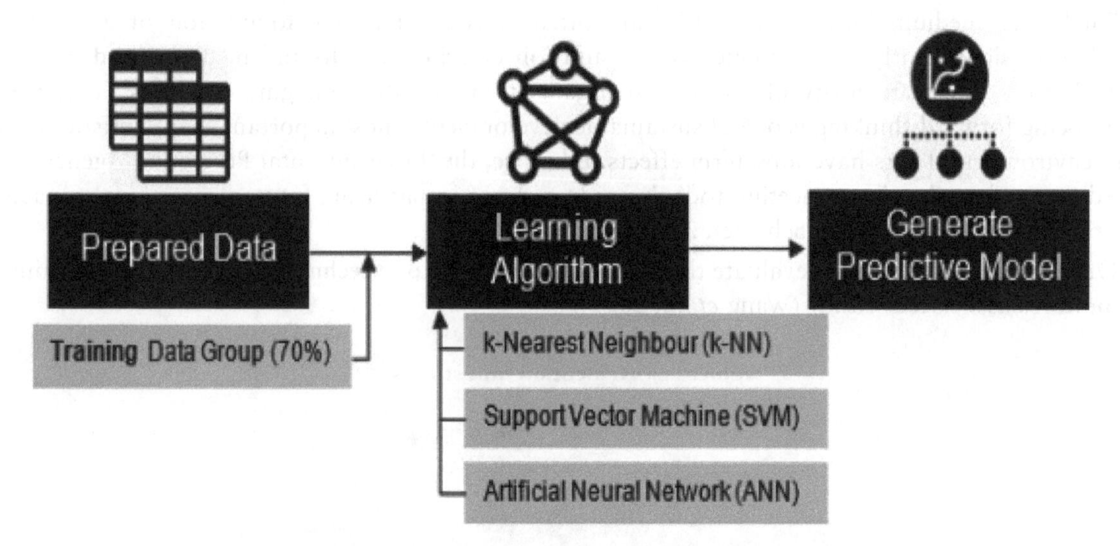

Figure 4. *Process of energy consumption prediction using machine learning. Source: Yin et al. (2021).*

Apart from that, the latest technological advancements can help in waste management and recycling by analysing historic data to optimise waste collection schedules and reduce greenhouse gas emissions. It can also increase the recycling rate by more than 60%. Regarding environmental monitoring and conservation, machine learning can detect deforestation patterns, predict water quality and identify endangered species. Last year, Microsoft used ML techniques to identify 1.9 square kilometres of degraded land for restoration (Wang *et al.*, 2021). A table has been illustrated below to compare different machine learning algorithms to optimise energy consumption.

Table 1: Comparison of different Machine Learning algorithms for energy consumption.

Machine Learning Algorithms	Optimisation Performance
LightGBM	40%
XGBOOST	25%
ANN	60%
LR	35%

5. DISCUSSION AND FINDINGS

Special function needs for green technology implementations are developed in the operation process or supplementary system in terms of substitute or special equipment. Existing operational techniques may appear to be more effective and dependable, yet they may be inconsistent with green technology. The modification in acceptance technique necessitates process reengineering as well as supporting facilities (Guo *et al.*, 2018). When replacing present manufacturing inputs with environmentally acceptable choices, companies must consider their processing characteristics. Modern machinery can handle materials such as metal, steel and cement in some circumstances, but not plastics or woods. Furthermore, according to Kuo and Smith (2018), the intended infrastructures may consume a lot of energy and generate a lot of trash during functioning.

6. CONCLUSION

Green technology is a wide concept that describes the use of engineering and innovation to produce environmentally-friendly products. Cleantech relates to items or services that boost operating effectiveness while cutting costs, decreasing energy use, minimising waste or mitigating negative effects on the environment.

Moreover, the combination of machine learning methods with green technology also has additional benefits. Large dataset analyses, energy efficiency improvements and environmental pattern prediction are all possible using machine learning algorithms, allowing for more precise resource management and decision-making. Opportunities to further improve the effectiveness and efficiency of environmental sustainability activities are presented by this integration. The results highlight how crucial it is to implement green technology practices, include machine learning approaches and deal with related issues if you want to achieve long-term environmental sustainability goals.

7. FUTURE SCOPE

This research paper has discussed the GT or green technology strategies and their impact on sustainable development. In this present era, green technological enhancement is widely used in multiple sectors to create ultimate and sustainable development. Through go green concept, organisations can use renewable things for further use. Researchers create a new global game-based model that incorporates these elements and investigates different factors that affect overall organisational development and acceptance decisions. Researchers also offer management advice to businesses.

REFERENCES

Awan, U., A. Kraslawski, J. Huiskonen, and N. Suleman (2020). "Exploring the locus of social sustainability implementation: a South Asian perspective on planning for sustainable development," *Universities and Sustainable Communities: Meeting the Goals of the Agenda 2030*. Springer: Cham, Switzerland.

Cohen, M. C., R. Lobel, and G. Perakis (2016). "The impact of demand uncertainty on consumer subsidies for green technology adoption," *Management Sci.*, 62(5): 1235–1258.

Geddes, A., T. S. Schmidt, and B. Steffen (2018), "The multiple roles of state investment banks in low-carbon energy finance: an analysis of Australia, the UK and Germany," *Energy Policy*, 115, 158–170.

Guo, Y., X. Xia, S. Zhang, and D. Zhang (2018). "Environmental regulation, government R&D funding and green technology innovation: evidence from China provincial data," *Sustainability*, 10(4), 940.

Hottenrott, H., S. Rexhäuser, and R. Veugelers (2016). "Organisational change and the productivity effects of green technology adoption," *Resour. Energy Econ.*, 43, 172–194.

Kuo, T. C., and S. Smith (2018). "A systematic review of technologies involving eco-innovation for enterprises moving towards sustainability," *Journal of Cleaner Production*, 192, 207–220.

Lisi, W., R. Zhu, and C. Yuan (2020). "Embracing green innovation via green supply chain learning: the moderating role of green technology turbulence," *Sustain. Dev.*, 28, 155–168.

Toyota Motor Corporation (2020). *Environmental Report 2018*, Available from: https://global.toyota/pages/global_toyota/sustainability/report/er/er18_full_en.pdf, Accessed on: 8 March, 2022.

Wakeford, J. J., M. Gebreeyesus, T. Ginbo, K. Yimer, O. Manzambi, C. Okereke, M. Black, and Y. Mulugetta (2017). "Innovation for green industrialisation: an empirical assessment of innovation in Ethiopia's cement, leather and textile sectors," *J. Clean. Prod.*, 166, 503–511.

Wang, X., S. H. Cho, and A. Scheller-Wolf (2021). "Green technology development and adoption: competition, regulation and uncertainty—a global game approach," *Management Science*, 67(1), 201–219.

Williams, S., and A. Schaefer (2013). "Small and medium-sized enterprises and sustainability: managers' values and engagement with environmental and climate change issues," *Business Strategy and the Environment*, 22(3), 173–186.

Yacob, P., L. S. Wong, and S. C. Khor (2019). "An empirical investigation of green initiatives and environmental sustainability for manufacturing SMEs," *Journal of Manufacturing Technology Management*.

Yin, S., N. Zhang, B. Li, and H. Dong (2021), Enhancing the effectiveness of multi-agent cooperation for green manufacturing: dynamic co-evolution mechanism of a green technology innovation system based on the innovation value chain," *Environ. Impact Assess. Rev.*, 86, 106475.

Zhu, L., J. Luo, Q. Dong, Y. Zhao, Y. Wang, and Y. Wang (2021). "Green technology innovation efficiency of energy-intensive industries in China from the perspective of shared resources: dynamic change and improvement path," *Technol. Forecast. Soc. Change*, 170, 120890.

17. An Empirical Analysis in Understanding the Role of Blockchain Technology in Supporting Collaborative Economy in Emerging Economies

Jeidy Panduro-Ramirez[1], Jenny Ruiz-Salazar[2], Jady Vargas-Tumaya[3], Freddy Ochoa-Tataje[1], and Dante De-la-Cruz-Cámaco[4]

[1]Universidad Cesar Vallejo

[2]Universidad Privada del Norte

[3]Universidad Nacional de Educación Enrique Guzmán y Valle

[4]Universidad Tecnológica del Perú

ABSTRACT: Although blockchain is still comparatively a latest innovation to the industry, the technology is not widely accepted in the global financial environment, but it has some distinguishing features that set it apart from other connected breakthroughs. Although the technology has gained considerable notoriety through Bitcoin and other cryptocurrencies, its use in the marketplace is not regarded. The significance of blockchain technology on advanced markets has been recognised. The purpose of this study is to investigate and evaluate how blockchain technology can aid in the creation and expansion of cooperative economic models in developing nations. Secondary qualitative methods have been utilised by researchers to gather information. Blockchain mainly focuses on promoting a collaborative economy globally, which is the subject of a number of academic studies that have been gathered and examined. The research study advances knowledge of how blockchain technology might benefit emerging economies' collaborative economies. It highlights the opportunities, advantages, difficulties and challenges regarding the use of blockchain as well as the consequences for equitable economic growth.

KEYWORDS: Blockchain technology, cryptography, security, economy, innovation

1. INTRODUCTION

The blockchain technology is comparatively new and one of the most important technologies in today's world that is used in storing and recording information in a secured way. Blockchain technology is impossible to hack, and retrieving the data and information is very difficult through this technology. This technology is efficient in tracking as well as transferring information and is considered one of the effective technologies in the emerging economies of the world. The use of modern gadgets has reduced work pressure as well as errors; however, the digitalised equipments are very easy to hack and extract confidential information. Therefore, through blockchain technology, the security system can enhance the efficiency of this new emerging economies (Yaga *et al.*, 2019). Considering the intimate ties that exist between blockchain systems and the Internet in developed countries and the enabling of mentoring trading, few research have looked into how these two effects are linked (Cârlan, 2019).

As a result of these issues, international marine corporations are working to develop more appropriate technology operation techniques. The naval company has made note of blockchain based, which has been steadily applied towards the international container supply chain network and has created the opportunity

DOI: 10.1201/9781003532026-17

for revolutionary innovation in the nautical sphere. The available studies on the use of blockchain infrastructure and the encouragement of the conversion and strengthening of the coastal distribution network are still lacking (Liu *et al.*, 2021). Second, they identify the problem areas of coastal expansion and the synergy point of combining marine and blockchain via maritime studies.

1.1. Objectives and Hypothesis

- To comprehend how blockchain technology supports emerging economies' collaborative economies.
- To examine the advantages, difficulties and possibilities brought on by the use of blockchain technology in the collaborative economy.
- To investigate how blockchain technology may affect inclusive economic development and female empowerment in emerging economies.
- To pinpoint the crucial elements impacting the effective application and uptake of blockchain technology in the cooperative economies of developing nations.

Hypothesis 1: *Blockchain technology adoption in emerging economies' collaborative economies results in lower costs, better traceability and more transparent peer-to-peer transactions.*

Hypothesis 2: *Blockchain technology offers exceptional chances for equitable economic development, financial inclusion and the empowerment of disadvantaged communities in developing nations.*

Hypothesis 3: *The adoption of blockchain technology has a favorable impact on participant trust, efficiency and transparency in emerging economies' collaborative economies*

2. LITERATURE REVIEW

2.1. Concept of Blockchain

The blockchain is a decentralised public ledger that exist in various networks. The blockchain technology is mostly useful while dealing with cryptocurrencies. The blockchain technology is a process in which all the information about the transaction and tracking the assets of the business can be done.

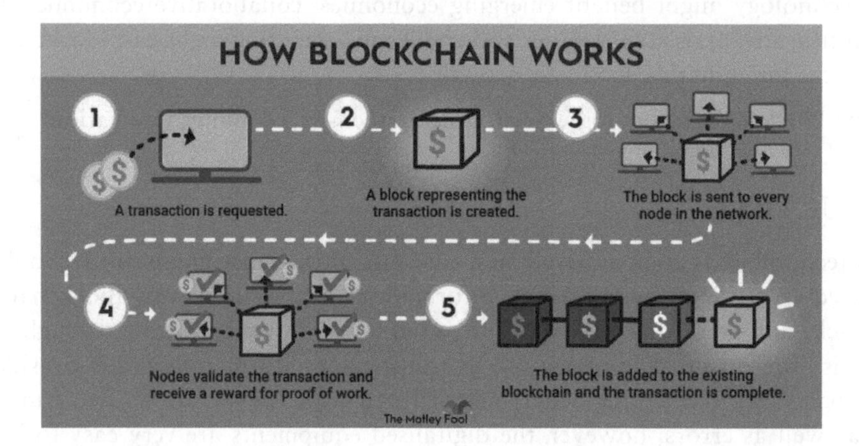

Figure 1. *Working of a blockchain. Source: Herweijer and Swanborough (2018).*

Here the assets can be tangible in nature. In the blockchain technology, the main concept is that an individual can record and distribute the information among the blocks but is unable to edit the information (Saha *et al.*, 2019).

2.2. Various Characteristics of Blockchain Technology

Although relatively new to the industry and yet not extensively recognised in the global economic climate, blockchain offers a number of unique characteristics that set it apart from other related technologies. The several characteristics of blockchain technology include:

1. Immutability: this feature is one of the effective features of blockchain technology as no information can be destroyed or hacked through this technology (Anwar, 2018).
2. Decentralised: the network does not have any specific governing body and is maintained by a group of members which makes it easier to access.
3. Security: A secure environment is created using cryptography, and therefore, it provides a layer of protection from the various malicious internet activities.
4. Faster settlement: the blockchain technology provides fast services unlike traditional systems.
5. Distributed ledgers: the ledgers are distributed publically; thus, everyone can see the transactions and, therefore, diminish the corruption process.

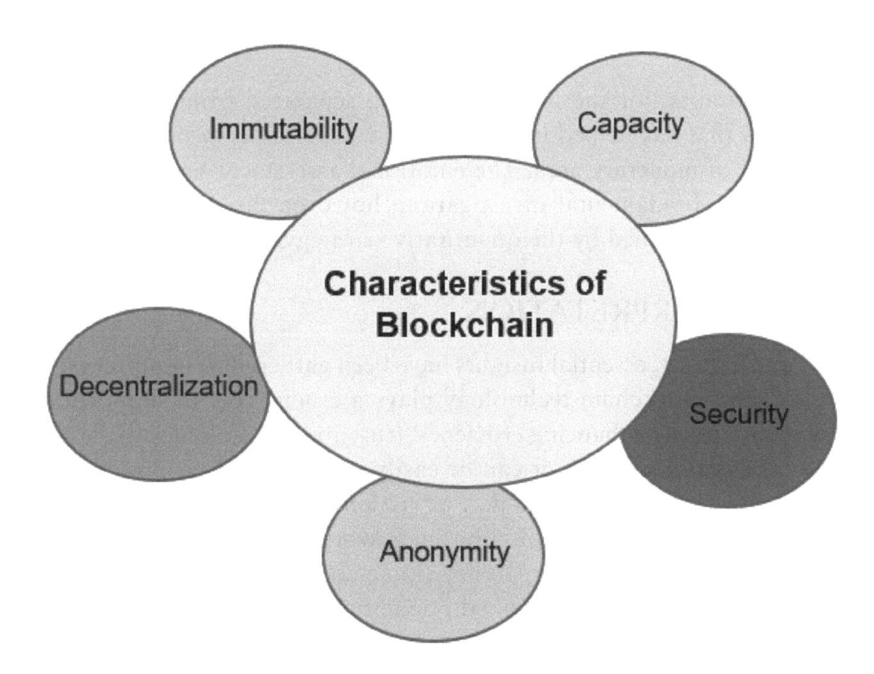

Figure 2. *Various features of blockchain. Source: Self-developed.*

2.3. Impact of Blockchain Technology in Various Service Systems

In manufacturing industry as well as in finance industry, the usefulness or the impact of blockchain has been a much discussed topic nowadays. The various improvised features in the blockchain technology have been proved to be very helpful in these industries. The use of cryptography in blockchain technology has enhanced the security system, which has a great impact over the financial service systems. The banking and insurance sector are much vulnerable to hacking and security crisis (Herweijer and Swanborough, 2018). Therefore, the use of blockchain technology has provided secured transaction records and better services to its customers.

2.4. Advantage and Disadvantage of Blockchain Technology

2.4.1. Advantage

Utilising blockchain in various service systems has several benefits since it improves security and fosters greater consumer, business partner and stakeholder trust promotes the working sector's transparency, which automatically boosts productivity and performance.

2.4.2. Disadvantage

Though the technology has various advantages, however, some disadvantages can also be seen while using this technology. The blockchain technology becomes slower to process when many users use the network at a same time.

3. RESEARCH METHODOLOGY

In this examination program, an efficient way has been utilised to direct the exploration to come by the right consequences of the effect and convenience of blockchain technology in different assistance industry like banking and manufacturing. The examination approach has been finished utilising the fitting techniques that can give a credible information to the detailing and legitimacy of the exploration work. Hence in this examination work, both essential and auxiliary information are being utilised to obtain the expected outcomes in light of the blockchain innovation in arising economies. It is critical to comprehend the ease of use of the innovation and its effectivity in view of safety worries as it can assist with diminishing the fake exercises and can assist in making a powerful correspondence with the different business components. This aides in getting the genuine information on the effect and productivity of blockchain technology (Saberi *et al.*, 2019).

The auxiliary information is, for the most part, being separated from the diary, articles as well as different organisation sites that has helped in acquiring the general perspective on the impacts of AI and its productivity in the field of monetary area. The emotional assessment has helped with separating data that was assembled from the fundamental investigation; however, the data that are accumulated in the discretionary procedure was examined by the quantitative strategy.

4. ANALYSIS AND INTERPRETATION

Based on analysing a large data set, essential insights have been gathered to interpret related to the research topic. It has been found that blockchain technology plays a crucial role in supporting the collaborative economy in emerging economies by enhancing efficiency, trust and transparency while making transactions. Based on the characteristics of blockchain, it can be easily identified that it provides a decentralised and transparent ledger to ensure secure and verifiable transactions. It has significantly improved trust among participants in emerging countries. In the global business world, 75% of business leaders consider using blockchain technology due to trust and transparency (Liu *et al.*, 2021). Moreover, blockchain technology can also provide financial services to the unbanked population in emerging economies. It has been found that blockchain technology has significantly bridged the gap in financial services by providing access to more than 1.9 million adults. Apart from that, peer-to-peer sharing platforms powered by blockchain facilitate collaborative consumption by providing accommodation, services and sharing resources. It has been found that the global sharing economy will reach a valuation of 330 billion by 2026.

Figure 3. *Sharing economy platforms and total funds raised.*

Furthermore, blockchain technology also enables safe, secure and efficient microfinance services to provide small loans and other essential financial services to entrepreneurs in emerging economies. Although the global cost of remittance is about 7%, using blockchain technology this cost can be reduced. It also helps in developing and providing access to decentralised applications as the total number of blockchain users will be increasing to over 100 million in 2022 (Darlington, 2021). However, for building an effective blockchain network some steps need to be followed that has been illustrated in the below table.

Table 1: Steps for building a blockchain network.

Stages for Development	Processes Involved
Step 1	Every node collects information on transactions
Step 2	Each node collects new transactions into a new block
Step 3	Findings a difficult proof of work by the nodes for the new block
Step 4	Broadcasted the identified block to all nodes
Step 5	All blocks are reviewed for acceptance for every valid transaction
Step 6	Expressing the acceptance by creating a new block in the chain using the hash of the accepted block

5. DISCUSSION AND FINDINGS

The above data and various discussions about blockchain and its influence on the emerging economy provides effective information of the technology. The use of secondary data has shown various advantages as well as disadvantages of the technology. As no information can be destroyed or edited, therefore, the information in the network cannot be misused. One other factor that has put a positive impact over the technology is the transparency (Dubey *et al.*, 2020). This reduces corruption and proves to be effective in manufacturing, financial as well as other economic sectors of the economy. The use of cryptography makes the technology much more valuable and trusted by various business organisation. However, it also has various disadvantages and still requires improvements to be more effective in the use of this industries. The high energy consumption and the delayed process makes it difficult to operate in the fast-moving world.

6. CONCLUSION

Based on the above discussion done over the impact of blockchain technology, it can be concluded that the innovation has helped various emerging economies to conduct a safe and secure business process in various sectors of the economy. The enhancement in security of data transactions is one of the major impacts of the blockchain technology in the various industries of the economies. The faster settlement factors in the blockchain technology makes it much more favourable for the emerging economies as it does not take much time and provides quick and easy solutions to the various business process. The research has provided valuable insights to governments, companies and scholars on useful information about how to use blockchain technology to create cooperative economic models in poor nations. Emerging economies have the chance to promote inclusive growth, improve transparency and empower communities while removing historical barriers to economic participation by embracing blockchain technology.

7. FUTURE SCOPE

Blockchain technology is creating an inventive perspective that is keeping its engraving in most of the parts of cutting-edge lifestyle. It is obvious that the real execution of this can lessen the complexities of present-day life in an organized manner. It is not only productive for the monetary area but it can loosen up its arm to the transportation and arranged activities system and clinical benefits of space. Also, its extensive plan is truly prepared for upgrading the step-by-step lifestyle of people as it can reduce the work strain and security breach in each space of the economy.

REFERENCES

Anwar, H. (2018). *6 Key Blockchain Features You Need to Know about! 101 Blockchains.* Available at: https://101blockchains.com/introduction-to-blockchain-features/. (Accessed on: 1 March 2022).

Cârlan, V. (2019). *Maritime Supply Chain Innovation: Costs, Benefits and Cost-effectiveness of ICT Introduction.* University of Antwerp.

Darlington, N. (2021). *What is Blockchain Technology? A Step-by-Step Guide For Beginners.* Blockgeeks. Available at: https://blockgeeks.com/guides/what-is-blockchain-technology/. [Accessed on: 1st March, 2022]

Dubey, R., A. Gunasekaran, D. J. Bryde, Y. K. Dwivedi, and T. Papadopoulos (2020). "Blockchain technology for enhancing swift-trust, collaboration and resilience within a humanitarian supply chain setting," *International Journal of Production Research*, 58(11), 3381–3398.

Golosova, J., and A. Romanovs (2018). "The advantages and disadvantages of the blockchain technology," *2018 IEEE 6th Workshop on Advances in Information, Eelectronic and Electrical Engineering (AIEEE)* (pp. 1–6). IEEE.

Herweijer C., and J. Swanborough (2018). "8 ways blockchain can be an environmental game-changer," *World Economic Forum.* Available at: https://www.weforum.org/agenda/2018/09/8-ways-blockchain-can-be-an-environmental-game-changer/. (Accessed on: 1 March 2022).

Liu, J., H. Zhang, and L. Zhen (2021). "Blockchain technology in maritime supply chains: applications, architecture and challenges," *International Journal of Production Research*, 1–17.

Saberi, S., M. Kouhizadeh, J. Sarkis, and L. Shen (2019). "Blockchain technology and its relationships to sustainable supply chain management," *International Journal of Production Research*, 57(7), 2117–2135.

Saha, A., R. Amin, S. Kunal, S. Vollala, and S. K. Dwivedi (2019). "Review on 'Blockchain technology based medical healthcare system with privacy issues'," *Security and Privacy*, 2(5), e83.

Yaga, D., P. Mell, N. Roby, and K. Scarfone (2019). "Blockchain technology overview," *arXiv preprint arXiv:1906.11078.*

18. Investigation of Machine Learning Approaches in the Financial Services

Jesús Padilla-Caballero[1], Obed Matías-Cristóbal[1], Kathy Flores-Cabrera-De-Ruiz[2], and María Mamani-Ticona[3]

[1]Universidad César Vallejo

[2]Universidad Nacional de Ucayali

[3]Instituto Superior Pedagógico Privado Nueva Esperanza

ABSTRACT: Over the past few years, there have been significant technological breakthroughs that have impacted everyone and every part of life. One such development that has the potential to alter the course of human history is machine learning (ML) technology.

Artificially intelligent computer software may imitate and behave like a person. In most corporate operations nowadays, ML and artificial intelligence (AI) are applied. The financial services sector is one where the sending of AI and ML is extending rapidly. AI and ML are supplanting the main part of fundamental monetary capabilities like risk appraisal, exchanging stocks and application for credits.

ML will not entirely replace the need for finance specialists, but it will give finance managers more time to concentrate on the crucial, strategic components of their business rather than tedious, repetitive duties. The significant target of this study is to assess the utilization of AI and ML in different financial processes as well as the impacts of these forward leaps on undertakings, laborers and monetary subject matter experts.

KEYWORDS: Machine learning, artificial intelligence, financial services, risk management, stock trading

1. INTRODUCTION

Machine learning (ML) is enabling advances in robotics, computer vision and natural language processing. There is a significant deal of interest in using ML in other wide sectors where there is an excess of data as a result of these excellent applications. Because of ongoing specialized improvements in the disciplines of artificial intelligence (AI), especially ML, this age is in an exceptionally lucky position. Recently performed truly, an errand is currently delivered through programming and robotized innovations (Cavalcante *et al.*, 2016). As indicated by eminent American PC expert John McCarthy, AI alludes to the science and designing part of making clever robots. Cognitive intelligence, which is the capacity to handle problem-solving, manage thinking, assimilate incredibly complicated ideas, and learn quickly via experience, is a key component of how modern robots function and perform. Natural level intelligence, which most people possess, is distinct from cognitive intelligence. In the upcoming, profound upheaval of the finance industry, AI is a supporting force. One of these rapidly growing fields is the identification of financial fraud. According to McAfee, financial fraud-related cybercrime costs the world economy 600 billion dollars annually, or 0.8% of GDP (Dyzma, 2019).

AI has been demonstrated to be quite successful in detecting financial fraud. Reportedly to a current Forbes magazine article, AI systems may significantly decrease payment and money-related frauds, according to 81% of financial fraud detection specialists. Payments related to fraud are becoming less common as a result of AI's capability of interpreting trend-based insights from algorithms that use ML

DOI: 10.1201/9781003532026-18

(Dyzma, 2019). AI employs ML and predictive analytics to quickly detect any abnormality in very large datasets. A ML algorithm's prognostic value increases with the amount of data it is fed. In most cases, ML methods choose to increase prediction and efficiency by carefully analysing historical data from a large data network. As a portion of the payment processing analysis, big banks may also deploy fraud detection tools based on predictive analytics (Columbus, 2019). Predictive analytics is yet another tool banks can employ to spot fraud in mobile programs used for banking or for placing orders and making payments for products and services.

2. LITERATURE REVIEW

According to McCarthy *et al.* (2006), a 1955 Dartmouth Summer study proposal is where the idea of AI first appeared. More advanced versions of this theory claim that 'every aspect that develops or another component of intelligence is able in principle, be defined with such accuracy that a machine may be made to imitate it'. To 'find how to make machines utilize language itself, form concepts and abstractions, solve problems currently left to humans, and improve them', was done (DeepMind, 2016). Since then, AI has developed into a subject of study whose objective is to give machines the abilities required to tackle difficult jobs. A trend of success has consequently gradually formed. When it comes to sophisticated judgment-related challenges, such as selecting decision variables from a wide pool of candidates who must fit into a high-dimensional space, machines routinely outperform humans (Saygin *et al.*, 2000). This is demonstrated through Turing tests, which assess whether machines can display intelligent conduct. In reality, when deep learning is used as a guide, ML has excelled at several tasks (Dixon *et al.*, 2020).

Though many people believe that AI, ML and related ideas like deep learning as well as data science are incomprehensible (Wall, 2018), this is not always the case. In essence, ML is a kind of AI that creates methods that let computers find patterns in datasets. Data science is a unique area of research that blends deep learning, ML and AI to produce insightful results. On the other hand, deep learning is a subset of ML that gives robots the information they need to address complex issues.

In past academic evaluations, the expanding corpus of studies on AI and ML in finance was noted. The study of text mining and predictive analytics in banking, for instance, is examined by Das (2014). While reviewing papers on credit risk and bankruptcy that employed hybrid models, which combine classical modelling with AI, artificial neural networks and other ML methods, De Prado et al. (2016) saw an increasing tendency in finance research. In their extensive analysis of the available research, West and Bhattacharya (2016) divided the body of work on fraud in finance detection into categories depending on the types of fraud dealt with, the algorithms used, and the effectiveness of the detection approaches.

3. RESEARCH METHODOLOGY

ML is a method used by computers to improve performance or generate precise predictions. These techniques usually depend on improving a reward or loss function. The potential of ML is vast, as it offers the prospect of reducing product and service expenses, expediting business operations and providing improved customer service. Acknowledged as a critical domain in this era of unparalleled technological advancement, ML is rapidly gaining traction across nearly all industries (Lee and Shin, 2020).

3.1. Supervised Learning

On tagged samples, this group of algorithms receives training. Predictions can be produced for unlabelled examples using the trained model. The most frequent issues with the supervised learning task are those involving classification, regression and ranking. Credit rating and bankruptcy prediction tasks both fall within the large multi-class classification category in FRM. Numerous studies have shown that using a single classifier to predict bankruptcy or determine credit scores is possible. The most popular two individual classifiers for this are neural networks and support vector machines.

3.2. Unsupervised Learning

The technique used to uncover similarities in unlabelled data is known as unsupervised learning. In this instance, there are no labelled data accessible. With unlabelled data, specific problems (such as anomaly detection, dimensionality reduction, outlier identification, clustering and outlier detection) can be solved using unsupervised learning approaches.

3.3. Reinforcement Learning

In reinforcement learning environments, the student can interact actively with the surroundings. The learner's objective is to maximize reward across all of its environmental interactions. For risk-optimized dynamic portfolio allocation, researchers and practitioners have embraced reinforcement learning algorithms because of their dependability in studying the environment and selecting the most appropriate strategy.

3.4. Semi-supervised Learning

When label access is possible but expensive, semi-supervised learning issues are common. These methods provide predictions on the unlabelled data by analysing both the labelled and unlabelled data. The topic is very pertinent to applied ML research since many real-world problems (such as nance and healthcare) have a tough time collecting labels.

4. RESULT AND DISCUSSION

ML is becoming increasingly prevalent in business applications, with numerous solutions already in place and many more under exploration. Risk management in banks has received a lot of attention since the global financial crisis, which has led to a persistent focus on identifying, evaluating, disclosing and managing risks (Leo *et al.*, 2019). Giving a business (or client) a score to determine whether or not they are likely to default is known as credit scoring. The majority of the research has focused on identifying potential consumers as 'good' or 'bad' to aid in credit choices and credit risk management. As a result, algorithms that deal with classification predominate. ML techniques have been shown to perform better in categorization and forecast accuracy than traditional statistical methods.

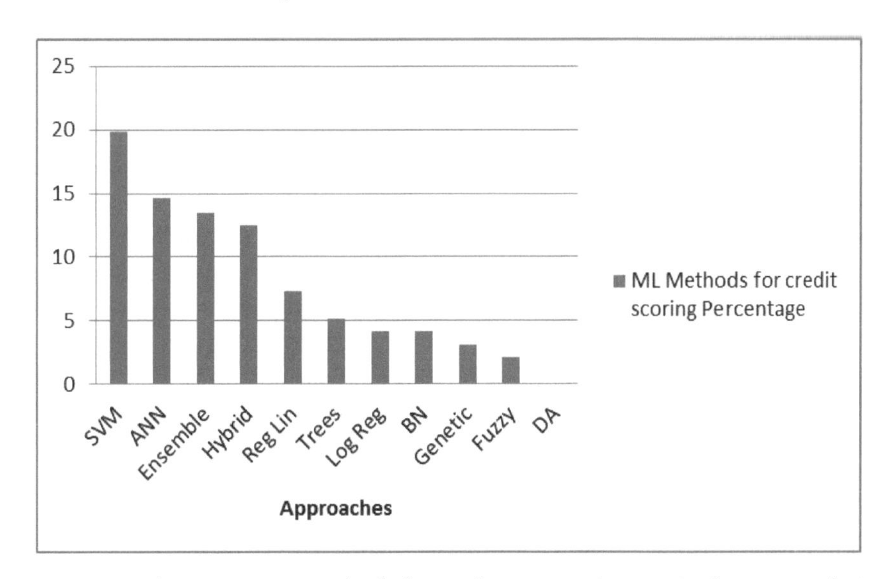

Figure 1. *Machine Learning methods for credit scoring. Source: Author's compilation.*

We conducted an experimental investigation of ML techniques for categorizing credit scoring. Figure 1 shows the numerous ways that ML can be used to determine credit scoring. SVM and ANN, however, are thought to be a commonly used and well-proven ML strategy.

Further research on the applicant of ML to financial risk management does not seem to have been done adequately. Numerous studies have looked at market risk or volatility from the perspective of managing investments, portfolios and interest rate risk. Liquidity risk, which has drawn considerable attention from regulators ever since the financial crisis, has been investigated in various use cases.

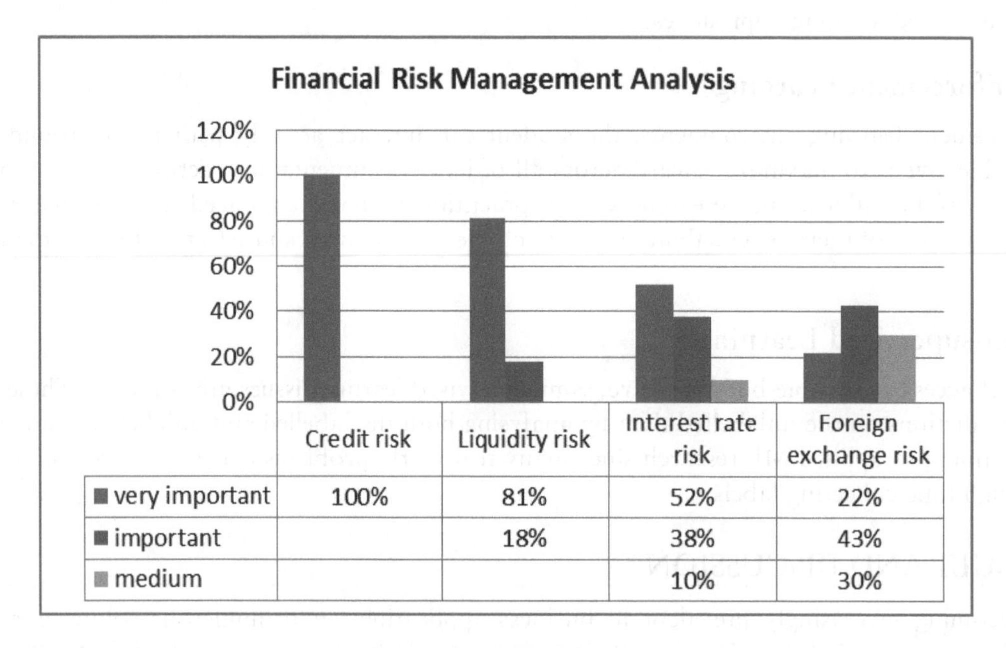

Figure 2. Machine Learning for managing financial risk analysis. Source: Author's compilation.

Figure 2 portrays the financial risks according to credit risk, liquidity risk, interest risk and foreign exchange risk. Albeit each of the previously mentioned financial perils is significant, credit risk is the most significant. To perceive each of these, we are utilizing AI calculations.

AI strategies break down the components behind the spread and effect of foundational risk inside the monetary organization, prompting progressions in the current guidelines of the financial market (Kou *et al.*, 2019).

5. CONCLUSION

It is anticipated that risk management will work to use machine learning strategies to enhance their capabilities. The banking and finance sector is commonly regarded as having a promising future for ML. This study has examined, investigated and evaluated the literature on the use of ML for risk management in the banking industry. The primary focus of the study seems to be credit risk management. In some studies on market risk, ML has been used to forecast risk, interest rate curves and market regime shifts. Little study has been done in this area, despite the growing industry focus on liquidity risk as a result of regulators' concerns. We have examined the most current applications of ML for risk management in the financial sector. We highlighted issues that have been thoroughly investigated as well as those that require additional inquiry. The well-researched fields include fraud detection, credit rating, bankruptcy prediction and volatility forecasting. Modern ML models, particularly deep learning models, have been extensively used to solve these difficulties. On the other hand, subjects like claims modelling, loss reserving and mortality projection have not drawn as much attention.

REFERENCES

Cavalcante, R. C., R. C. Brasileiro, V. L. Souza, J. P. Nobrega, , and Oliveira, A. L. (2016). Computational intelligence and financial markets: A survey and future directions. *Expert Systems with Applications*, 55, 194–211.

Das, S. R (2014). "Text and context: language analytics in finance," *Foundations Trends(R) in Finance*, 8(3), 145–261.

de Prado, J. W., V. de Castro Alcântara, F. de Melo Carvalho, K. C. Vieira, L. K. C. Machado, and D. F. Tonelli (2016). "Multivariate analysis of credit risk and bankruptcy research data: a bibliometric study involving different knowledge fields (1968–2014)," *Scientometrics*, 106(3), 1007–1029.

DeepMind (2016). *The Story of AlphaGo—Barbican Centre*. Retrieved May 10, 2021, from Google Arts & Culture https://artsandculture.google.com/exhibit/thestory- of-alpha/0wLCk0X1qEe5KA.

Dixon, M. F., I. Halperin, and P. Bilokon (2020). *Machine Learning in Finance: From Theory to Practice*. Springer International Publishing.

Dyzma, M. (2019). *Fraud Detection with Machine Learning: How Banks and Financial Institutions Leverage AI*. Retrieved 19 March 2020, from https://www.netguru.com/blog/fraud-detection-withmachine-learning-how-banks-and-financial-institutions leverage-ai.

Columbus, L. (2019). *How AI is Protecting Against Payments Fraud*. Retrieved 19 March 2020, from https://www.forbes.com/sites/louiscolumbus/2019/09 /05/how-ai-is-protecting-against-paymentsfraud/# 305acba44d29.

Kou, G., X. Chao, Y. Peng, F. E. Alsaadi, and E. Herrera Viedma (2019). "Machine learning methods for systemic risk analysis in financial sectors."

Lee, I., and Y. J. Shin (2020). "Machine learning for enterprises: applications, algorithm selection, and challenges,". *Business Horizons*, 63(2), 157–170.

Leo, M., S. Sharma, and K. Maddulety (2019). "Machine learning in banking risk management: a literature review," *Risks*, 7(1), 29.

McCarthy, J., M. L. Minsky, N. Rochester, and C. E. Shannon (2006). "A proposal for the Dartmouth summer research project on artificial intelligence," *AI Magazine*, 27(4), 12.

Saygin, A. S., I. Cicekli, and V. Akman (2000). "Turing test: 50 years later," *Minds and Machines*, 10(4), 463–518.

Wall, L. D. (2018). "Some financial regulatory implications of artificial intelligence," *Journal of Economics and Business*, 100, 55–63.

West, J., and M. Bhattacharya (2016). "Intelligent financial fraud detection: a comprehensive review," *Computer Security*, 57, 47–66.

19. An Overview of Exploring the Potential of Machine Learning Approaches on Financial Services Industry

Priti Gupta[1], Gunjan Shamra[2], M. Vishnuvardhan Reddy[3], Sudarshana Sharma[4], Pranjal Rawat[5], T. Ch. Anil Kumar[6], and Shabana Faisal[7]

[1]Assistant Professor, P.G. Department of Economics, Bhupendra Narayan Mandal University (West Campus) P.G. Centre, Saharsa, Bihar

[2]Institute of Business Management, GLA University, Mathura, India

[3]Assistant Professor, St. Martin's Engineering College, Secunderabad – 500100

[4]Assistant Professor, Amity Business School, Amity University, Madhya Pradesh, Gwalior

[5]Research Scholar, School of Management, Graphic Era Hill University Devi Road Sitabpur, Near Khushi Hotel Kotdwar, Uttrakhand

[6]Assistant Professor, Department of Mechanical Engineering, Vignan's Foundation for Science Technology and Research, Vadlamudi, Guntur Dt., Andhra Pradesh, India – 522213

[7]University of technology, Bahrain

ABSTRACT: The expansion of its use in numerous service industries is a result of technological advancements and people' growing reliance on modern technology and artificial intelligence. Nevertheless, the use of current technologies and 'artificial intelligence' influence significantly on the financial service industries, one of the key industrial sectors, as it has helped to reduce errors and improve security. The potential risk is substantially higher in the financial sector because it interacts with a global network. To provide a precise and useful outcome, secondary research has been carried out. Furthermore, it has been observed that a lot more work needs to be done to increase the prominence of mechanical learning and the usage of artificial intelligence in the finance industry. The results show how machine learning (ML) techniques have a wide range of applications in the financial services sector, from risk assessment to consumer interaction. The report focuses on the advantages, difficulties and possibilities linked to ML adoption in financial services, offering insightful information for financial organizations, regulators and researchers hoping to use ML to its full potential in this field.

KEYWORDS: Machine learning, artificial intelligence, finance and banking, technology, big data

1. INTRODUCTION

In this world of digitalisation, machine learning or ML has achieved a successful 'Human Level Performance' in different domains such as speech recognition, image classification as well as machine translation. Munkhdalai *et al.* (2019) have stated that ML technology is considered as a subset of 'Artificial Intelligence' or AI that is primarily concerned with optimising business processes with minimal or no human participation. In today's large-scale banking organisations, 'systematic risk' is always present, so perceptive and automatic ML methods have become an important tool for assessing and detecting risks arising from '***global networks in financial system***', '***market sentiments***', '***big data of financial transactions***' and '***risk proclivity***', among other things (Oyewola *et al.*, 2021).

It has been observed that the capability of 'ML' in better decision making in financial aspects in businesses was investigated by focusing on '***neural networks***' for '***economic decision making***'. In the year 2010, the '***Journal of Banking & Finance***' published a preliminary research series evaluating whether ML techniques may improve financial decision as well as 'credit risk management' in the banking sector.

DOI: 10.1201/9781003532026-19

Besides, Sujith *et al.* (2022) have identified that financial firms would be better equipped to handle the financial threats of 'artificial intelligence' and 'ML'.

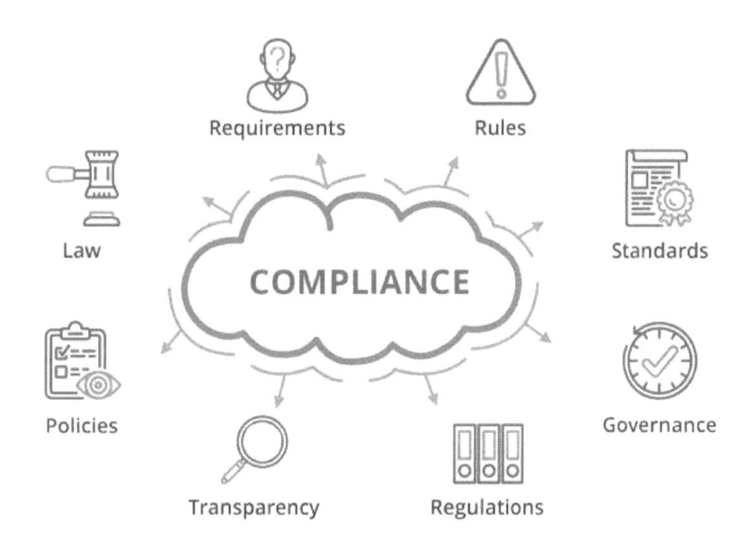

Figure 1. *Use of machine learning in banking sector. Source: Munkhdalai et al. (2019).*

From the above image, it has been observed that ML technology helps to enhance '***security***', '***process automation***' and '***underwriting and credit scoring***' that may enhance financial services among organisations. ML algorithms typically generate tailored reports based on available data, providing basic and straightforward insights to different management levels.

1.1. Objectives and Hypothesis

- To examine the advantages and benefits of implementing ML techniques in a number of financial services areas, including risk assessment, fraud detection, customer sentiment analysis, investment portfolio optimisation and algorithmic trading.
- To identify the difficulties and factors that must be taken into account when applying ML in the financial services sector, including data quality, security and privacy issues, model interpretability and legal compliance.
- To evaluate the potential for innovation and the creation of individualized financial solutions in the financial services industry.

Hypothesis 1: *Implementing ML in financial sector can change a number of fields by enhancing process automation, accuracy and efficiency.*

Hypothesis 2: *For the ethical and responsible use of ML in financial services, issues including data quality, privacy and security worries, model interpretability and regulatory compliance must be resolved.*

Hypothesis 3: *Utilising ML in financial services can lead to enhanced consumer interaction, customised financial solutions, real-time risk assessment and fraud protection.*

2. LITERATURE REVIEW

In recent decades, experts and government leaders have employed the idea of systemic financial risk to assess the likelihood of disruption to customers, financial markets and even the economy. As per the notation of Hargreaves *et al.* (2017), mainly, 'systematic financial risk' would be a major risk factor for financial system stability, negative effect on economic growth, along with information failure in the same (Leo *et al.*, 2019).

2.1. More Accurate Forecasts

Data analysis and reporting are the two important roles that are needed for performing financial accounting. According to Sadgali *et al.* (2019), there is a high risk of managing large financial data manually. In order to avoid data risk, ML application is implemented which can handle all financial data efficiently. Moreover, banking companies need to access financial data, cash flow and online transactions all the time for predicting inventory management.

2.2. Developing Innovative and Collaborative Workplace

ML technique that recommends or completes accountancy codes automatically reduces errors and time saving. ML will evaluate and evaluate the data, identifying inconsistencies and provide a list of errors for auditors to review (Gu, 2020).

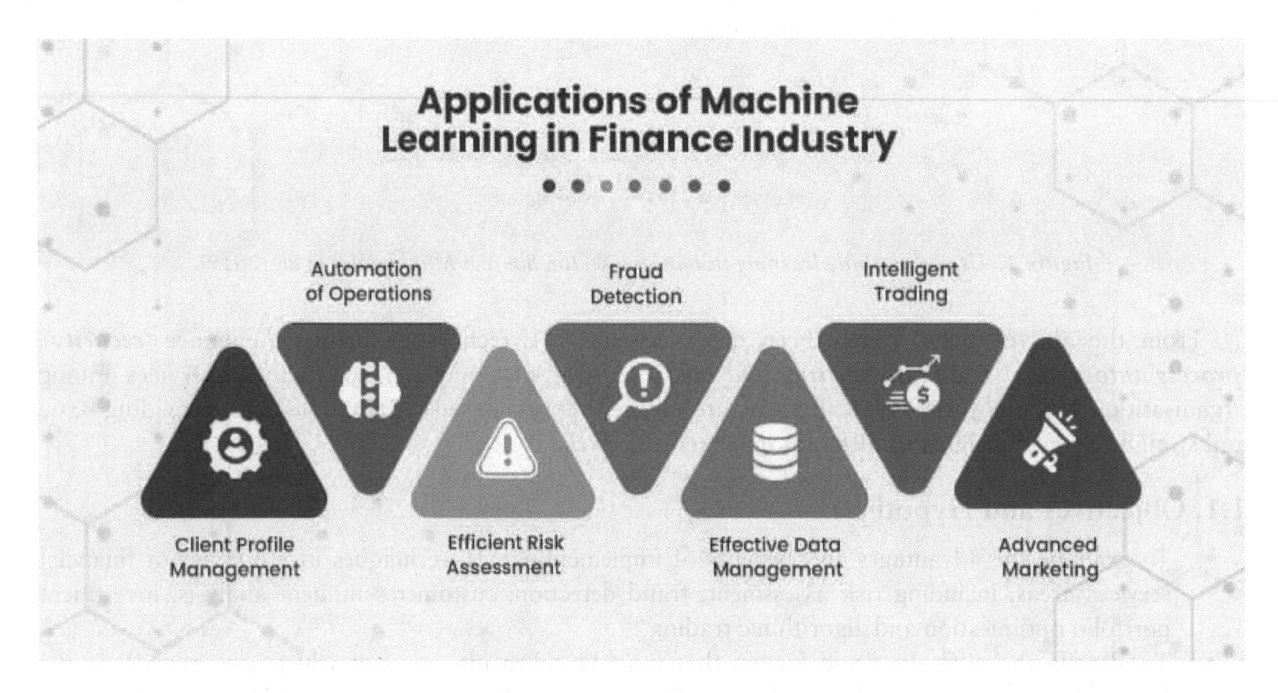

Figure 2. Benefits of Machine Learning in finance industry. Source: Self developed.

2.3. Develop Effective Communications with Business Partners

By adapting Natural Language Processing or NLP, banking organisations can conduct effective and powerful conversations with business partners including trading partners. Many accountants get engaged in routine communication activities, such as 'contacting customers' on a regular basis.

2.4. Detection and Prevention of Fraudulent Activities

In financial industry like banks, fraudulent activities pose a bigger threat as it holds billions of accounts. The banking sector requires great security in order to safeguard the accounts and the data. The financial sector or industry, therefore, is always in a great threat of security breach, and therefore requires effective technology and MLin order to make the security system more robust. In the financial industry, fraud is considered to be one of the major threats to the sector as it can damage several economic activities (Aziz *et al.*, 2019). Therefore, with the advancement in technology and artificial intelligence the security systems are made stronger. However, the modern fraudsters have technical knowledge; therefore, the banking companies emphasises in ML in order to combat the fraudulent activities or transactions of cash.

2.5. Loan Underwriting

The ML in banking and insurance industry is very useful as the companies have data and information of millions of users, and thus have huge underwriting process.

Hence, using ML techniques in financial industry is very much useful as it saves times as well as expense of the organisation. The data scientists working in various financial sector can use the algorithms of ML to process big data and information of the customers to make proper decisions for granting loan or not. There are numerous ways to train the algorithms based on income, occupation and loan history as well as customers credit behaviour.

3. RESEARCH METHODOLOGY

In this research program, a systematic way has been used in order to conduct the research to get the correct results of the impact and usefulness of ML in financial industry like banking and insurance. The research methodology has been done using the appropriate methods that can provide an authentic data for the formulation and validity of the research work. Therefore, in this research work, secondary data is being used in order to get the required results. It is important to understand the usability of ML and its effectivity in regard of security concerns as it can help decrease the fraudulent activities and can help in creating an effective communication with the various business elements.

Secondary data analysis can also be beneficial for cross-cultural research, as researchers can access datasets from different countries or cultures to compare and contrast findings. This can help to identify similarities and differences across different cultures, which can be important for understanding global issues and developing effective policies and interventions. Besides, use of secondary data is also prominent in this research work. The secondary data are mostly being extracted from the journal; articles as well as various company websites that has helped in gaining the overall view of the effects of ML and its efficiency in the field of financial sector.

The use of mix method is, therefore, regarded as appropriate in this research paper as it helps in getting wide range of opinion about the researched topic. The research project has been directed in a consecutive system. In such manner, it is worth focusing that the investigation should likewise be possible in turn around design. The subjective examination has assisted with breaking down information that were gathered from the essential exploration though the information that are gathered in the optional strategy were investigated by the quantitative technique. There were a few preventions that arose in the approach to directing this exploration. Be that as it may, this multitude of blocks was figured out to direct this exploration venture and take a general thought regarding the subject of the significance of ML these days in upgrading the development, security and manageability of financial enterprises.

4. DISCUSSIONS AND FINDINGS

According to 'McKinsey & Co.,' banks' 'risk functions' will need to be very different from how they are presently by 2025. The expected changes in 'risk management' are due to the deepening and broadening of legislation, shifting customer expectations and shifting risk types. New products, services and risk management techniques are being made possible by the utilisation of developing technology and advanced analytics. By identifying complex, nonlinear patterns in enormous datasets, ML, one of the cutting-edge technologies with substantial implications for risk management, can enable the development of risk models that are more accurate. Every new piece of data that is provided has the potential to improve these models' ability to anticipate the future, which will eventually increase. ML is projected to be employed in a range of settings inside a bank's risk organisation. Additionally, it has been argued that a project including ML could help update the risk management divisions of banks.

Table 1: Different ML methods.

Classification of Methods	Description
Network model	Finding and regulating systemically important financial institutions is one of the key responsibilities of macroprudential regulatory agencies. The modern banking system and the global financial market have evolved into a sophisticated network with components like debt relations and securitisation, therefore it is crucial to keep that in mind. Some of the essential methods include centrality degree analysis, dual networks, network architectures, Bayesian graphical models, etc. These techniques are consistently employed to determine crucial nodes in intricate networks or to investigate risk exposure in financial networks.
Big data analysis	Big data has become a well-liked subject in both business and academics. The related methods are employed to uncover the relationships between a wide range of variables and vast amounts of data, including interbank liquidity and global capital flow. For instance, there is a lot of interest in the connection between online contracts and the market for stocks as well as the fact that stock prices are influenced by public sentiment (Malali and Gopalakrishnan, 2020). In order to recognise and distinguish systemic risk, big data analysis aims to combine divergent financial data and expand expertise. Big data analysis therefore shares characteristics with other application domains that include heterogeneous data with conventional data mining approaches, such as classification methods.
Text mining	Financial managers may be able to identify risk characteristics or market attitudes by using text mining to analyse massive amounts of financial data, such as news, financial data, tweets and other circumstances from social networks. The development of a word base, the extraction of popular words and sentiment analysis and other topics are covered in existing literature. Financial text mining has been done using a variety of techniques. For instance, the ranking method of categorisation, the examination of the percentage of good and negative words and the classification of emotional words
Statistical methods and Econometrics	Financial evidence research most frequently employs traditional techniques like statistics and economics. Causation analysis, correlation analysis and regression analysis can all be used to study the relationships between the variables that make up financial items (Sadgali *et al.*, 2019). Additionally, more econometric techniques have always been used to assess systemic risk using values of risk exposure, such as conditional value as risk, conditional-risk and systemic expected loss (SES), among others.

Source: Self-developed

5. CONCLUSION

Based on the above discussion, it can be concluded that ML and artificial intelligence are becoming an important approach in every sector of the economy. The increasing dependency and advancement in technology not only makes the job easier but also makes it productive and effective. The future decisions and the accurate predictions are very effective in not only financial sector but in every business process. Therefore, the use of this technologies will help in individuals as well as economic growth. The decreasing number of errors and better communication between stakeholders makes the approach much more effective. Though some improvements are required to be made to ensure positive growth and better security in the financial sector. The security breaching and misuse of information and data is a huge crisis in the financial sector. Therefore, proper measures and improvements are required in terms of ML and its engagement in banking sector. The report also finds chances for innovation in the financial services sector by incorporating

ML. Real-time risk assessment, fraud protection, the creation of customised financial products and increased consumer involvement are all examples of this. ML can help financial organisations become more competitive overall by revealing new market trends, enhancing fraud detection systems and more.

6. FUTURE SCOPE

ML is a developing innovative viewpoint that is keeping its imprint in the majority of the fragments of advanced way of life. It is apparent that the legitimate execution of this can diminish the intricacies of present-day life in a coordinated way. It is not just gainful for the financial sector, but it can stretch out its arm to the transportation and planned operations framework and medical services space. Moreover, its comprehensive arrangement is genuinely equipped for redesigning the day-by-day way of life of individuals as it can lessen the work strain and errors in every area of the economy. In this way, one can say that the eventual fate of AI is colossally prosperous as it can improve the effectiveness of different areas of present-day life.

REFERENCES

Aziz, S., M. Dowling, H. Hammami, and A. Piepenbrink (2019). "Machine learning in finance: a topic modeling approach," *European Financial Management*.

Gu, N. (2020). "Digital financial inclusion risk prevention based on machine learning and neural network algorithms," *Journal of Intelligent & Fuzzy Systems*, (Preprint), 1–16.

Hargreaves, C. A., V. Reddy, and R. V. Reddy (2017). "Machine learning application in the financial markets industry," *Indian Journal of Scientific Research*, 17(1), 253–256.

Joshi, N. (2022). *How Machine Learning is Impacting the Finance Industry*. www.bbntimes.com. Available at: https://www.bbntimes.com/financial/how-machine-learning-is-impacting-the-finance-industry. [Accessed on: 28th February, 2022].

Kou, G., X. Chao, Y. Peng, F. E. Alsaadi, and E. Herrera-Viedma (2019). "Machine learning methods for systemic risk analysis in financial sectors," *Technological and Economic Development of Economy*, 25(5), 716–742.

Leo, M., S. Sharma, and K. Maddulety (2019). "Machine learning in banking risk management: a literature review," *Risks*, 7(1), 29.

Malali, A. B., and S. Gopalakrishnan (2020). "Application of artificial intelligence and its powered technologies in the Indian banking and financial industry: an overview," *IOSR Journal of Humanities And Social Science*, 25(4), 55–60.

Munkhdalai, L., T. Munkhdalai, O. E. Namsrai, J. Y. Lee, and K. H. Ryu (2019). "An empirical comparison of machine-learning methods on bank client credit assessments," *Sustainability*, 11(3), 699.

Oyewola, D. O., E. G. Dada, J. N. Ndunagu, T. A. Umar, and S. A. Akinwunmi (2021). "COVID-19 risk factors, economic factors, and epidemiological factors nexus on economic impact: machine learning and structural equation modelling approaches," *Journal of the Nigerian Society of Physical Sciences*, 395–405.

Sadgali, I., N. Sael, and F. Benabbou (2019). "Performance of machine learning techniques in the detection of financial frauds," *Procedia Computer Science*, 148, 45–54.

Sujith, A. V. L. N., N. I. Qureshi, V. H. R. Dornadula, A. Rath, K. B. Prakash, and S. K. Singh (2022). "A comparative analysis of business machine learning in making effective financial decisions using structural equation model (SEM)," *Journal of Food Quality*, 2022, 6382839, 7 p.

20. An Analysis of Artificial Intelligence Contributing the Performance Management of Professors in Educational

Priti Gupta[1], Roop Raj[2], Ambarish Ghosh[3], Kushagra Kulshreshtha[4], N. Krishna Kumar[5], Arpan Shrivastava[6], and László Pataki[7]

[1]Assistant Professor, P.G. Department of Economics, Bhupendra Narayan Mandal University (West Campus) P.G.Centre, Saharsa, Bihar

[2]Assistant Professor in Economics, Economics Department, RRM Group of Institute, Kurukshetra, Haryana

[3]Assistant Professor, FMS ICFAI University, Raipur C.G.

[4]Institute of Business Management, GLA University, Mathura, India

[5]Associate Professor, UICSA, Guru Nanak

[6]Assistant Professor, Department of Management, Prestige Institute of Management and Research, Indore

[7]Associate Professor, John von Neumann University, Hungary

ABSTRACT: Performance management is a method that primarily aids in assessing both individual and group performance and can then assist in the adoption of efficient strategies to eventually attain performance excellence. When taking this into account, performance management is necessary across all industries. Explore the pertinent study for further information on how to conduct logical and scientific research, meet expectations, and further combine positivist philosophy, deductive method and descriptive design. Additionally, this study combined a quantitative data analysis technique with primary data collection in the form of a survey. A survey of 50 professionals in the educational sector was conducted for this study. The findings of this study show how using artificial intelligence (AI) to control professor performance in educational institutions is possible. They highlight the advantages, difficulties and opportunities involved with using AI and offer guidance for academic institutions, administrators, and researchers looking to use AI for efficient professor performance management.

KEYWORDS: Performance, evaluation, enhancement, learning, progress, assessment, management, experience, educational sector

1. INTRODUCTION

Performance management is primarily a corporate management technique that aids in monitoring and evaluating workers' performance so that management may gauge the organization's success based on it. Employees and managers can collaborate on planning, reviewing, and evaluating individual employee performance as well as quantifying their contribution to the advancement of the company through the performance management process. For example, active learning in the form of practical learning experience emerged as the significant factor that essentially impacts performance management for professors' performance in the educational sector.

Validity and reliability along with identification of effective and ineffective performance emerged as other significant criteria that essentially help determine eventual performance management for professors.

DOI: 10.1201/9781003532026-20

1.1. Objectives and Hypothesis

- To investigate how artificial intelligence (AI) may improve how academics in educational institutions are managed in terms of performance.
- To examine the pros and disadvantages of using AI to manage academics' performance, taking into account factors like objectivity, consistency, and real-time feedback.
- To discover the difficulties and factors, such as data protection, ethics and algorithm transparency, that come with integrating AI into professors' performance management procedures.
- To evaluate the potential for AI integration to improve performance management procedures, including thorough data analysis, individualised feedback and focused help and training.

Hypothesis 1: *Artificial intelligence is being incorporated into performance management processes to improve the efficiency, impartiality and consistency of grading professors at educational institutions.*

Hypothesis 2: *By providing thorough data analysis, individualised feedback, and the ability to pinpoint areas that need specific help and training, the incorporation of AI creates chances for improving performance management processes.*

Hypothesis 3: *By removing prejudice and giving real-time feedback to help professors' professional development and advancement, the use of AI in performance management provides fair evaluations.*

2. LITERATURE REVIEW

The concept of performance management lies in helping organisational managers to assess employee performance and provide them necessary feedback based on that. On the other hand, the main aim of a performance management system is to create an appropriate working environment within an organisation which essentially allows workers to learn, grow and deliver their best performance. This in turn also boosts employee morale by making employees feel comfortable about the learning and development atmosphere within the organisation through which they can grow constantly. Moreover, effective application of performance management also helps with improving accountability and thus, eventually enhancing employee performance level.

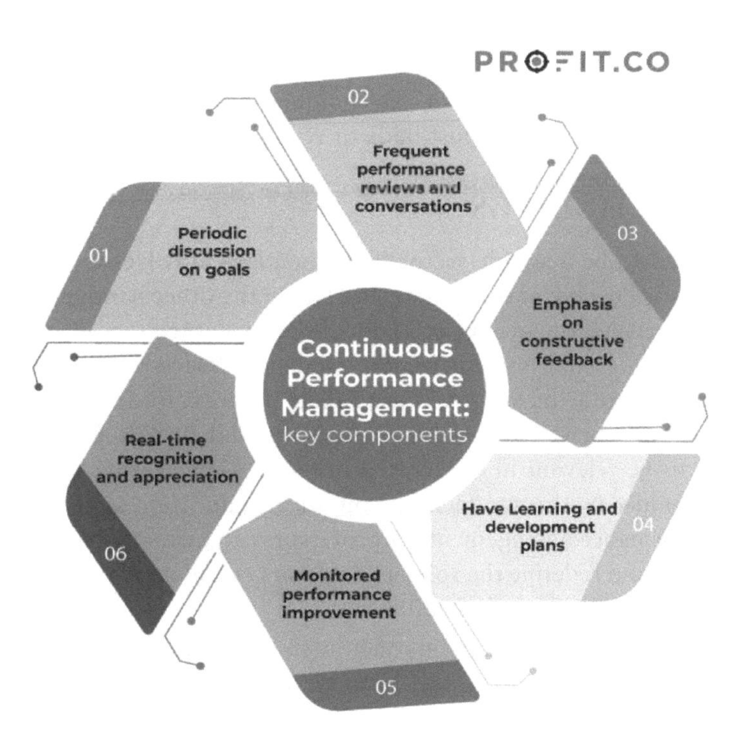

Figure 1. Key benefits of performance management.

The need for performance management in the educational sector lies in enhancing the overall system and ensuring students are provided with best and necessary guidance from the teachers (Audenaert *et al.*, 2019). This further helps understand the eventual impact of teaching skills in the educational institutions by measuring student progress. Performance management in the education sector essentially includes looking at evidence of the impact created by other teachers and later use it as an instance for teachers to learn and understand the ideal teaching process. Upon considering this, the need for constant learning and development emerged as one of the major processes of enhanced performance management.

Punctuality in class appears to be one of the fundamental reasons for performance management in higher education. Therefore, for professors it is important to maintain class punctuality to motivate students and ensure they attend classes. This in turn significantly helps professors establish an effective communication with students and help them achieve academic objectives (Cadez *et al.*, 2017). On the contrary, aptitude and attitude emerged as the basic factors that make essential contributions to professors' performance and enhance the outcome for students. Subject mastery and effective teaching methodology appear to be fundamental factors that make significant contributions to professors' performance in the educational sector. It is important for the professors to have adequate knowledge on the subject they teach while application of specific and effective teaching methodology helps professors enhance students' performance (Paweloszek, 2015).

3. RESEARCH METHODOLOGY

Research methodology is essentially defined by the specific procedures used to identify, choose, process and analyse necessary data related to any topic. Effective application of research methodology essentially helps evaluating validity and reliability of a study, which also ensures maintenance throughout consistency of a study. The relevant study uses the positivist research philosophy to conduct a detailed examination of the implementation of performance management and how it affects academics' performance. Value freedom and the ability to evaluate quantitative data are essential benefits of positivism philosophy which certainly helped this research distinguish different factors contributing to professors' performance management and their specific impact on the same (Park *et al.*, 2020). One of the main reasons behind integration of positivism philosophy is its ability to conduct objective-driven research. This significantly helped the concerned study to maintain data relevance and enhance the eventual outcome. On the other hand, the ability to highlight necessary data patterns and predicting trends emerged as significant benefits of the concerned research philosophy. Secondary data analysis can be used to replicate and validate research findings from previous studies. By using the same dataset and analytical techniques, researchers can confirm or challenge prior findings, which can help to build a more robust body of research in a particular field (Pearse, 2019).

4. DISCUSSION AND FINDINGS

Educational sector happens to be a simple sector that essentially involves only a few stakeholders upon which the overall progress is dependent (Camilleri, 2021). On the other hand, professors possibly play the most significant role in enhancing the overall progress of the concerned sector that essentially helps create a strong and effective learning environment for the students. Professors are responsible for providing appropriate education to the students and measuring their progress to determine the impact of effective teaching and ensuring personal as well as career growth of the students. However, professors' performance evaluation also appears to be relevant in this regard in terms of measuring the effectiveness of teaching and learning guidance provided by them to facilitate the ensure education system (Hung, 2017). Therefore, integration of proper performance management for professors can certainly be a significant way to address core educational challenges and redefine the role of professors in enhancing personal and long-term career progress of the students (Frederiksen *et al.*, 2020).

Learning theories provide precise instructions on how students should learn; this indicates a connection between the traits and instructional strategies that can be used to develop a strategy for learning. Table 1 lists the methods advised by the various learning theories as well as the personal experiences of the subject-matter experts from the various fields. These insights can be used to create learning environments that are incredibly successful.

Table 1: "Relationship between student patterns and learning theories".

'Learning theory'	'Instruction Characteristics'	'Associated Instructional Patterns'	'Recommended Strategies'
'Conductivism Learn by imitation'	'Use experimental procedures Reactive Educating is passive'	'Tutorial Training Observation'	"Repetition Stimulus and response association Feedback Reading"
"Constructivism Learn by experience"	'Consider previous learning Flexible and modifiable knowledge Build knowledge actively'	"Research/construction exploration Scientific method Objective based scenario"	"Discussion panels Puzzle Design of collaboration experiences"

It is crucial to identify the participants in education in any method of instruction once the ideas underpinning learning have been well understood. This definition aims to make the participants and their functions in education obvious. the student who is in charge of his education and the teacher who directs him or her towards the learning.

Besides, computer programmes called chatbots or conversational assistants are able to communicate with users using language-based user experiences. A chatbot's primary goal is to replicate intelligent human interaction such that the user's experience resembles that of speaking with a real person as closely as feasible. One of the services a chatbot may provide for a user is information on an item, event, or action. The first stage in how chatbots typically work is the use of natural language. It also relies on discussions with a predetermined flow and planned exchanges that offer minimal room for interpretational ambiguity.

On the other hand, effective integration of best teachers in higher education is the only way to enhance collective progress of the concerned sector. It is to be seen that there are different methods to evaluate a teacher's performance. Considering this, the style of teaching appears to be a significant process to determine if the teacher is suitable for providing appropriate education to students (Cho *et al.*, 2017). However, teachers' personality and cognitive abilities also matter in the education sector in terms of providing effective learning to students and ensuring they achieve a good education. It is observed that personality traits such as attitude and attributes play an integral role in determining teachers' capability to connect with the students at the higher level and provide them with effective education (Shams, 2019). On the other hand, professors often consider the most effective and suitable teaching styles to communicate with students and facilitate their learning experience. Upon considering this aspect, the need and importance of attitudes and attributes of professors can be understood in this regard. Moreover, it is the personality of the professor that attracts the most to students in terms of developing a strong relationship and seeking their help whenever needed. Based on this observation, it can be said that effective integration of experienced professors with charming personalities can certainly be a key requirement in the education sector itself. Furthermore, it is also observed that professors with attractive personalities connect with students easily for natural reasons which highlight the significance of attitude and aptitude in this regard. Complex personality of the professors would create a distance between them and the students which further can make a direct impact on student progress and thus professors' performance can be evaluated. Upon considering this particular factor, a clear need for an effective and attractive personality is needed for the professors in terms of facilitating student experience. On the other hand, based on the analysis the role of attitude and attribute in determining professors' performance and work as key performance management tools can also be identified in this regard. This also set a specific and basic requirement for the professors in terms of addressing student needs and ensuring collective growth of the concerned sector. Upon this, if the performance management system is well backed with AI technology, it would help to make the education sector stronger. Since the world is advancing, like every sector, the education sector should also adopt the AI system in their performance management.

Ensuring effective student progress and constant growth are considered to be the eventual goals for professors in the education sector. A lot is dependent on the style of teaching adopted by the professors which essentially connects with students and ensures their active participation in academics. Simple yet effective style of trenching emerged as one of the necessary requirements for the professors which certainly helps them establish a strong bond with students and facilitate the overall learning process. Different teaching methods have different impacts on the students which essentially play a crucial role in enhancing the eventual outcome of student progress. Easy and effective learning makes it easy for the students to relate with the professor's perspective and essentially helps measure the impact of teaching, which clearly defines the significance of the teaching method in this context.

5. CONCLUSION

Performance management is considered as an effective evaluation tool to determine the progress at the individual and collective levels. On the other hand, professor performance management seems to be a pertinent idea at the higher education level that greatly contributes to student advancement in both academics and careers as well as personal development. This factor makes it clear that academics require an efficient performance evaluation system. Strong methodological principles are incorporated into this research through the integration of positivist philosophy, the deductive technique and the descriptive design. On the other hand, this study also incorporates primary data gathering in the form of a survey and a quantitative data analysis technique. The professor's disposition and qualities were apparent as the crucial performance evaluation tools after taking into account the data analysis and fundamental results. On the other hand, teaching methods emerged as another significant consideration for professors' performance management in the educational sector. According to the research, incorporating AI tools like ML algorithms and natural language processing into performance monitoring in educational settings has many advantages. AI makes it possible to evaluate students in an unbiased and consistent manner, do away with bias and offer immediate feedback to help educators improve. By improving the effectiveness of data collection, processing and feedback production, it makes it possible to evaluate performance in a more precise manner.

6. FUTURE SCOPE

This study essentially discusses basic factors for the professors to put effective performance and facilitate student knowledge. On the other hand, the concerned research also discusses the need and effectiveness of professors' performance management that essentially helps incorporate specific practises in the educational sector and ensure uninterrupted student progress. Upon considering this aspect, this research can certainly be used for specific educational institutions such as colleges and universities to outline necessary guidelines while appointing professors. On the other hand, the concerned research can further be used to highlight different personal traits of professors in terms of facilitating educational experience to students.

REFERENCES

Audenaert, M., A. Decramer, B. George, B. Verschuere, and T. Van Waeyenberg, (2019). "When employee performance management affects individual innovation in public organizations: the role of consistency and LMX," *The International Journal of Human Resource Management*, 30(5), 815–834.

Cadez, S., V. Dimovski, and M. Zaman Groff (2017). "Research, teaching and performance evaluation in academia: the salience of quality," *Studies in Higher Education*, 42(8), 1455–1473.

Camilleri, M. A. (2021). "Using the balanced scorecard as a performance management tool in higher education," *Management in Education*, 35(1), 10–21.

Cho, C. H., J. H. Jung, B. Kwak, J. Lee, and C. Y. Yoo (2017). "Professors on the board: Do they contribute to society outside the classroom?" *Journal of Business Ethics*, 141(2), 393–409.

Frederiksen, A., L. B. Kahn, and F. Lange (2020). "Supervisors and performance management systems," *Journal of Political Economy*, 128(6), 2123–2187.

Hung, C. L. (2017). "Social networks, technology ties, and gatekeeper functionality: implications for the performance management of R&D projects," *Research Policy*, 46(1), 305–315.

Park, Y. S., L. Konge, and A. R. Artino (2020). "The positivism paradigm of research," *Academic Medicine*, 95(5), 690–694.

Paweloszek I. (2015). "Approach to Analysis and Assessment of ERP System. A Software Vendor's Perspective, Proceedings of the 2015 Federated Conference on Computer Science and Information Systems 2015, M. Ganzha, L. Maciaszek, M. Paprzycki, Annals of Computer Science and Information Systems,", 1415–1426, IEEE, doi: 10.15439/2015F251.

Pearse, N. (2019. "An illustration of deductive analysis in qualitative research," *18th European Conference on Research Methodology for Business and Management Studies* (p. 264).

Shams, F. (2019). Managing academic identity tensions in a Canadian public university: The role of identity work in coping with managerialism. *Journal of Higher Education Policy and Management*, *41*(6), 619–632.

21. Impact of Quality Management and the Internet of Things on Success Factors in a Large Enterprise

Ahmad Y. A. Bani Ahmad[1], Nidhi Chaturvedi[2], Thimmiaraja J[3], Brijesh Goswami[4], Arun M[5], Arif Hasan[6], and Viktor Fórián-Szabó[7]

[1]Associate Professor, Department of Financial and Accounting Science, Middle East University, Amman 11121, Jordan

[2]Senior Lecturer, School of Business Management, Emirates Aviation University, Dubai, UAE

[3]Department of Information Technology, Dr. Mahalingam College of Engineering and Technology, Pollachi

[4]Assistant Professor, Institute of Business Management, GLA University, Mathura

[5]Department of MCA, School of Computing, Kalasalingam Academy of Research and Education, Kriahnanlovil, Arun Vincent

[6]Amity Business School, Amity University Madhya Pradesh

[7]Economist, Hungary

ABSTRACT: The 'Internet of Things (IoT)' is regarded as a special network that allows for the configuration and connection of various systems, processes, and individuals in order to facilitate the free information and data flow for making key decisions. The highest unique selling proposition for improving the quality of service in the hotel sector is the quality of services and managing quality. The use of IoT helps to ask customers for more self-assistance checks, which makes it easier for them to book hotels and other services, respond to their wants and expectations, and estimate patterns so that good quality assurance can be effectively maintained. The employment of smart tagging and other processes concentrates on using the offerings and helps draw in more clients. IoT is anticipated to expand more quickly and is ready to unleash greater possibilities. The study's goal is to understand how IoT can improve quality management in the top hotel industry. The study is being prepared by the researchers using secondary data. Secondary data analysis is particularly useful for longitudinal research studies, which involve collecting data over an extended period. By using existing longitudinal datasets, researchers can track changes and trends over time, which can be especially important for studying the effects of policies or interventions. Among primary and secondary data, only secondary qualitative data has been chosen because it is comparatively less time-consuming, and the researchers will leverage large data sets from previous studies. Apart from that, this research paper has found that integration of IoT and quality management can significantly contribute towards the success of enterprises. Organisations can emphasise robust quality management, real-time monitoring, and data analysis to improve service quality and ensure customer satisfaction.

KEYWORDS: Internet of Things, service quality, quality management, frequency analysis.

1. INTRODUCTION

The application of 'Internet of Things (IoT)' has been introduced by Ashton, which enables in describing the ability of the sensors to connect with the internet and offer enhanced services which benefits the stakeholders. It has been stated that IoTis a network which supports in connecting with the different systems and subsystems with the identifiable address so as to offer better and intelligent services to meet the needs of the users. The current digitisation of the business environment has enabled in scalability and adaptability of IoT as it supports in performing the operation better, enables the individuals to complete the task at the quick

DOI: 10.1201/9781003532026-21

span of time and is also involve in better quality of the services delivered. The application of IoT supports in reducing the errors related to the business process, and hence, quality can be enhanced at ease (Buhalis and Leung, 2018). The application of IoT systems support in collecting the data and information from different sources and types of sensors, this information is then collated and organised for making critical analysis for better decision making. The management can also state the data precision, ranges, device specification, etc and thereby can enable in gathering real-time information about the necessary inputs.

The IoT supports the organisation without making any constraints or limitations covering geographical locations, time, individuals etc. and offers the necessary tools which enable the people to interact with others on real real-time basis, connect with different systems and gather the data for performing high-quality services and support in realising the goals of the organisation. IoT in the current decade is gaining more attention as the critical technology which supports businesses to realise their objectives in a sustainable manner. Furthermore, with the spread of the internet around the world, business enterprises and management are focused in connecting with their employees, vendors, suppliers, customers and others for better delivery of services. The competitive landscape has been increasing with more players offering similar goods and services, hence, it is critical to use digital technologies for delivering quality services at zero error, reduced cost and less time.

In service-related organisations like hospitality and hotels, the quality management is highly important as it enables the customers to visit the place more often. The management in the hospitality industry enables in using these tools to understand the demand forecasting, address the growing needs of the customers and deliver services with high quality so that the customers are delighted always. The application of IoT in the hotels supports the management in understanding the room booking trend, helps the visitors and customers to book the rooms in advance, check quick civility of various other services like food orders, site seating, booking cabs and other services, these aspects support in enhancing the quality management and other related aspects in the hospitality industry.

In addition, it is noted that the IoT enable in changing the manner in which the data and information are transmitted, and analysis are made for effective quality management in the hospitality industry. These technologies support in understanding the sensor, user devices and other systems for effective understanding of the data, forecast the pattern and manage the operations effectively so that better quality of services can be offered to the customers. In hospitality industry, the quality of services and managing them is the critical success factor; hence IoT are being applied for better performance management and support in understanding the various quality measures for effective realisation of the organisation's goals.

The hotel business will be used by the researchers to analyse the essential impact of employing IoT technologies for improved quality maintenance, which is a key component of the article in understanding the total influence of quality assurance and the internet of things on performance indicators in a major enterprise. The IoT tools enable in enquiring the overall understanding of various circumstances related to the customer and devices so that better services can be offered with the highest quality standards (Torres, 2018).

1.1. Objectives and Hypothesis

- To investigate the impact of quality management practices concerning organisational success in a large enterprise
- To explore the role and significance of IoT regarding success factors within large organisational enterprises
- To evaluate the potential impacts and the effects of integration of IoT technologies with quality management in the context of large organisations

Hypothesis 1: *There is a significant relationship between the implementation of quality management practices and success factors in a large enterprise.*

Hypothesis 2: *There is a significant interaction effect between quality management practices and IoT technologies on success factors in a large enterprise*

Hypothesis 3: *The adoption of IoT technologies significantly impacts success factors in a large enterprise.*

2. LITERATURE REVIEW

In order to enhance, co-create, and adjust the hospitality experience, technology is crucial. A few things tourists may do to enhance their experience include scheduling your days, researching statistics, and finding local attractions (Kansakar *et al.*, 2019). Among the IoT uses in a hotel are smart room service, automated hotel rooms, seamless check-in and check-out, smart washing and seamless check-in and check-out. The electronic key card that the hotel delivers to the customer's phone to access the room is another sort of key tag used in hotels. With the help of their past medical history, the visitor's current health state may be monitored via sensors and transmitted to the hospital in an emergency (Lee, 2019). The security of all hotels is an important part and requires great care and investment for the safety of our customers. IoT Smart Video can filter out suspicious behaviour in a surveillance camera file and can detect and reduce the risk of theft or intrusion. The hotel can implement tracking systems to carry out clever activities and services, such as locating visitors and sending them a personalised greeting. Many hotel establishments, including Hong Kong Hotel Icon, give visitors a 'Handy' cell phone to use while they're there, adding further functionality (Zhang *et al.*, 2021). The Smart Room, which gives a personalised room feel and a voice-controlled widget that adapts and works with the visitor, is now being tested at Hilton and Marriott hotel offices. The visitor can engage with the virtual help interfaces to enhance their value proposition. Kitchen personnel can correctly arrange the utilisation of available commodities by using sensors set there that can identify the expiration dates of food and drink (Torres, 2018). A built-in artificial intelligence sensor can recommend recipes to kitchen staff with available products.

Smart resorts can personalise the ambiance of a guest's room using information gathered from a prior visit. The IoT can be used by the smart hotel to manage garbage locally in an eco-friendly way, save energy, provide better customer service, and maintain the building. The terminal gathers a lot of internal and external data when IoT enters the hotel and the city, such as the customer's position, the existence of the resources they need, the weather, the state of the roads, and traffic levels. This information may not directly impact the user's value encounter, but it may have an impact on the travellers' perception in general (Wu and Cheng, 2018).

3. RESEARCH METHODOLOGY

Secondary data analysis can significantly increase the efficiency of research projects by saving time and resources that would otherwise be spent on primary data collection. Researchers can use existing datasets to explore research questions, identify patterns, and generate new hypotheses without having to conduct their own studies. Large datasets that would be difficult or impossible to obtain using primary data-collecting techniques are also accessible through secondary data analysis. Researchers investigating uncommon or difficult-to-reach people or phenomena might particularly benefit from this since they can get data from other researchers or governmental organisations. Thus, researchers have employed a supplementary way of data analysis to acquire pertinent and factual data. Considering solely secondary qualitative research is the important emphasis of this study because to its dependability, capacity to give thorough knowledge, and accessibility. Concerning the data collection process, researchers have collected data from literature reviews, case studies, and industry reports. Various reputable data were considered to collect information from academic journals and industry resources. However, researchers have not critically focused on one particular geographic location. They aimed to collect resources from a broad range of sources to provide a comprehensive understanding of the significant impact of quality management and IoT across different industries. Moreover, all the collected data was processed and thoroughly analysed using thematic analysis.

4. DISCUSSION AND FINDINGS

According to the European Commission, small and medium-sized businesses (SMEs) are companies with fewer than 250 workers and a revenue or turnover of less than €50 million. Even though it is possible to distinguish between the three groups (micro, small, and medium), we will refer to them all collectively in this study as 'SMEs', even if Table 1 gives a more thorough definition.

Table 1: The European Commission definition of SMEs.

Category	Staff	Turnover	Total balance
'Medium'	'<250'	'≤e 50 m'	'≤ e 43 m'
'Small'	'<50'	'≤e 10 m'	'≤ e 10 m'
'Micro'	'<10'	'≤e 2 m'	'≤ e 2 m'

According to the 'European Commission', '99% of all EU workers are employed by SMEs, making them the engine that drives the continent's economic growth'. As a result, they intend to encourage entrepreneurship and enhance the business climate for SMEs. Additionally, the US government emphasises the significance of its SMEs, which account for two-thirds of all newly created jobs.

The technologies of Industry 4.0 are underutilised and occasionally downright neglected by industrial SMEs, according to previous literature evaluations. The characteristics of SMEs have been discovered and reported in many studies. Specific SME traits that are advantageous in a digital transformation have been described. Entrepreneurship, creativity, a focus on learning and power centralisation were the traits found. In order for SMEs to maintain their competitiveness, investments must be made in and changes made to processes, systems, and technology. These investments and continual improvements are still necessary for SMEs to compete in Industry 4.0. Particularly cloud solutions have been found to be widely accepted, in part because of their simplicity in comparison to other Industry 4.0 technologies and the advantages of adopting them. They also learned that SMEs cannot use other Industry 4.0 technologies, such as autonomous robots, cyber-physical systems, machine-to-machine communication, etc., because of their relatively high cost. Regarding the aforementioned traits of SMEs and the significance of IoT and AI in SMEs. Previous research on the relationship between technology and organisations has mainly focused on the theories that either technology is an objective, external force that has a predetermined impact on organisational characteristics like structure or that technology is a result of human action, with technology being a result associated with strategic choice and social action. To understand technology as a social and physical construct that is always changing, it is vital to distinguish between human behaviour that influences technology and human activity that is influenced by technology. It also assumes structural features as the result of human intervention. In addition, actors acting in a particular social environment physically produce technology, and players also socially construct it by giving it various meanings.

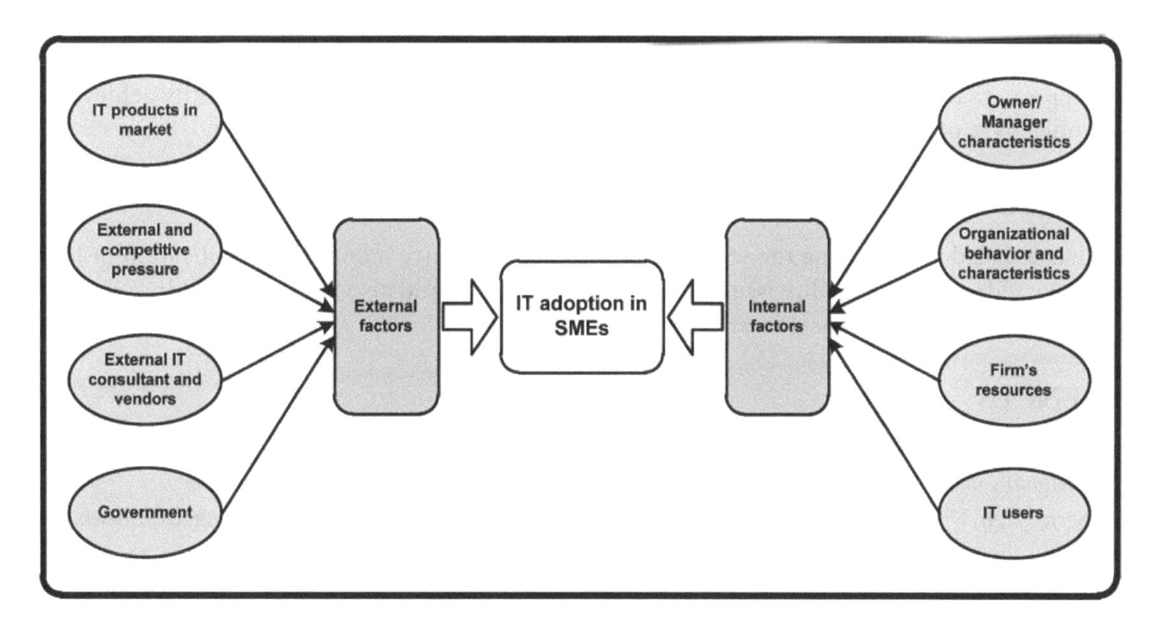

Figure 1. *Technological concept with SME business*

Analytical techniques like machine learning can be used on IoT device data. Machine learning, often known as AI, is becoming more and more popular in businesses as a part of Industry 4.0, notably software firms like Google and Facebook. With the availability of larger datasets and improved processing power, deep learning has gained popularity and, for example, outperformed traditional picture classification techniques. Deep learning techniques are always improving and outperforming their forerunners, which points to the intense research activity in the area. A 1997 study examined the use of 'artificial intelligence (AI)' techniques as an emerging technology for industry maintenance.

The 'IoT' is seen as a single network that allows different systems, processes and people to be configured and interconnected to allow free 'flow of data' and information for critical decision-making. Service quality and quality management are the most original sales proposals to offer superior service in the hotel sector. To employ IoT support to enable consumers to book rooms and other services more independently, to satisfy demands and expectations, and to anticipate standards in order to maintain superior quality control. Smart beacon technology and other methods put the emphasis on using services, which draws in additional clients (Lee, 2019). The hotel industry management uses these tools to understand demand forecasts, meet growing customer needs and deliver high-quality services to keep customers happy. Using IoT in hotels helps management understand the trend of booking a room, helps guests and customers make a reservation in advance to quickly check various other services such as food reservation, seating, taxi booking and other services. Supports development aspects of quality management and other related aspects in the hotel industry. In the hospitality industry, the quality of services and their management are critical success factors, so IoT is used to support better performance management and understanding of the various quality indicators to effectively achieve organisational goals (Hassija *et al.*, 2019).

5. CONCLUSION

It has already been said that IoT is a network that supports the interconnection of different systems and subsystems with identifiable addresses to provide better and smarter services according to users' needs. The current digitalisation of the business environment has made IoT scalable and adaptable because it facilitates better performance of functions, enables individuals to complete tasks quickly and also means better quality of services. The IoT application supports the reduction of business process errors so that quality can be easily improved. Using IoT systems supports the collection of data and information from different sources and sensor types and then the collection and organisation of this information to perform critical analysis for better decision-making. The handling can indicate data accuracy, bandwidth, device specifications, etc. and thus enables the collection of real-time information about the necessary inputs. Moreover, this research has followed a structured approach while collecting and analysing information. The secondary qualitative research method was utilised by reviewing the literature, case studies, and industry reports that are relevant to the research topic. The researchers have successfully achieved all the research objectives of this study and met the hypothesis made at the start of the study.

6. FUTURE SCOPE

The IoT has enabled in enhancing the service offerings in an effective manner, the application of IoT tends to support in addressing the quality issues in an effective manner and also involves in addressing the needs and requirements of the customers in an efficient manner.

REFERENCES

Buhalis, D., and R. Leung (2018). "Smart hospitality—Interconnectivity and interoperability towards an ecosystem," *International Journal of Hospitality Management*, 71, 41–50.

Femenia-Serra, F., B. Neuhofer, and J. A. Ivars-Baidal (2019). "Towards a conceptualisation of smart tourists and their role within the smart destination scenario," *The Service Industries Journal*, 39(2), 109–133.

Hassija, V., V. Chamola, V. Saxena, D. Jain, P. Goyal, and B. Sikdar (2019). "A survey on IoT security: application areas, security threats, and solution architectures," *IEEE Access*, 7, 82721–82743.

Kansakar, P., A. Munir, and N. Shabani (2019). "Technology in the hospitality industry: prospects and challenges," *IEEE Consumer Electronics Magazine*, 8(3), 60–65.

Lee, I. (2019). "The Internet of Things for enterprises: an ecosystem, architecture, and IoT service business model," *Internet of Things*, 7, 100078.

Torres, A. M. (2018). "Using a smartphone application as a digital key for hotel guest room and its other app features," *International Journal of Advanced Science and Technology*, 113, 103–112.

Wu, H. C., and C. C. Cheng (2018). "Relationships between technology attachment, experiential relationship quality, experiential risk and experiential sharing intentions in a smart hotel," *Journal of Hospitality and Tourism Management*, 37, 42–58.

Zhang, L., D. Jcong, and S. Lcc (2021). "Data quality management in the Internet of Things," *Sensors (Basel)*, 21(17), 5834.

22. A Comparison of the Effects of Robotics and Artificial Intelligence on Business Management and Economics

Ahmad Y. A. Bani Ahmad[1], Priti Gupta[2], Thimmiaraja J[3], Brijesh Goswami[4], Arun M[5], Geetha Manoharan[6], and Dalia Younis[7]

[1]Associate Professor, Department of Financial and Accounting Science, Middle East University, Amman 11121, Jordan

[2]Assistant Professor, P.G. Department of Economics, Bhupendra Narayan Mandal University (West Campus), P.G. Centre, Saharsa, Bihar

[3]Department of Information Technology, Dr. Mahalingam College of Engineering and Technology, Pollachi

[4]Assistant Professor, Institute of Business Management, GLA University, Mathura

[5]Department of MCA, School of Computing, Kalasalingam Academy of Research and Education, Kriahnanlovil, Arun Vincent

[6]School of Business, SR University, Warangal, Telangana

[7]College of International Transport and Logistics, AASTMT University, Egypt

ABSTRACT: The firm must adapt to incorporating new and innovative technology in the current digital world to increase production and output. A better administration of many business processes, including finance, advertising, supply chain management, human resource management, and economic elements like cost, realising higher income, GDP, etc., is made possible by the application of AI. Nearly 91 replies were obtained, and these data were examined utilising percentage rate calculations by the investigator using both primary and secondary data. The secondary data is obtained from a variety of online libraries, including EBSCO and others, in order to comprehend earlier research done in the subject. On the other hand, the implementation of robotics into business organisations and robotics has been transforming the world. Robotics plays a crucial role in business management as well as economics by reshaping and revolutionising current business practices. This research study will also explore the effects of robotics by focusing on the challenges and opportunities associated with the implementation process in various sectors. Regarding gathering information about the implementation of robotics and artificial intelligence, researchers have followed a secondary qualitative research approach. This research method has been chosen as it can provide significant information from broad perspectives. Based on the data collection, this research has compared the significant effects of 'Robotics and Artificial intelligence' in business management and economics.

Keywords: Artificial intelligence, robotics, management, sustainable development, economics

1. INTRODUCTION

It has been noted that the world is changing at a rapid pace, the advent of advanced tools like robotics, process automation and 'Artificial Intelligence (AI)' has led to an increase in enhancing the business management process and support in realising better benefits for different stakeholders. The technological advancements have made the world a small place, and businesses are trying to make critical connection with their stakeholders for achieving sustainable growth and development (Choi, 2019). The world is evolving quickly. There has never been a period in history when the abrupt changes brought on by the

DOI: 10.1201/9781003532026-22

advancement of information technology have not had an effect on nearly all spheres, including political, economic, and social life (Moloi and Marwala, 2020).

Artificial intelligence is applied in order to create computer intelligence. The three subcategories of artificial intelligence are machine learning, intelligent systems, and agile computations. Computational intelligence is the use of intelligent natural systems, such as ant colonies or bird flocks, as the building blocks of artificial intelligence. Systems that calculate the shortest path between two points, like Google Maps, have effectively been built using evolutionary algorithms (Ruiz-Rea *et al.*, 2020). Governments are being forced to establish their e-government, and financial institutions are being established as a result of the 'digital age' that started with the Internet and mobile technologies. In this era, businesses opened their cloud and web stores to interact with their client base. On tablets, smartphones and social media sites. Although other sciences like biotechnology, nanotechnology, biology, and so on are interwoven with the digital era (Chopra, 2019). Overall, it is a step towards the 'spatial economy', while other innovations have a greater direct or indirect impact on businesses and the economy. Artificial intelligence and robotics are the names of these advances. The industrialisation and mechanisation, primarily in Britain and by automakers, marked the beginning of the 'industrial age'. Early in the 20th century, business and the economy were significantly impacted by the production and supply side of the emerging economy. The advancements of the industrial period had an impact on the factors that drove production, including capital, entrepreneurship, labour and land. Mechanisation, culture, learning, finance and agribusiness are partly to blame for these consequences (Wong, 2020).

Along with artificial intelligence, Robotics has also emerged as one of the most significant technological advancements in recent years. It has become one of the most powerful sources of business management as well as economics. It usually offers a wide range of solutions to business problems by embracing automation, increasing operational efficiency, and improving precision. Nowadays, many organisations have already adopted and integrated robotics into their business operations, from Logistics to healthcare. Robotics has significantly reduced human errors. Moreover, this significant transformation also has profound implications related to profitability, productivity, and overall economic growth (Marwala, 2018).

1.1. Objectives and Hypotheses

- To explore the effects of 'robotics and AI' on business management to improve efficiency, productivity, and business performance
- To analyse the implications of these two emerging technologies on economics
- To identify the challenges and opportunities associated with the implementation of AI and robotics into business management
- To examine the economic effects of implementing AI and robotics in this current business world

Hypothesis 1: *The integration of AI and Robotics can significantly influence business management practices*

Hypothesis 2: *The adoption of AI and robotics can affect job opportunities for human*

Hypothesis 3: *The Implementation of AI and robotics has significant effects on economic growth and market dynamics*

2. LITERATURE REVIEW

'AI' is capturing every element of human life, and this is causing a fundamental shift in how people perceive and behave. Generally speaking, people have looked to economic and financial theories to help them grasp ideas like competitiveness, long-term economic expansion, knowledge gaps or information distortions, and the importance of failure as a component in manufacturing and decision-making. among other hypotheses, the industrial process allocates resources (Wirth, 2018).

These theories' fundamental characteristic is their attempt to reduce the impact of uncertainty by attempting to return the future to the present. These ideas have a strong foundation in risk and risk management. Risk is defined as the influence of uncertainty on a target in '***International Organization for Standardization (ISO)*** Standard 31000'. In other words, uncertainty has a significant role in the deviation from anticipated results.

Economics as a science has always offered a solid foundation for comprehending uncertainty and what decision-making implies in the field of behavioural science. The economy has achieved this via a variety of future-prediction models (Rampersad, 2020).

The major characteristic of economic models, as was previously said, is that they aim to reverse the impacts of uncertainty by attempting to bring the coming back into the present. Artificial intelligence, according to scientists, 'does not genuinely provide us intelligence, but a crucial aspect of intelligence: forecasting' (at least in its current state). Humans have previously used this to develop monetary and financial ideas. Artificial intelligence impacts how economic and finance theories are delivered as it provides us with fundamental intelligence (Rampersad, 2016). In order to remove ambiguity and enable agents to make educated decisions, this study focused on how artificial intelligence will reinterpret several crucial economic and financial theories (Lee and Park, 2018).

Apart from AI, one of the most promising technological advancements is Robotics. It has significant capabilities to lead transformative changes within organisations. In the last few years, Robotics has become a disruptive force in the business world that offers many benefits to organisations from efficiency to precision (Chopra, 2019). More or less, every organisation has started adopting robotics to carry out human works. Robotics can be integrated into business management. It has many significant impacts on various aspects, such as process automation, supply chain management, organisational structure and decision-making processes.

3. RESEARCH METHODOLOGY

Secondary data analysis allows for comparative research, comparing data from different sources, regions or time periods. Researchers can explore variations, similarities, or differences across populations, cultures or contexts, enhancing the generalizability and applicability of their findings. Since secondary data analysis involves working with pre-existing data, it does not require direct interaction with human participants. This can alleviate ethical concerns related to informed consent, privacy or potential harm to participants. Researchers can focus on data analysis rather than data collection, reducing the burden on participants and ethical complexities associated with primary data collection. Secondary data analysis allows for the replication and validation of previous research findings. By reanalysing existing datasets, researchers can confirm or challenge prior conclusions, enhancing the robustness and reliability of research outcomes. Replication studies are essential for building scientific knowledge and verifying the generalizability of findings. As the researchers have considered a secondary qualitative research method, all the information has been collected from academic datasets, scholarly articles, government publications, and industry reports. Moreover, this research study also considered a wide range of datasets to gather information and provide essential insights about the effects of Robotics and AI on business management and economics.

4. DISCUSSION AND FINDINGS

Today's demands for quicker service delivery to a large number of clients at once have inadvertently increased the complexity of the business process. Manually handling such scenarios can lead to blunders that are not only costly in time and effort but also have a substantial impact on how a corporation is run. The AI-powered smart system can be integrated with the current business management area system in order to prevent such errors and the associated losses, allowing for the safe storage of all crucial documents in a single location. Based on the above figure it can be inferred that AI comprises several functions from speech recognition to expert systems. These make significant effects on business performance and economic growth. On the contrary, the integrated machine learning algorithm assists the company in obtaining an error-free output with virtually little waiting time for inexpensive operations (Ruiz-Rea *et al.*,

2020). Robots are one of the most productive innovations in the history of technical mechanisms because they can do a variety of tasks more precisely and effectively than people can. In addition to this, a number of industrial and medical errors can be virtually eliminated by utilising the robot's engagement facilities. The robot's ability to help the business with significant cost savings and real job changes that are cost-effective is both its most lucrative and interesting feature. A single robot can now effectively replace 1,000 workers thanks to the top-notch advancements in robotics that have been made in recent years.

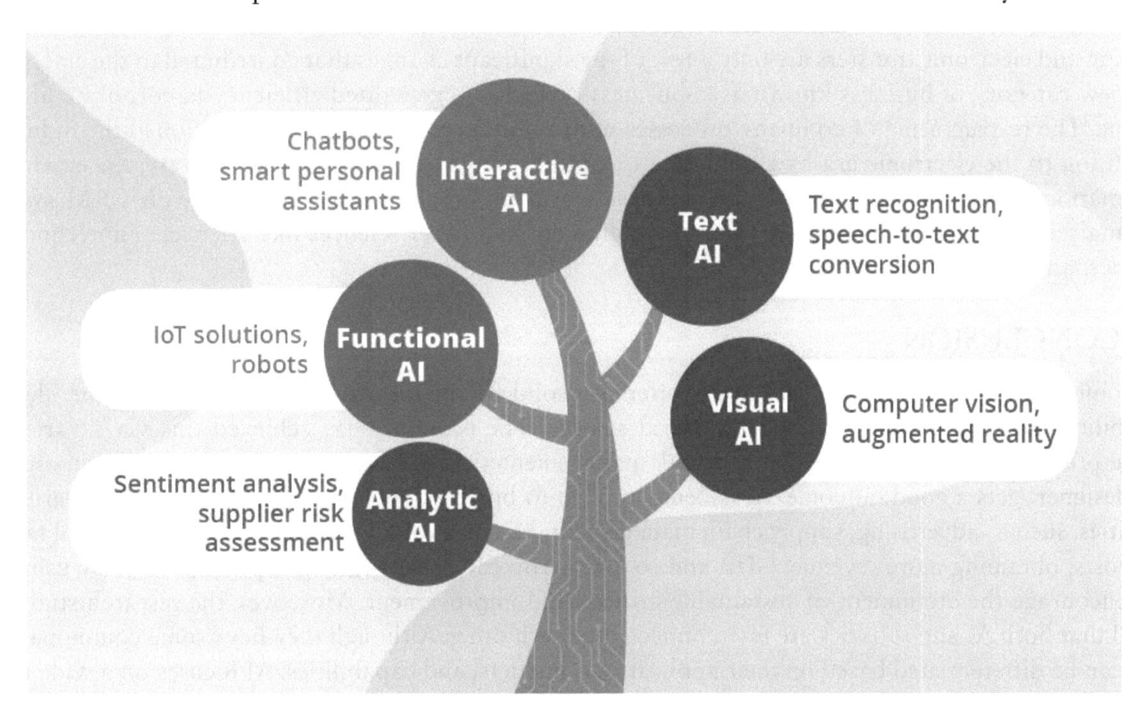

Figure 1. AI functions.

Table 1: Overview of applications and algorithms.

Subtopic	Algorithms	Applications
'Quality control management'	'Decision tree, SVM, NN'	'Quality cost reduction'
'Predictive maintenance'	'KNN, decision tree, PCA'	'Remaining the useful life'
'AI training and concept generation'	NN	'Object- recognition'

The employed algorithm and its application on the basis of the business management subtask are shown in the above Table 1. This table discusses the diverse algorithm that was used in the analysis section to produce sufficient output in accordance with the topic criteria.

NN models replicate the learning process used by the human brain and typically include weights, nodes and layers. On the contrary, the tree algorithm is a branching technical architecture that inherits many types of item information (Moloi, 2016). By combining these methods, AI technology becomes intelligent and aids the organisation by making key decisions regarding net company management by examining the state of the market and economic aspects. Additionally, AI employs models from 'supervised and unsupervised learning methodologies' that support singular value decomposition and effective principal component analysis. AI primarily uses Q-learning, a well-known 'reinforcement learning algorithm', for business management objectives.

The total findings lead to the conclusion that AI helps to enhance corporate management procedures and significantly boost economic metrics. Artificial intelligence is applied in order to create computer intelligence. The three subcategories of artificial intelligence are machine learning, computer science and flexible computer use. Computational intelligence is the use of intelligent natural systems, such ant colonies or bird flocks, to build intelligent machines. Systems that calculate the shortest distance between two points, like Google Maps, have successfully exploited computational intelligence. The 'digital age', which started with the internet and mobile phone technology, forced governments to establish financial firms on tablets, mobile phones, and social networking sites, and firms to open their online stores and cloud retailers to work with their clients. Agreements, electronic invoices, e-commerce, the online, mobile banking and electronic transfers are only a few of the significant changes that contributed to the emergence of a new category of business known as 'e-business', which has generated efficiency in corporate life and Unique. The restructuring of company processes with e-commerce settings has transitioned the industrial revolution to the electronic era by eliminating or optimising procedures. On the contrary, the expanding information environment compelled the business world to simultaneously respond with CRM systems and analyse big data. Despite the fact that the digital era and other sciences like science, nanotechnology, genetics, and so on are closely related.

5. CONCLUSION

Economics as a science has almost always offered a solid foundation for comprehension of the ideas of instability and decision-making in behavioural science. The economy has achieved this via a variety of future-prediction models. On the contrary, risk management aims to lessen or reduce these variables so that the 'designer' gets a good outcome. AI is being applied to business operations to help manage a variety of activities, such as advertising, supply chain management, human capital, finance, and other financial factors like costs, obtaining more revenue, GDP, and so forth. You can successfully grasp the process by using AI. and encourage the attainment of sustainable growth and improvement. Moreover, the research study has found that both AI and robotics are interconnected to each other. Although they have some commonalities, they can be differentiated based on their applications, impacts, and capabilities. AI focuses on a wide range of technologies, excels in performing complex tasks, and can analyse a large amount of data. Whereas, robotics focus on the use of physical machines that can perform tasks with precision and speed.

6. FUTURE SCOPE

This article is intended to understand the role of AI in the business management domain and economic aspects, the researcher has made a conceptual understanding on the implementation of AI in the related field, based on considering it can be stated that the study can be extended to specific type of industries covering banking, financial services, retail sector, manufacturing companies etc. this will enable in creating more understanding on the emerging role of AI in the domains and can explore more opportunities for the business.

REFERENCES

Choi, J. J., and Ozkan, B. (2019). "Innovation and disruption: Industry practices and conceptual bases," In J. J. Choi & B. Ozkan (Eds.), *Disruptive Innovation in Business and Finance in the Digital World* (vol. 20, pp. 3–13). Emerald Publishing Limited.

Chopra, K. (2019). "Indian shopper motivation to use artificial intelligence: Generating Vroom's ex-pectancy theory of motivation using grounded theory approach," *International Journal of Retail & Distribution Management*, 47(3), 331–347.

Lee, Y. K., and D. W. Park (2018). "Design of internet of things business model with deep learning artificial intelligence," *International Journal of Grid and Distributed Computing*, 11(7), 11–22.

Marwala, T (2018). *Handbook of Machine Learning: Foundation of Artificial Intelligence*. World Scientific Publication.

Moloi, T. (2016). "A cross sectoral comparison of risk management practices in South African organizations," *Problems and Perspectives in Management*, 14(3), 99–106.

Moloi, T., and T. Marwala (2020). "Introduction to artificial intelligence in economics and finance theories," *Artificial Intelligence in Economics and Finance Theories*, 1–12.

Rampersad, G. (2020). "Robot will take your job: Innovation for an era of artificial intelligence," *Journal of Business Research*, 116, 68–74.

Ruiz-Rea, J. L., J. Uribe-Toril, J. A. Torres, and J. De Pablo (2020). "Artificial intelligence in business and economics research: trends and future," *Journal of Business Economics and Management*.

Wirth, N. (2018). "Hello marketing, what can artificial intelligence help you with?" *International Journal of Market Research*, 60(5), 435–438. https://doi.org/10.1177/1470785318776841.

Wong, K. K. L., G. Fortino, and D. Abbott (2020). "Deep learning-based cardiovascular image diagnosis: a promising challenge," *Future Generation Computer Systems*, 110, 802–811.

23. A Resource-Based View Assessment of Artificial Intelligence and its Impact on Strategic Human Resources Quality Management Systems

Dhruva Sreenivasa Chakravarthi[1], Shaziya Islam[2], Sangram Singh[3], Richa Gupta[4], Melanie Lourens[5], M. Ravichand[6], and Ivanenko Liudmyla[7]

[1]Research Scholar, KL Business School, Koneru Lakshmaiah Education Foundation, Deemed to be University, Vaddeswaram Guntur District (A.P), India

[2]Associate Professor, Rungta College of Engineering and Technology, Bhilai

[3]Professor, Department of Business Sudies, Gulzar Group of Institutions, Khanna, Ludhiana

[4]Assistant Professor, Department of Computer Science and Engineering, Graphic Era Hill University, Dehradun, Uttarakhand

[5]Deputy Dean Faculty of Management Sciences, Durban University of Technology, South Africa

[6]Professor of English, Mohan Babu University, Tirupathi, (Erstwhile Sree Vidyanikethan Engineering College)

[7]Chernigiv Post Graduate Pedagogic Institute of K.D.Ushinsky, Ukraine

ABSTRACT: Artificial intelligence (AI) is advancing quickly and imitating human cognitive abilities. However, during the past few years, AI has made considerable strides and drawn a lot of funding and academic interest. Our study has chosen this prospective field of study in consideration of how human resources functions or jobs will change in the future and how they will react to artificial intelligence. The automotive industry has witnessed significant advancements in recent years, and the utilisation of data collection techniques is one such innovation. This research paper explores the novel application of data collection methods in the automotive industry for the first time. This study uses a mixed-methods strategy that combines qualitative and quantitative approaches. Artificial intelligence was shown to have a considerable beneficial effect on human resources.

KEYWORDS: Human resources, artificial intelligence, quality management system

1. INTRODUCTION

As a disruptive technology, artificial intelligence has the potential to automate and enhance a number of HR processes, including recruiting, overseeing performance, instruction, and retention of staff. By analysing the unique resources and capabilities that AI brings to HR, organisations can optimise their HR practices, improve decision-making, and foster a culture of continuous improvement. Explore the use of AI in automating and optimising the recruitment and selection processes in the automotive industry. This includes AI-powered resume screening, candidate sourcing, and automated interview scheduling. Investigate the ways in which AI can help HR departments forecast future talent requirements, detect skills gaps, and create specialised development and training initiatives to match the changing needs of the automotive sector. Humans must put more effort into increasing productivity, developing their skills and raising efficiency despite the risks and difficulties they face. Figure 1 vividly illustrates the advent of an era of automated business models by demonstrating the degree to which innovation and human resource management have been impacted by the level of digital penetration.

DOI: 10.1201/9781003532026-23

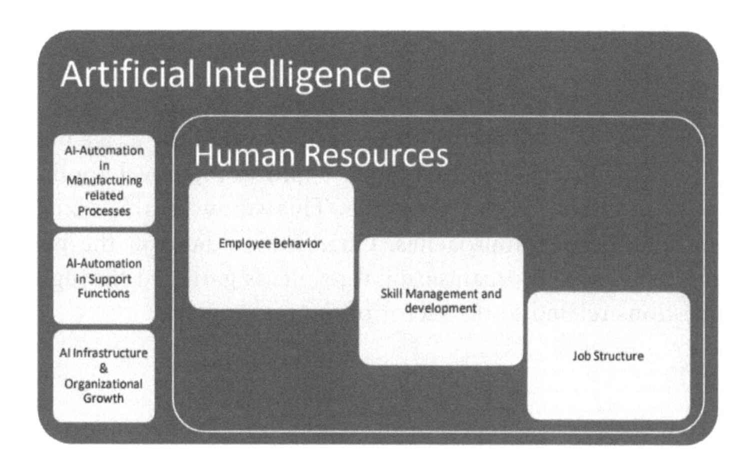

Figure 1.

1.1. Problem Statement

Artificial intelligence is posing a challenge to workers by assuming their cognitive and regular tasks and requiring newer skills. Because AI poses a danger to some jobs, eventually, all human resources will be replaced by machine resources.

1.2. Research Objectives

- To evaluate the status of AI deployment in strategic human resources (HR) quality management solutions at this time.
- Evaluating the impact of AI on workforce dynamics, job roles, and skill requirements within HR quality management systems.

2. LITERATURE REVIEW

In their study titled 'Impact of Artificial Intelligence on Human Resources', Khatri *et al.* (2020) discuss the effects of integrating AI-based systems and processes into businesses on talent management, including both technical and non-technical resources. Enterprises are drawn to the AI-based eco-system and environment as part of industry 4.0 ambitions. With a specific spotlight on the Indian subcontinent, Premnath and Chully (2020) distributed their subjective concentrate on Artificial intelligence in the Human Resource Management function. That's what the creators found in spite of the fact that artificial intelligence is being utilised in the enrolment and preparing processes, it hasn't saturated the HR capability in that frame of mind, however much it has in different areas (Liu *et al.*, 2021). The HR capability keeps on being a feasible region for artificial intelligence organisation. Vrontis *et al.* (2021) undertook a peer journal article study. According to the authors, organisations' primary goal should be to live with technology while also reskilling their human resources for the future. Technology shouldn't be permitted to take over and replace the role of people, invade their privacy, violate their rights, or utilise their data inappropriately (Bag *et al.*, 2021).

Organisations can use artificial intelligence to improve the client care insight by giving more significant proposals and practical choices (Manser Payne *et al.*, 2021). As indicated by the asset-based view (RBV; Majhi *et al.*, 2021), the utilisation of AI incorporates a blend of verifiable assets (Pack *et al.*, 2021c). These assets incorporate strong resources, work abilities, and authoritative coordination (Selz, 2020). At the point when an organisation effectively coordinates assets that are not handily repeated, it acquires an upper hand (Yasmin *et al.*, 2020) and works on, generally speaking firm execution (Chen and Lin, 2021). Subsequently, it is significant to investigate the components and key factors that impact the effect of AI on firm execution (Mikalef *et al.*, 2021), especially in the web based business industry where direct client communication is predominant (Wang and Fan, 2021).

3. RESEARCH METHODOLOGY

3.1. Research Design

The automotive business has seen critical progressions as of late, and the usage of information assortment strategies is one such advancement. This research paper explores the novel application of data collection methods in the automotive industry for the first time. This study uses a mixed-methods strategy that combines qualitative and quantitative approaches. Quantitative data on the perceived influence of AI on HR procedures, staff efficiency, and organisational results is gathered through survey questionnaires. Questionnaire had 10 questions related to the sector of the respondents.

3.2. Data Collection

The basic data was gathered via a field survey. For the research study, information was gathered from 230 respondents who represented 8 automotive firms in western India. Secondary data is gathered from books, journals (both online and printed), research papers, company reports, websites, and newspapers.

3.3. Sampling Method

To gather the responses, a stratified random sampling technique is used. The researcher divided the population into strata based on the organisational employee strength of several passenger vehicle manufacturers because the population for this study is relatively large.

3.4. Tools for Data Collection and Analysis

The main data from a selected sample is collected via a survey using a 5-point Likert scale. The IBM SPSS 21 programme is used to analyse the data using a variety of statistical tools.

4. RESULTS AND DISCUSSION

The original schedule of interviews and survey from strategic management were used to interview 230 respondents across eight sectors.

4.1. Demographic Information and Descriptive Analysis

For this study, one of the fundamental demographic characteristics is the composition of the population, which is crucial for any meaningful socio-economic analysis (Vannucci, and Colla, 2019). According to their position in their separate functions, industry officials filled out open-ended designations and functions. Percentage distribution of respondents is shown in Table 1.

Table 1: Percentage distribution of respondents.

Variables	Response Options	Percentage
Gender	Male	94.20%
	Female	5.80%
	Prefer not to say	
Qualification	Graduate	55.50%
	Post graduate	44.50%
Work Experience	Less than 15 years	35.10%
	16–20 years	25.90%
	More than 20 years	39.0%

Source: Author's compilation.

Out of 230 responders, Table 1 above shows that 94.2% are men and 5.8% are women. As a result, men make up the majority of the study's responses. Out of 230 responses, it can be seen from the same table's qualification makeup that 55.5% are graduates and 44.5% are postgraduates. Therefore, it is appropriate to assume that the majority of respondents have relevant awareness and understanding of the subject. In the same table above, it can be observed that out of 230 respondents, 35.1% have less than 15 years of work experience, 25.9% have between 16 and 20 years, and 39% have more than 20 years. As a result, according to their work experiences, the majority of the respondents to the survey are seeing changes in the phases or stages of technological growth.

The frequency distribution for functions is shown in Table 2 and Figure 1.

Table 2: Frequency distribution for functions.

Response Options	Frequency	Percentage
HR	23	10.00%
IT	10	4%
Manufacturing	24	10.43%
Project Management	26	11.30%
Quality Management	30	13%
R & D	42	18.26%
Sales and Marketing	15	6.52%
SCM	60	26.00%
Total	230	100%

Source: Author's compilation.

Out of 230 respondents, Table 2 above shows the AI percentage used in various sector that 10% represent the human resources function, 4% the information technology function, 10.43% the manufacturing function, 11.3% the project management function, 13% the quality function, 18.26% the research and development function, 6.52% the sales and marketing function, and 26% the supply chain management function. Hence, it is showing in figure 2, that in the bulk of organisational functions, AI is used (Manser Payne *et al.*, 2021).

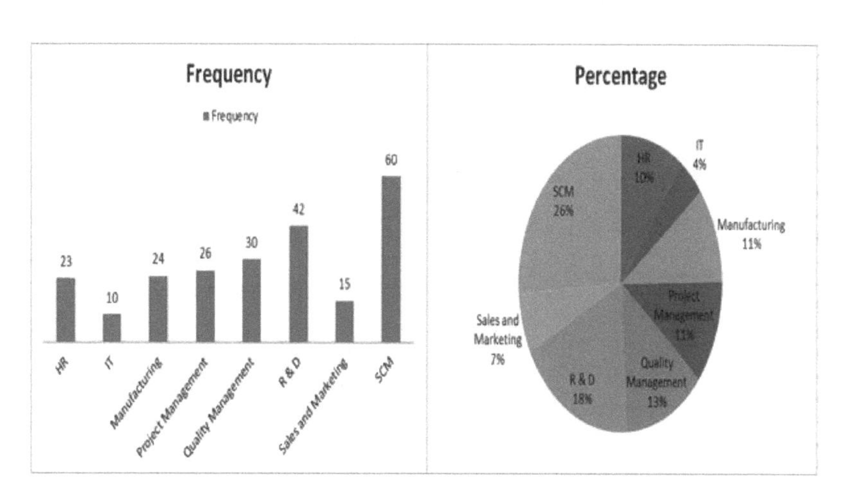

Figure 2. *Frequency distribution of AI for functions. Source: Author's compilation.*

5. CONCLUSION

The conclusions, which provided insights into organisational strategy and preparedness to develop human talent, led to recommendations. The findings demonstrated that cooperation with technology is essential for the successful adoption of artificial intelligence. The particular effects of AI on numerous HR processes, including hiring and selection, performance management, development and training, engagement among workers, and talent management, may be explored in more detail. This may help uncover possible problems and ethical issues while also shedding light on how AI can improve these procedures' efficiency and efficacy.

REFERENCES

Bag, S., J. H. C. Pretorius, S. Gupta, and Y. K. Dwivedi (2021). "Role of institutional pressures and resources in the adoption of big data analytics powered artificial intelligence, sustainable manufacturing practices and circular economy capabilities," *Technological Forecasting and Social Change*, 163, 120420.

Chen, Y., and Z. Lin (2021). "Business intelligence capabilities and firm performance: a study in China," *International Journal of Information Management*, 57, 102232.

Khatri, S., D. K. Pandey, D. Penkar, and J. Ramani (2020). "Impact of artificial intelligence on human resources," *Data Management, Analytics and Innovation: Proceedings of ICDMAI 2019, Volume 2* (pp. 365–376). Springer: Singapore.

Liu, Y., D. Vrontis, M. Visser, P. Stokes, S. Smith, N. Moore, A. Thrassou, and A. Ashta (2021). "Talent management and the HR function in cross-cultural mergers and acquisitions: the role and impact of bi-cultural identity," *Human Resource Management Review*, 31(3), 100744.

Majhi, S. G., A. Mukherjee, and A. Anand (2021). "Business value of cognitive analytics technology: a dynamic capabilities perspective," *VINE Journal of Information and Knowledge Management Systems*.

Manser Payne, E. H., J. Peltier, and V. A. Barger (2021). "Enhancing the value co-creation process: artificial intelligence and mobile banking service platforms," *Journal of Research in Interactive Marketing*, 15(1), 68–85.

Mikalef, P., K. Conboy, and J. Krogstie (2021). "Artificial intelligence as an enabler of B2B marketing: a dynamic capabilities micro-foundations approach," *Industrial Marketing Management*, 98, 80–92.

Premnath, E., and A. A. Chully (2020). "Artificial intelligence in human resource management: a qualitative study in the Indian context," *Journal of Xi'an University of Architecture & Technology*, XI, 1193–1205.

Selz, D. (2020). "From electronic markets to data driven insights," *Electronic Markets*, 30(1), 57–59.

Vannucci, M., and V. Colla (2019). "Quality improvement through the preventive detection of potentially defective products in the automotive industry by means of advanced artificial intelligence techniques," *Intelligent Decision Technologies 2019: Proceedings of the 11th KES International Conference on Intelligent Decision Technologies (KES-IDT 2019), Volume 2* (pp. 3–12). Springer Singapore.

Wang, M., and X. Fan (2021). "An empirical study on how livestreaming can contribute to the sustainability of green agri-food entrepreneurial firms," *Sustainability*, 13(22), 12627.

Yasmin, M., E. Tatoglu, H. S. Kilic, S. Zaim, and D. Delen (2020). "Big data analytics capabilities and firm performance: an integrated MCDM approach," *Journal of Business Research*, 114, 1–15.

24. Role of Machine Learning in the Fields of Marketing Management and Strategic Human Resource Management

Ganesh Waghmare[1], Swati Manoj Yeole[2], Nidhi Tandon[3], Amit Gupta[4], Melanie Lourens[5], Ravinder Goyal[6], and Leszek Ziora[7]

[1]Associate Professor, Business Administration Academics, Lexicon Management Institute of Leadership and Excellence, Pune, Maharashtra

[2]Associate Professor, Business Administration Academics, MIMA Institute of Management, Balewadi, Pune, Maharashtra – 411045

[3]Associate Professor, Department of Commerce, Manav Rachna International Institute of Research and Studies (Deemed University), Faridabad, Haryana

[4]Assistant Professor, Department of Computer Science and Engineering, Graphic Era Hill University, Dehradun, Uttarakhand

[5]Deputy Dean Faculty of Management Sciences, Durban University of Technology, South Africa

[6]Assistant Professor, Mechanical Engineering, Gulzar Group of Institutions Khanna, Ludhiana

[7]Faculty of Management, Czestochowa University of Technology, Poland

ABSTRACT: Similar objectives addressed at different audiences are shared by marketing and human resources. The branding of the business and its consumer communication are the responsibility of marketing. Employment branding is the responsibility of strategic human resource management (SHRM), which makes sure that both internal employees and external prospects have favourable perceptions of the company. However, the researchers have found that SHRM, marketing management and Machine learning are directly linked with each other. SHRM and marketing management focus on similar goals. Whereas machine learning helps build connections between SHRM and marketing management. Hence, machine learning has significantly helped organisations improve their SHRM and marketing efforts.

KEYWORDS: Marketing management, strategic human resource management, machine learning

1. INTRODUCTION

All organisations must, in fact, align their human resources strategy with their marketing efforts. While the marketing division communicates the company's brand to customers, the relationship between HR and marketing establishes a connection with workers. The staff of the company is the HR department's main clientele. Due of its connection to marketing, the industry uses training sessions, evaluations, and worker orientations to spread the brand messaging created through marketing to its employees (Antons *et al.*, 2018).

Additionally, machine learning uses computer algorithms that can analyse data and advance organically by knowledge. It is powered by artificial intelligence. The ML algorithm continuously analyses past data, predicts the future with excellent internal validity, and can carry out routine tasks by itself.

1.1. Objectives and Hypotheses

Researchers have followed some research objectives and Hypotheses to reach meaningful conclusions. The objectives of this research study are,

DOI: 10.1201/9781003532026-24

- To examine the role of machine learning in marketing management and its impacts on various marketing activities
- To evaluate the application of machine learning in SHRM and its impacts of various SHRM activities
- To identify and analyse the challenges and opportunities associated with integrating machine learning in marketing and SHRM
- To investigate the key success factors and best practices of machine learning to improve the efficiency of SHRM and marketing management

Hypothesis 1: *Machine learning has the capability to improve the marketing communication of an organisation*

Hypothesis 2: *The implementation of machine learning can have significant effects on SHRM*

Hypothesis 3: *Organizations face benefits as well as challenges in adopting machine learning in marketing and SHRM*

2. LITERATURE REVIEW

Machine learning differs from other kinds of software due to three key components: high-speed processing, a massive volume of high-quality data, and robust algorithms. As a result, machine learning technologies, at their core, deliver superior outcomes. In addition, machine learning technologies boost the accuracy and dependability of daily tasks by employing an algorithm that connects high-quality data with quick computing resources.

As per the views of Jain and Pandey (Antons and Breidbach, 2018), If marketers utilise machine learning (ML) properly, they can use it to transform their whole marketing campaign by extracting the most important insights from their datasets and applying them in real-time. In order to steadily gain consumers and maximise campaign value, ML assists marketers in making judgements that are dependable regarding the optimal distribution of funds across media channels or determining the most effective ad placement.

Samarasinghe and Ajith (2021), argues that Machine learning technologies do not automatically carry out activities to achieve marketing objectives; as Machines cannot learn themselves, they require the assistance of a human force. In order to study business goals, customer preferences, historical trends, the overall framework, and acquire competence, they require time and training. This requires patience, as well as trustworthy information and quality promises.

Delery and Roumpi (2017) reports that ML enables HR managers to reduce challenges and lead to well-informed decisions that impact worker and organisational success. For example, HR executives utilise ML algorithms to anticipate worker potential, tiredness, flight risk, and even inclusive engagement, resulting in more productive dialogues that increase worker experience, retention, and performance.

3. METHODOLOGY

In this review, a systematic method was utilised. It enables a more in-depth examination of the literature, which comes from various fields and employs a variety of methodologies and theoretical frameworks. Secondary data sources were also employed to compile responses to queries and analyse research methodologies in order to develop crucial information that would assist answer the issues posed by the study's objectives. The researchers have simply taken into consideration a secondary qualitative technique when it comes to the research study on the function of machine learning in the context of marketing and strategic human resource management. In order to enable in-depth exploration, this procedure was centred on assessing and analysing already-existing qualitative data. In terms of data collection, the researchers have collected data from reputable sources that include academic databases, government publications, journals, online articles, industry reports, business publications, and other relevant websites. The data has been gathered relevant to the research topic only. Whereas, thematic analysis was use to extract information from secondary qualitative data.

4. ANALYSIS AND INTERPRETATION

Machine learning may benefit businesses in a variety of ways (Marketingevolution.com, 2021). There are several ways for businesses to employ machine learning to create an all-encompassing marketing plan. For instance, marketing teams may have trouble deciding where to place message and advertising. Based on user choices, marketing teams may create informed strategies. However, they lack the adaptability and agility to alter their plan in response to the most recent customer data (Jiang and Messersmith, 2018).

Additionally, ML may track which messages users have responded to and build a more thorough user profile. Depending on the clients' preferences, marketing teams may then send them communications that are better tailored to them. For instance, Netflix utilises ML to identify the genres that a certain user is interested in. The preferences that the handler sees are then customised to suit their preferences. Additionally, when machine learning is used, businesses may gather consumer data, enabling marketing teams to increase conversion rates and enhance the customer experience (Paweloszek, 2015).

Additionally, ML have a significant influence on an organisation's human resources department. In order to enhance the efficiency and efficacy of SHRM activities, ML applications increase worker experience and facilitate business performance. All firms have been concerned in analysing data linked to Human Resources and have focussed on human capital. It is viewed as the most significant component impacting the business's growth and processes at all levels of human resource policy (Paweloszek, 2015).

It has been observed that ML allows HR sectors to progress the applicant and worker experience by mechanising repetitive, low-value tasks and freeing up time to emphasise the more tactical, inventive work. Many of these time-consuming tasks that HR staff must perform may be automated by machine learning (Sabbeh, 2018). By better comprehending the data, HR managers may predict attrition trends, communication issues, project development, worker engagement, and a variety of other essential events and concerns. This will enable them to identify any issues early and take corrective action before they cause major problems. More importantly, by understanding the realities around employee turnover and implementing the necessary changes to lessen the issue. Moreover, ML can increase working efficiency by assisting workers in matching the company's strategic and operational needs.

Table 1: Relationship of ML with SHRM.

S/no.	SHRM Practices	Benefits of ML
1.	Employee involvement	ML helps HR managers in structured initiatives for the involvement and empowerment of employees.
2.	Talent management	ML aids in identifying and develop talent by analysing employee' data
3.	Employee development	Machine learning helps in mapping employee's competency and effectiveness.
4.	Performance management	ML tools review and examine the performance of the employees.
5.	The efficiency of the HR process	HR functions have been electronically enabling to improve efficiency and competencies.

Source: Self-made.

Hence, this shows the relationship between machine learning and marketing management, ultimately benefiting a business's growth and progression.

5. RESEARCH/FINDINGS

Marketing Management, SHRM and Machine Learning are all closely intertwined, which ultimately drives the growth of the business. This is an essential finding in the understanding of this research paper.

The study concludes that marketers need to have access to a lot of data. As a result, the ML tool is successfully utilised to manage huge amounts of data regarding customer preferences, market trends, and other factors that will affect the success of ML-enabled marketing. This information is obtained via the company's CRM, marketing initiatives, and website data. This may be improved by marketers using second- and third-party data. As a result, marketers pick the appropriate media channels, the appropriate pricing, and the appropriate timing for product promotion using accurate data and other outside factors that may affect a purchase decision. Additionally, marketing teams must demonstrate marketing value and return on investment to executive stakeholders. Marketing strategies may advance these objectives with the aid of ML technologies, distribute funds to profitable campaigns in an efficient manner, and provide marketing statistics that demonstrate the worth of a campaign. As a result, using ML to marketing management aids in achieving qualitative objectives like improving the client experience.

Additionally, it has been shown that machine learning will provide crucial insights into these factors, enabling HR managers to handle them more quickly and efficiently. These perceptions may be very beneficial for increasing productivity and reducing staff turnover rates.

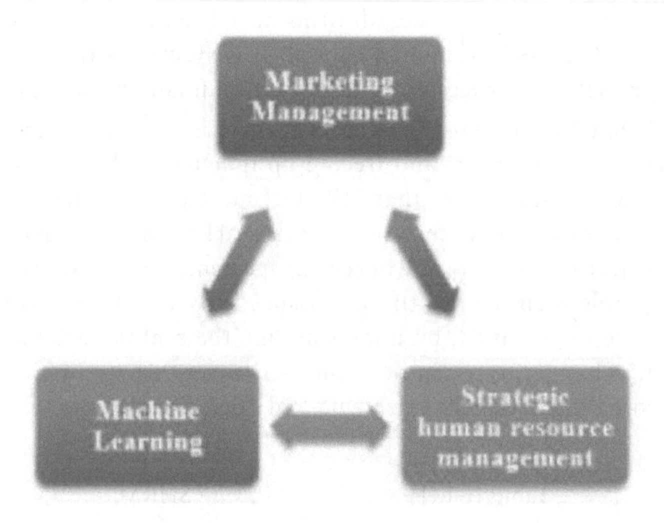

Figure 1. Marketing management, SHRM and machine learning. Source: Self-made.

It has been found that ML technology developments in business are bringing machines and employees closer together and researching methods to use it to enhance efficiency, convenience, and proficiency.

Based on analysing and interpreting the collected data, it has been found that there is a significant relationship between marketing management, machine learning, and human resource management. Their mutual contribution of them helps businesses to enhance performance and improve decision-making. In the context of the day-to-day relationship between marketing management and machine learning, various machine learning algorithms, techniques, and data analytics help in marketing. A large amount of customer data can be analysed easily using machine learning algorithms. It can be estimated that within 2024, more than 90% of marketing organisations will be implementing machine learning and AI. On the contrary, the implementation of machine learning can significantly enhance sales by 15–20% and reduce customer churn by 10–15% . As the machine learning market will reach $30 billion by 2026, more than 80% of marketing organisations believe that AI and machine learning are extremely essential for marketing. Apart from that, machine learning can also impact human resources management significantly by analysing a wide range of data sets of employees, including employee recruitment, performance, recruitment, engagement, and retention. Moreover, by adopting machine learning, HR managers can identify patterns and trends that help them optimise workforce planning and make data-driven decisions. It has been found that around 50% of organisations have also implemented machine learning to transform their HR process (Jiang and Messersmith, 2018). More than 60% of organisations will adopt machine learning to optimise workforce planning and talent acquisition. It can help HR managers to reduce 50% of their time while performing administrative tasks (Delery and Roumpi, 2017). Furthermore, the application of machine learning has been illustrated below using a table.

Table 2: Application of machine learning in marketing and SHRM.

Marketing	Strategic Human Resource Management
Customer segmentation	Talent acquisition
Demand forecasting	Employee engagement and retention
Social media analysis	Chatbots
Ad optimization	HR analytics and reporting
Marketing campaign optimization	Predictive analytics for workforce planning
Fraud detection	Performance management
Personal recommendation system	Sentiment analysis

6. CONCLUSION

The study has concluded that ML, SHRM and marketing management are significantly connected with each other. The main reason the SHRM and marketing departments are linked is that they have similar goals that help both the business development and ML assist both departments in their functions and operations. Therefore, the proper application of AI can result in time savings and improved efficiency in businesses.

Machine learning is a method for swiftly labelling and examining large data sets. People can accomplish something independently, but a machine can do it quickly and on a far better scale. Numerous marketing leaders believe that automation and machine intelligence will permit their team to emphasise strategic marketing initiatives. As customer expectations for more tailored, appropriate, and cooperative experiences rise, machine learning is an essential tool in meeting business goals. It enables marketers to progress consumer segments, offer more relevant creative campaigns, and more efficiently assess success.

Furthermore, ML can help HR by streamlining the procedure, reducing mistakes, and streamlining the department's process. ML is better furnished to identify the precise requirements of numerous employees and provide tailored training, rewards and incentive programmes for workers.

REFERENCES

Antons, D., and C. F. Breidbach (2018). "Big data, big insights? Advancing service innovation and design with machine learning," *Journal of Service Research*, 21(1), 17–39.

Delery, J. E., and D. Roumpi (2017). "Strategic human resource management, human capital and competitive advantage: is the field going in circles?" *Human Resource Management Journal*, 27(1), 1–21.

Jiang, K., and J. Messersmith, "On the shoulders of giants: a meta-review of strategic human resource management," *The International Journal of Human Resource Management*, 29(1), 6–33.

Marketingevolution.com (2021). *AI Marketing: Components, Benefits, and Challenges | Marketing Evolution [Internet]*. Marketingevolution.com. 2021 [cited 8 September 2021]. Available from: https://www.marketingevolution.com/marketing-essentials/ai-markeitng.

Sabbeh, S. F. (2018). "Machine-learning techniques for customer retention: a comparative study," *International Journal of Advanced Computer Science and Applications*, 9(2).

Samarasinghe, K. R., and Dr. A. Medis (2021). *Artificial Intelligence based Strategic Human Resource Management [Internet]*. [cited 8 September 2021]. Available from: http://file:///C:/Users/HP/Downloads/3059-1-3041-1-10-20200327.pdf.

25. Impact of Internet of Things on Cyber Security Risks and Its Solutions in Banking and Finance Sector

Mohanad Fayiz Al-Dweikat[1], Priti Gupta[2], Akash Bag[3], Dinesh Gupta[4], Rishika Yadav[5], Mohit Tiwari[6], and Tomasz Turek[7]

[1]Faculty of Business, Isra University Amman-Jordan

[2]Assistant Professor, P.G. Department of Economics, Bhupendra Narayan Mandal University (West Campus), P.G.Centre, Saharsa, Bihar

[3]Ph.D Scholar, Amity University, Raipur, Chhattisgarh

[4]Assistant Professor, Amity Business School, Amity University Gwalior, Madhya Pradesh

[5]Assistant Professor, Department of Computer Science and Engineering, Graphic Era Hill University, Dehradun, Uttarakhand

[6]Assistant Professor, Department of Computer Science and Engineering, Bharati Vidyapeeth's College of Engineering, Delhi A-4, Rohtak Road, Paschim Vihar, Delhi

[7]Czestochowa University of Technology, Poland

ABSTRACT: IoT and digital technology have had an influence on most bank systems in developing countries, and that's how quality management of technology concerns and cyber security management are expanding. The Internet of Things (IoT) influences cybersecurity threats and IT solutions in the banking and finance industry, and this impact is being thoroughly explored thanks to the mixed-methods approach. As the significant financial sector adjusts to expanded distant operations, there is a chance for increasing vulnerabilities and a possible magnified impact of cybersecurity attacks. IoT device growth in the banking industry has several advantages but also raises novel cybersecurity dangers. To safeguard their networks, their statistics, and users' financial details in the IoT age, financial institutions must establish strong cybersecurity safeguards, remain watchful, and react to evolving threats.

KEYWORDS: Internet of things, cyber security, banking, finance sector

1. INTRODUCTION

Smart devices, which associate gadgets functioning from home, public spaces, offices, colleges, and schools, are all around the current generation. Using the IoT, customers take advantage of this chance to work more intelligently. Companies in the banking, finance, and insurance sectors can quickly adopt IoT technology. The banking and financial industries have developed a new method for gathering crucial client data utilising IoT sensor devices and mobile phones. To develop new sectors, all banks have started interacting with their customers through social media, mobile devices, and other innovative sensors in recent years (Ioana, 2020). Due to an increase in the number of online, net banking accounts and a corresponding decrease in the usage of the conventional in-person services offered by branches of banks, financial institutions providing electronic banking additionally claim reduced operating expenses and increased revenue (Suseendran *et al.*, 2020). However, these are not the only advantages for banks and their customers who use online banking. Based on information from financial institutions, it may be deduced that as electronic online banking develops, there are increasingly more instances of cybercrime.

DOI: 10.1201/9781003532026-25

In recent years, all facets of the economy have developed significantly thanks to the development of the commercial banking systems. The advancement of technology and the use of digital technology in banking systems are two key factors in their expansion (Polymeni *et al.*, 2022).

1.1. IoT-Assisted Banking Services

The linked to the internet, when linked to the relevant bank, IoT can exchange data that has been stored in the cloud. There are billions of interconnected gadgets in use worldwide. All clients may see account details and convey access via mobile devices since the bank has granted such devices permission to exchange data on the cloud (Choithani *et al.*, 2022). The main benefit of IoT in banking services is the availability of credit cards and debit cards for simple access to bank services. Additionally, the bank can assess how frequently ATMs are used in a given location to decide whether to increase or decrease ATM installations. All client information is saved in IoT devices while being used (Polymeni *et al.*, 2022).

1.2. Research Objectives

- To determine how IoT affects cybersecurity risk in banking and finance
- To add IoT cyber risk to the highest-level cyber security transition in the banking industry by amending present cyber safety rules and standards.

2. LITERATURE REVIEW

2.1. Cyber-attacks in Finance and Banking Sector

The primary target of cyber-attacks is the banking and financial industry. In today's networked society, an attacker can quickly obtain personal data. As a result, cybercrime is currently dreaded all across the world. Figure 1 shows how the Internet of Things (IoT) joins devices to FinTech, cloud computing, ML, and modern throughput. A deliberate abuse of the web, digitalisation, networks, and frameworks for processing is known as a cyber-attack. To change the framework and hide code that could think twice about information, the programmer utilises malevolent code or programming to online violations, like taking financial record data and getting delicate well-being data. The risk of cyber-attacks is ascending alongside the arrangement of IoT-associated devices in the financial business. The IoT connected device communicates, analyses and provides new technological approaches. Other types of private information, besides just data, are also shared over the IoT. As a result, the dangers are extremely significant (Wang *et al.*, 2019).

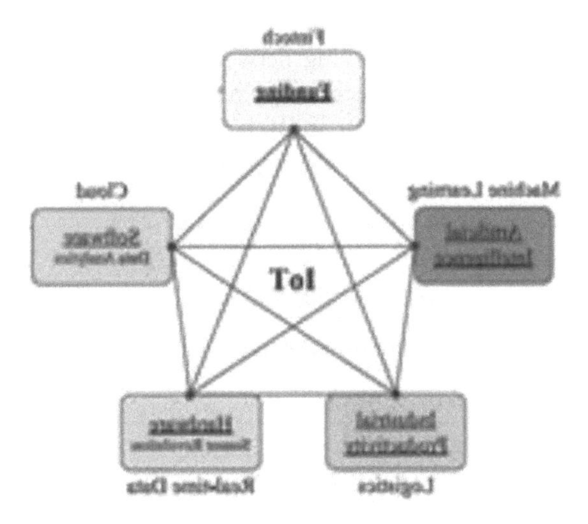

Figure 1. Internet of things. Source: Suseendran et al. (2020).

2.2. IT Solutions for IoT Cybersecurity in the Banking and Finance Sector

Robust authentication mechanisms, such as two-factor authentication and biometrics, can enhance IoT device security and prevent unauthorised access (Thach *et al.*, 2021). Executing network division and firewalls disengages IoT devices from basic frameworks, limiting the expected effect of compromised devices (Ali *et al.*, 2019). IDPS solutions can detect and prevent IoT-related attacks by monitoring network traffic, analysing device behaviour and raising alerts for potential threats. Developing comprehensive security governance frameworks, risk assessment methodologies and incident response plans is crucial for effective IoT security management in the banking and finance sector (Mhaskar *et al.*, 2021).

3. RESEARCH METHODOLOGY

3.1. Research Design

The investigation's approach used a mix of qualitative and quantitative techniques. This mixed-methods approach made it possible to fully explore how the IoT has affected cybersecurity threats and IT solutions in the banking and finance industry.

3.2. Sampling

Purposive sampling was utilized to select participants for interviews and focus group discussions. Stratified random sampling was employed for surveys to ensure a diverse representation of professionals from various job roles, levels and organisations within the banking and finance sector.

3.3. Data Analysis

Qualitative data coding and categorisation were conducted to facilitate the interpretation of participants' perspectives and experiences. For quantitative data, statistical analysis was performed on survey data using appropriate techniques related to IoT-related cybersecurity risks and IT solutions.

4. RESULT AND DISCUSSION

4.1. Overview of IoT and Banking System

The banks in our country have given innovation and IT arrangements that apply to banking exchange frameworks a ton of consideration starting around 2000 up to now. The utilisation of POS, credit and debit cards, and store ATMs has helped with monetary movement (Alghamdi *et al.*, 2020). In this IoT era, banks now have more technical connections with both commercial and individual customers. These connections are password-protected and very secure. Foreign banks create a friendly and trustworthy atmosphere for their clients by establishing their security banking networks. Local banks, however, are also gradually upgrading their Internet banking system. IoT technology has a variety of advantages, including data safety and greater protection levels. Then, based on our study, let's take a look at some of the banking-related IT application targets in our nation in Table 1 below (Long *et al.*, 2020).

Table 1: Bank targets survey 2020.

Target	2020	2019
Banks acquire certificates for information security	45.50%	28.60%
The core banking ratio is used by banks	100%	100%
Automation core banking	80.20%	77.40%
Risk management system	68.40%	62.50%
Online deposits for individuals	95.50%	92.30%

Target	2020	2019
Online salary payments for employees	90.60%	88.50%
Online bill payments for individuals	96.80%	96.40%
Mobile banking	100%	94.40%
SMS banking	94.80%	98.30%

Source: Long et al. (2020).

In Figure 2, the graph describes the targets achieved in one year after following the IoT in Banking and its impact on cyber security(Longet al., 2020). It is visible in the graph, that the influence of IoT on cyber-security can be the solution for finance management but still, it is moderate and can be improved.

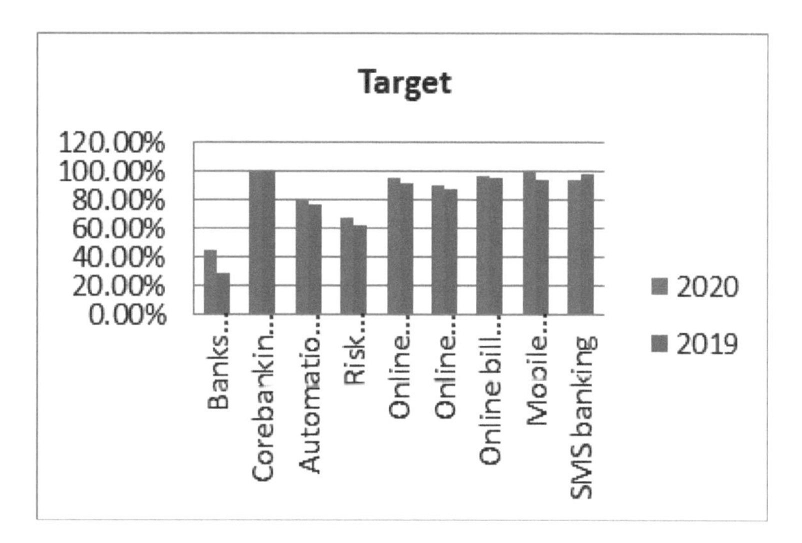

Figure 2. *ICT reports of banking and its target in the years 2020 and 2019. Source: Long et al. (2020).*

4.2. Opportunities and Threats of IoT on Cyber Security

In reality, we also do the following SWOT analysis on cyber security and digital banking.

- All bank branches are becoming more interconnected as a result of the rapid development of digital banking and core banking. Additionally, both data transfer and transaction speeds have increased. It gives liquidity to the economy more quickly;
- The digital technological Banks can gain from evolution thanks to cloud computing, saving large data, and IoT. It successfully provides efficient online water payment channels for consumers, energy, smartphones, internet, and utility bills.
- A high joblessness rate welcomed on by expanded financial mechanisation; An expansion in cyber-attacks and cyber safety that could bring about the deficiency of thousands of information records.
- The risk that clients will lose their record access keys; also, huge risks from programmers breaking into stages and taking web-based information.

5. CONCLUSION

This article has covered the general functions of modern banking technologies, challenges faced by the financial and banking sectors, and Baking's cyber-security when connecting IoT with the internet to store all data on the cloud. Risks associated with digital technology in banking include the potential for data loss owing to hackers, terrorists, and viruses; an increase in the amount of time spent on PCs and various workplace illnesses including headaches, stress, and a sensation of loneliness.

REFERENCES

Al Ghamdi, S., K. T. Win, and E. Vlahu-Gjorgievska (2020). "Information security governance challenges and critical success factors: systematic review," *Computers &Security*, 99, 102030.

Ali, I., S. Sabir, and Z. Ullah (2019). "Internet of Things security, device authentication, and access control: a review," *arXiv preprint arXiv:1901.07309*.

Choithani, T., A. Chowdhury, S. Patel, P. Patel, D. Patel, and M. Shah (2022). "A comprehensive study of artificial intelligence and cybersecurity on Bitcoin, crypto currency and banking system," *Annals of Data Science*, 1–33.

Ioana, C. P. R. (2020). "An inquiry on top banks by tier 1 ranking from central and Eastern Europe," *Annals of the University of Oradea, Economic Science Series*, 29(2).

Long, G., Y. Tan, J. Jiang, and C. Zhang (2020). "Federated learning for open banking," *Federated Learning: Privacy and Incentive* (pp. 240–254). Cham: Springer International Publishing.

Mhaskar, N., M. Alabbad, and R. Khedri (2021). "A formal approach to network segmentation," *Computers & Security*, 103, 102162.

Polymeni, S., D. N. Skoutas, G. Kormentzas, and C. Skianis (2022). "Findeas: A fintech-based approach to designing and assessing IoT systems," *IEEE Internet of Things Journal*, 9(24), 25196–25206.

Suseendran, G., E. Chandrasekaran, D. Akila, and A. Sasi Kumar (2020). "Banking and FinTech (financial technology) are embraced with IoT devices," *Data Management, Analytics and Innovation: Proceedings of ICDMAI 2019, Volume 1* (pp. 197–211). Springer: Singapore.

Thach, N. N., H. T. Hanh, D. T. N. Huy, and Q. N. Vu (2021). "Technology quality management of the industry 4.0 and cybersecurity risk management on current banking activities in emerging markets-the case in Vietnam," *International Journal for Quality Research*, 15(3), 845.

Wang, R., C. Yu, and J. Wang (2019). "Construction of supply chain financial risk management mode based on the Internet of Things," *IEEE Access*, 7, 110323–110332.

26. A Privacy-Preserving Healthcare IoT Framework Using Blockchain Technology

Satish Kumar Kalhotra[1], M. Siva Sangari[2], Ansuman Samal[3], Bibhuti Bhusan Pradhan[4], Astha Bhanot[5], R. Umamaheswari[6], and Sudheer S. Marar[7]

[1]Professor, Department of Education, Rajiv Gandhi University, Rono Hills Doimukh Arunachal Pradesh, India

[2]Associate Professor, Department of CSE, KPR Institute of Engineering and Technology, Coimbatore, India

[3]Professor, Faculty of Hospitality and Tourism Management, Siksha O Anusandhan Deemed to be University, Bhubaneswar, Odisha, India

[4]Pro-VC cum Registrar, Siksha O Anusandhan Deemed to be University, Bhubaneswar, Odisha, India

[5]PNBA University, Riyadh, KSA

[6]Professor, Department of Computer Science and Engineering, Gnanamani College of Technology, Tamil Nadu, India

[7]Professor & HOD, Department of MCA, Nehru College of Engineering and Research Centre, Kerala, India

ABSTRACT: The significance of clinical consideration has expanded emphatically, making it perhaps life's most fundamental part. Medical services specialists are currently using wearable innovation in light of the Web of Things to accelerate the symptomatic and therapy process. The Web has been utilised to associate billions of sensors, contraptions and autos. For the treatment and care of patients, one such technology is RPM. In this research study, the researchers have adopted the secondary qualitative research methodology by utilising existing literature, academic resources, case studies and industry reports for collecting information. Whereas, the outcomes of this study revealed that the integration of blockchain technology and IoT can significantly enhance security, privacy protection, and data integrity. This study also highlighted the potential benefits as well as challenges of various IoT frameworks. The way forward includes performing future studies about the research study and evaluating practical implementations of IoT and blockchain technology in terms of feasibility and effectiveness.

KEYWORDS: Healthcare, disease, machine learning, blockchain technology, IoT

1. INTRODUCTION

The number of medical patients is skyrocketing in many nations, making it increasingly challenging for patients to find primary physicians or carers. Through distant patient observation, the rise of IoT and wearable innovation lately has enhanced patient quality of care. Additionally, it enables doctors to see more patients. Patients are monitored and cared for via remote patient monitoring (RPM), which takes place outside of the typical clinical setting. In the beginning, it offers patients an inherent convenience of service. As needed, patients can maintain contact with medical professionals. Additionally, it lowers medical expenses furthermore, increases the expectation of care (Stamatellis *et al.*, 2020).

2. LITERATURE REVIEW

The gathering, storage, and management of sensitive patient data makes the healthcare industry one of the most sensitive industries. The conveyance of clinical benefits to patients has gone through a worldview change since the Web of Things (IoT) was acquainted with the medical services area.

DOI: 10.1201/9781003532026-26

The expression 'Web of Things' (IoT) portrays an organisation of genuine items like hardware, vehicles, structures, and other actual items that have networks, programming, and sensors worked in (Miyachi and Mackey, 2021).

2.1. IoT Framework for Healthcare Using Blockchain

A permissioned blockchain is used in Alaba *et al.* (2021)'s proposed blockchain-based IoT system for medical services to ensure the classification and security of patient information. The IoT layer, the information layer, the blockchain layer, and the application layer make up the system's four layers. The Internet of Things (IoT) layer is made up of IoT-enabled wearables, sensors, and medical equipment that gathers patient data. The data layer is in charge of handling and archiving the information gathered by IoT devices (El Majdoubi *et al.*, 2021).

2.2. Blockchain-based Structure for Far-off Medical Services Monitoring with Privacy Protection

Zhang *et al.* presented a blockchain-based security-saving architecture for remote medical care monitoring in 2020. The framework's blockchain architecture combines a public blockchain with a permissioned blockchain. While the patient's public data is preserved on the public blockchain, the patient's private information is stored on the permissioned blockchain.

2.3. Blockchain-based Healthcare Framework for Privacy Protection

Blockchain technology was used to provide a hybrid blockchain architecture-based healthcare paradigm (Shen *et al.*, 2019). The three levels of the framework are the information layer, the blockchain layer, and the application layer. The information layer is in charge of gathering and managing patient information, whilst the blockchain layer is utilised to store patient data and ensure its integrity and confidentiality.

2.4. IoT Framework Based on Blockchain for Personalised Healthcare Services

In 2020, Yao *et al.* suggested a blockchain-based IoT infrastructure for individualised healthcare services. The system uses a permissioned blockchain to guarantee the categorisation and security of patient information. The three levels of the framework are the information layer, the blockchain layer, and the application layer. Numerous studies have explored using blockchain-based solutions to improve the security and protection of IoT devices used in the healthcare industry. In order to assure safe and private data exchange in the healthcare IoT ecosystem, Kouicem *et al.*, (2021) suggested a framework that incorporates blockchain technology, smart contracts, and machine learning algorithms (Li *et al.*, 2022).

A blockchain-based healthcare IoT system was suggested in a different study by Ghanbari *et al.*, (2020) to enable secure and privacy-preserving data sharing across healthcare providers. Only parties with permission to access patient data can access it thanks to the framework's utilisation of savvy agreements to administer access control and information-sharing agreements. Patients may also keep tabs on who has access to their data thanks to the proposed framework, which increases accountability and transparency.

3. RESEARCH METHODOLOGY

Secondary data analysis enables researchers to conduct large-scale studies that would be impractical or impossible using primary data collection methods. By accessing comprehensive datasets, researchers can examine phenomena on a broader scale, explore trends over time, and analyse relationships between variables that may require a significant sample size for meaningful results. As the researchers have only relied on secondary qualitative data, all the data relevant to this research topic was collected from academic resources, journal papers, books, industry reports, and online articles. These collected data helped the researchers to gather a wide variety of data to provide essential insights about the challenges and benefits of IoT and blockchain technology. On the other hand, thematic analysis was used to extract information from large data sets by identifying patterns and themes. Moreover, Secondary data analysis can be particularly useful for exploratory research or hypothesis generation. Researchers can analyse existing data to identify patterns, formulate new research questions, or generate hypotheses for further investigation.

4. ANALYSIS AND DISCUSSION

The protection and security of touchy patient information have become a significant problem as IoT devices proliferate in the medical services industry. The improvement of blockchain innovation offers a potential remedy for these issues. Studies are conducted in this research and debate to examine the advantages and hardships of using blockchain innovation in medical services IoT systems to protect privacy.

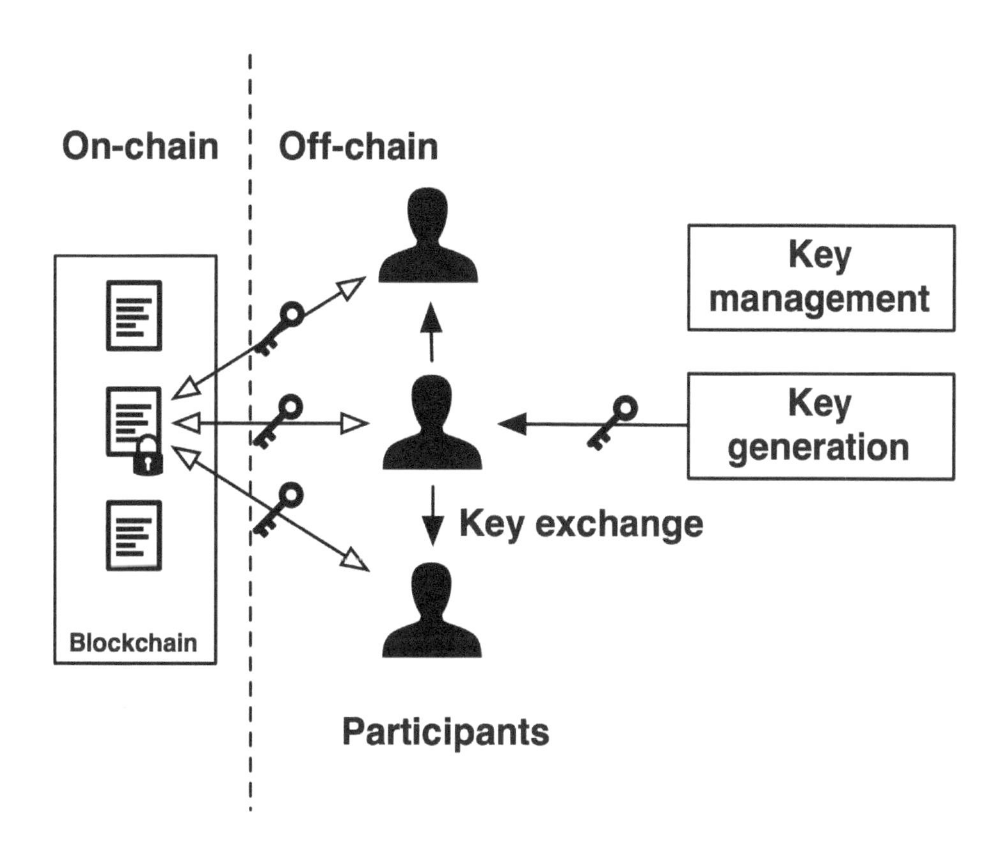

Figure 1. The user chain and doc chain: data encryption.

The Web of Things (IoT) in medical services has been expanding quickly because it makes it possible to monitor patient information progressively, empowering the early recognition and anticipation of sicknesses. There are rising worries about the security and protection of this information, though, as more delicate patient data is amassed and transmitted. The use of blockchain technology has been suggested as a potential remedy for these issues (El Azzaoui *et al.*, 2022). Researchers have looked into the advantages and difficulties of implementing a blockchain-based structure for IoT frameworks in medical care that safeguard patient protection. A hash function is used in the context of blockchain technology to construct a hash of the information in a block. The new block contains the previous block's hash, resulting in a chain of blocks that is encrypted. An illustration of a hash function is as follows:

$$h(x) = \text{SHA256}(x)$$

In this illustration, the widely-used hash algorithm SHA256 produces a 256-bit hash from the input data x. Any modification to the input data will produce a different hash because the resulting hash is specific to the input data. As a result, it is highly challenging for hackers to alter the data contained in the blockchain.

Table 1: Example of healthcare data permission structure.

Stakeholders	Role	Implementation
Patients	They are the data owner of information through the process that is dictated by smart contacts	It enables to provision of information about others' relationships with the stakeholders. Patients can engage with anyone to access the data
System administrators	They are responsible for mapping identification codes to blockchain wallets to allow integration with existing healthcare ID systems	They can connect with the smart contacts of patients that will help enable verification and identification
Healthcare providers	Their role is associated with the licensed healthcare providers verified from external data sources. They set relationship status in their patient's smart contacts	Through this, a PPR relationship can be established between two nodes in the system. One node store and manage information while the other provides care.

4.1. Benefits of Blockchain Technology for IoT in Healthcare

One of the key advantages of using blockchain technology in healthcare IoT installations is its decentralised nature. The decentralised network of nodes that makes up a blockchain provides a distributed ledger that keeps track of all transactions. Decentralisation makes it possible for the data recorded on the blockchain to be immutable because there isn't a single point of failure or control (Kasyap and Tripathy, 2021).

4.1.1. Data Security

The potential of blockchain technology to offer a safe and tamper-proof platform for storing sensitive patient data is another important benefit of adopting it in healthcare IoT systems. Blockchain technology uses cryptographic methods to make sure that the data kept there is secure and cannot be changed or removed (Hossein *et al.*, 2019).

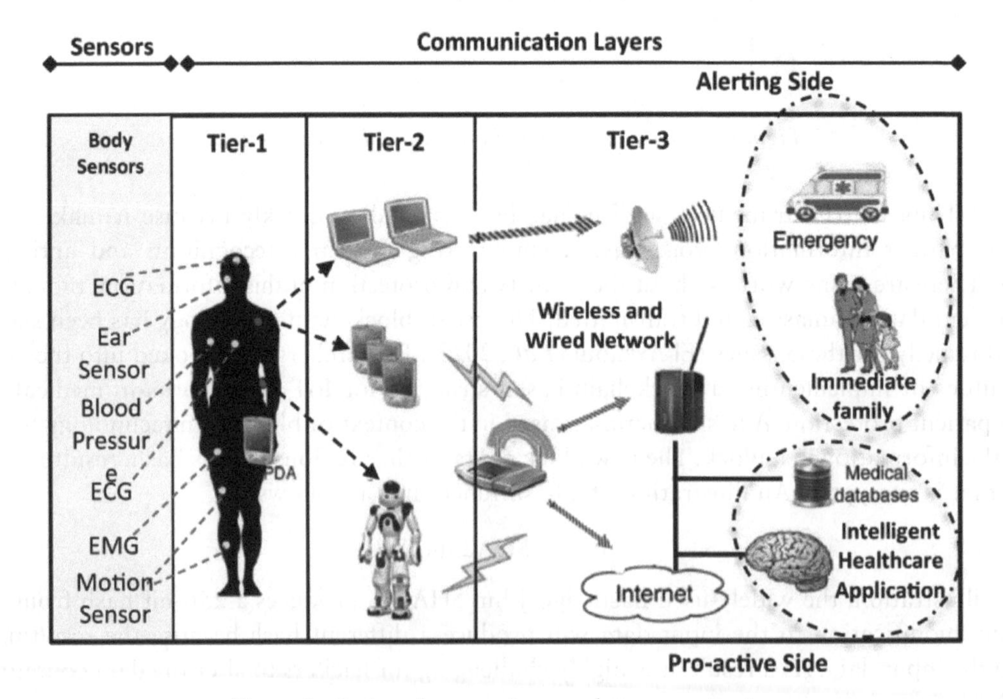

Figure 2. Patient data monitor: real-time monitoring.

4.1.2. Privacy Protection

Using blockchain technology in IoT frameworks for healthcare also provides a way to protect patient privacy. Because blockchain technology is decentralised, no single organisation has control over the patient data that is stored there.

4.1.3. Smart Contracts

Essentially self-executing contracts that are programmed to run when specific criteria are satisfied, are another feature of blockchain innovation that can be utilised. Shrewd agreements can be used to control who has access to patient data, limiting access to only those who are authorised.

The SHA-256 (Secure Hash Algorithm 256-bit) function, which accepts input messages of any length and generates a fixed-size output of 256 bits, is another widely used cryptographic hash function. The SHA-256 function's mathematical formula is:

$$\text{SHA-256}(\text{input message}) = \text{“}A\Big(H\Big(B\Big(C\Big(D\Big(E\Big(F\Big(G\big(\text{input message}\big)\Big)\Big)\Big)\Big)\Big)\Big)\Big)\text{”}$$

where:

A, B, C, D, E, F and G is a set of logical operations performed on the input message in a particular order.

The SHA-256 hash of the input message is the result of the final logical operation. This equation demonstrates how the SHA-256 function takes an input message and uses a number of logical operations to create a distinct hash that is a representation of the input message.

4.2. Issues with Using Blockchain Technology in IoT for Healthcare

Regulatory Compliance: Ensuring regulatory compliance is one of the main obstacles to the utilisation of blockchain innovation in medical services IoT platforms. Since the healthcare sector is highly regulated, using blockchain technology may be subject to regulatory standards and requirements. As a result, healthcare providers might need to make sure that their usage of blockchain technology complies with laws like HIPAA or GDPR.

Scalability is a problem when implementing blockchain technology in IoT frameworks for the healthcare industry. Because blockchain technology is decentralised, each node in the network needs to store a copy of the blockchain. Each node needs a growing quantity of storage as the blockchain gets bigger (Dwivedi *et al.*, 2019).

4.2.1. Interoperability

Healthcare IoT frameworks may involve a number of stakeholders, such as insurance providers, government organisations, and healthcare providers. It can be difficult to ensure interoperability across various parties, especially when employing blockchain technology.

4.2.2. Cost

Implementing blockchain technology in IoT frameworks for the healthcare industry may be expensive. A blockchain network needs a lot of resources, including staff, hardware, and software, to be implemented and maintained.

The difficulty of interoperability while adopting a blockchain-based infrastructure is another difficulty. Information can be challenging to share and access healthcare data among various healthcare institutions and providers since information is frequently kept in different formats and on multiple platforms (Dwivedi *et al.*, 2019). Using standardised data formats and protocols can assist ensure that data can be exchanged and accessed by various parties, which can help to increase interoperability. Concerns about patient data privacy also exist, particularly when it comes to disclosing information to researchers and healthcare professionals. Although blockchain technology can aid in protecting data privacy, it's still essential to ensure that patients have control over their own data and can determine who has access to it.

5. CONCLUSION

With more IoT devices being utilised in healthcare, protecting the privacy of sensitive patient data is becoming a serious issue. Blockchain technology has emerged as a viable remedy for privacy concerns in healthcare IoT because to its decentralised and secure nature. IoT devices that are connected to the internet communicate data with one another, leaving the network open to cyberattacks. Since privacy and security are of the highest significance in the healthcare sector, protecting patient data from unauthorised access is crucial. Blockchain technology has solved IoT security for the healthcare sector by providing a secure, decentralised, and transparent platform. The proposed frameworks' consensus method guarantees data integrity, which is essential in healthcare systems.

REFERENCES

Dwivedi, A. D., G. Srivastava, S. Dhar, and R. Singh (2019). "A decentralised privacy-preserving healthcare blockchain for IoT," *Sensors*, 19(2), 326.

El Azzaoui, A., H. Chen, S. H. Kim, Y. Pan, and J. H. Park (2022). "Blockchain-based distributed information hiding framework for data privacy preserving in medical supply chain systems," *Sensors*, 22(4), 1371.

El Majdoubi, D., H. El Bakkali, and S. Sadki (2021). "SmartMedChain: a blockchain-based privacy-preserving smart healthcare framework," *Journal of Healthcare Engineering*, 2021.

Hossein, K. M., M. E. Esmaeili, and T. Dargahi (2019). "Blockchain-based privacy-preserving healthcare architecture," *2019 IEEE Canadian Conference of Electrical and Computer Engineering (CCECE)* (pp. 1–4). IEEE.

Kasyap, H., and S. Tripathy (2021). "Privacy-preserving decentralized learning framework for healthcare system," *ACM Transactions on Multimedia Computing, Communications, and Applications (TOMM)*, 17(2s), 1–24.

Li, T., H. Wang, D. He, and J. Yu (2022). "Blockchain-based privacy-preserving and rewarding private data sharing for IoT," *IEEE Internet of Things Journal*, 9(16), 15138–15149.

Miyachi, K., and T. K. Mackey (2021). "hOCBS: a privacy-preserving blockchain framework for healthcare data leveraging an on-chain and off-chain system design," *Information Processing & Management*, 58(3), 102535.

Shen, M., Y. Deng, L. Zhu, X. Du, and N. Guizani (2019). "Privacy-preserving image retrieval for medical IoT systems: a blockchain-based approach," *IEEE Network*, 33(5), 27–33.

Stamatellis, C., P. Papadopoulos, N. Pitropakis, S. Katsikas, and W. J. Buchanan (2020). "A privacy-preserving healthcare framework using hyperledger fabric," *Sensors*, 20(22), 6587.

27. A Machine Learning Approach for Predicting Parkinson's Disease Using Big Data

Harish Reddy Gantla[1], P. N. Siva Jyothi[2], Ramesh Kumar[3], A. Sivaramakrishnan[4], Rania Mohy ElDin Nafea[5], T. K. Revathi[6], and Reshma V. K[7]

[1]Associate Professor, Department of Computer Science and Engineering, Vignan Institute of Technology and Science, India

[2]Associate Professor, Department of Information Technology, Sreenidhi Institute of Science and Technology, Hyderabad, India

[3]Subject Matter Specialist/Scientist (Agricultural Extension), Department of Agricultural Extension, Krishi Vigyan Kendra, Ambala, Haryana, India

[4]Associate Professor, Department of CSE, Koneru Lakshmaiah Education Foundation Vaddeswaram, Andra Pradesh, India

[5]College of Administrative and Financial Sciences, University of Technology Bahrain, Bahrain

[6]Assistant Professor, Department of Computer Science and Engineering, Sona College of Technology, Salem, India

[7]Assistant Professor, Department of Computer Science and Engineering, Sri Krishna College of Engineering and Technology, Coimbatore, India

ABSTRACT: The study offers a novel method for using medical imaging to diagnose Parkinson's disease that is based on deep learning. When trained on medical pictures like magnetic resonance imaging and dopamine transporter scans, deep segmentation and persistent neural networks extract knowledge that is then analysed and applied. Interior models of the prepared DNNs make up the extricated information that is applied through move obtaining and space variation to create a solitary design for Parkinson's disease prediction across various healthcare settings. Sizable experimental research is presented to demonstrate how the suggested method may accurately detect Parkinson's disease by utilising various medical image sets taken in actual settings.

KEYWORDS: Machine learning (ML), big data, Parkinson's, disease, DNN

1. INTRODUCTION

Medical imaging and other biomedical signal analyses now in use have long relied on feature extraction coupled with qualitative as well as quantitative processing. State-of-the-art performance in key computation duties, such as a vision for computers, speech recognition for interaction, and natural language processing, has been made possible by recent advancements in AI (ML) alongside profound brain organisations (DNNs). DNNs may be educated as end-to-end designs that incorporate several network types and produce outputs in the form of numbers or symbols. This is on the grounds that they are skilled at translating huge volumes of information, signals, pictures and picture groupings, recognising designs inside them, and afterwards really involving those examples for characterisation, relapse and expectation. In many different challenges, encouraging outcomes have been attained (Wingate *et al.*, 2020). One of the most prevalent neurodegenerative diseases in persons between the ages of 50 and 70 is Parkinson's disease (PD), particularly in nations with large older populations like the United States as well as the European Association.

DOI: 10.1201/9781003532026-27

2. LITERATURE REVIEW

Over 10 million individuals around the globe suffer with PD, a neurological condition. For patients to receive successful treatment, early diagnosis and precise illness progression prognosis are essential. By examining vast volumes of patient data, algorithms for machine learning, or ML, have the possibility to help in the early recognition and conclusion of Parkinson's sickness. The Writing Survey will go over late investigations that have been finished on the utilisation of enormous information and AI to anticipate Parkinson's sickness. The utilisation of help vector machines (SVM), irregular timberland displaying (RF), and ANN (counterfeit brain organisation) were three different AI (ML) techniques that were utilised in a concentrate by Arora *et al.* (2020) for anticipating the degree of Parkinson's illness (Salmanpour *et al.*, 2019). The MDS-UPDRS, created by the Development Issue Society, was utilised in the study to rate the severity of PD. The information was gathered from the PD patient's clinical, genetic, and imaging information stored in the 'Parkinson's Advancement Marker Initiative (PPMI) database'. The findings revealed that the SVM algorithm, which had an accuracy of 86.22%, outperformed RF (83.56%) as well as ANN (79.29%). The study came to the conclusion that ML algorithms could help clinicians anticipate the severity of PD.

In another exploration exertion, Joundi *et al.* (2019) fostered an original technique for utilising huge information to gauge the headway of Parkinson's infection. A profound learning model called the LSTM, or long momentary memory, brain network was utilised in the review. The information was gathered by the PPMI and comprised statistics on the population, medical evaluations, and imaging results. The study's findings demonstrated that the model using LSTM had a 76% accuracy rate in predicting the course of PD. The study came to the conclusion that the model developed by LSTM can offer insightful information on anticipating Parkinson's disease progression. Huge information was utilised in a concentrate by Prashanth *et al.* (2020) to utilise AI to gauge the probability of fostering Parkinson's sickness (Shahid and Singh, 2020). The study used five distinct 'machine learning (ML) algorithms': 'the choice tree, SVM, RF, for example, k-nearest neighbour (KNN), and logistic regression', and it used data from the 2007 National Health as well as Nutrition Examination Surveys (NHANES). According to the study's findings, the accuracy of the SVM algorithm was the highest at 89.1%, followed by those of the KNN algorithm (84.8%), RF (81.9%), tree of choices (78.6%), as well as logistic regression methods (76.4%). According to the study's findings, large data combined with ML algorithms can help predict the likelihood of developing Parkinson's disease.A profound learning model called the consideration-based present moment and long haul memory (ALSTM) counterfeit brain network was as of late utilised by Zhu *et al.* (2021) to anticipate the degree of Parkinson's illness. The study made use of demographic information, clinical evaluations, and imaging data gathered by the Parkinson's Phase Markers Initiative (PPMI). The study's findings demonstrated that, with a precision of 91.67%, the ALSTM model performed better than other conventional ML algorithms. According to the study's findings, deep learning models could offer useful information on estimating the impact of Parkinson's illness (Cao *et al.*, 2020).

The study made use of demographic information, clinical evaluations, and imaging data gathered by the PPMI. The review utilised the gradient enhancing machine (GBM), RF, & SVM as three different machine learning (ML) techniques. According to data from neuron imaging studies, Singh and Tawfik (2019) employed deep learning algorithms to predict Parkinson's illness. A long-transient memory, or LSTM, organisation was utilised in the study to learn seasonal trends in the data and a convolutional neural network (CNN) for obtaining features from brain images. Using the data set of 44 patients along with 44 healthy controls, the study was able to predict Parkinson's disease with an accuracy of 93.75%. In a different investigation, Makkieh *et al.* (2020) analysed gait data to foresee Parkinson's sickness utilising an AI approach. In light of stride information gathered from wearable sensors, a k-nearest neighbour (KNN) classifier was utilised in the study to predict Parkinson's disease. Using an information set of 40 patients along with 40 healthy controls, the study was able to predict Parkinson's disease with an accuracy of 96.25%. This research collectively shows the potential of algorithms based on AI for Parkinson's illness prediction using a variety of big data sources, including medical records, genetic data, image data, motion data, audio data and cell phone sensor data (Zeroual *et al.*, 2020).

3. RESEARCH METHODOLOGY

Secondary data sources often provide a vast amount of information collected over extended periods, covering diverse geographic regions, populations and time limits. This enables researchers to analyse a more extensive range of variables and draw broader conclusions from the data. Additionally, Secondary data collection is especially beneficial for long-distance or cross-cultural research. Researchers can access data from different regions or countries without the need for extensive travel or language proficiency, enabling broader and more diverse research investigations. This is the reason researchers have considered secondary data collection method to collect useful and relevant data related to research topic. In order to examine the effectiveness of a machine learning approach for the prediction of Parkinson's disease using big data, the researchers only focused on collecting secondary data only. Concerning data analysis, various academic databases, pieces of literature, industry reports, case studies, journal articles and other online website sources were used. These secondary sources have helped the researchers to gather information from a broad perspective. Whereas, in terms of data analysis, thematic analysis was used by analysing and identifying patterns to detect themes to draw an effective conclusion.

4. ANALYSIS AND DISCUSSION

In this study, PCA is combined in a deep learning framework to evaluate the progression of PD. The performance of the model may be enhanced by PCA's solution to the multi-co linearity issue in the data and its reduction of the overall dimension of the feature space (Bi *et al.*, 2021).

In high-dimensional data, deep networks may discover complicated structures. Furthermore, it doesn't need explicit selection and extraction of features like other ML algorithms.

'A CNN-RNN network, based on the ResNet-50 CNN structure, with one FC layer (1500 units) and two GRU RNN layers (with 128 units each); the V5 are extracted from the second GRU layer'

When there are several associated variables present in a multivariate study, PCA is utilised to extract the features. By breaking down the area of input into a collection of factors that are indistinguishable from one another, PCA effectively solves the issue of multicollinearity. Principal components (PCs) are the names for these uncorrelated variables. The majority of the variability in the dataset may typically be captured by the first few PCs. The greatest measure of changeability in the information is contained in the primary head part, and the leftover fluctuation is contained in the furthest degree practical in every part that follows (Song *et al.*, 2023). Therefore, it is possible to lower the dataset's dimensionality while keeping the most essential components of each variable by dropping the least significant PCs.Assume that the UPDRS score is influenced by an array X that comprises 'p' numbered variables, $X = (x_1, x_2, ..., x_p)$. The set of additional variables = $(1, 2, ..., p)$ that PCA produces is linearly connected to the input region but uncorrelated. These additional variables are referred to as PCs, with PC1 containing the largest amount of variation found in the dataset and decreasing as PC1 to PCp. Each PC's variance contribution is calculated as

$$\varepsilon_i = \alpha_{i1}x_1 + \alpha_{i2}x_2 + ... + \alpha_{ij}x_j + ... + \alpha_{ip}x_p$$

Profound brain networks are complex counterfeit brain organisations. These organisations might be prepared with a ton of information to advance consequently how to represent facts at different levels of abstraction. A back-propagation approach is used by DNNs to find intricate structures in huge datasets. The internal values of the network are changed by the backpropagation technique using the error. Convolution brain organisations, intermittent brain organisations, profound Boltzmann machines, profound conviction organisations, profound autoencoders and profound brain networks are a couple of instances of the different profound learning varieties (Kleanthous *et al.*, 2020).

Three factual appraisal measurements – mean complete blunder (MAE), the root-mean-square mistake (RMSE), and the proportion of assurance (R2) – have been applied to look at the precision of the expectation model. Eqs mirror the comparing assessment measurements. The 'hydroGOF' *R* package was used to compute these evaluation measures.

$$MAE = \frac{1}{n}\sum_{i=1}^{n}\left|\widehat{y_i} - y_i\right|$$

$$RMSE = \sqrt{\frac{\sum_{i=1}^{n}\left|\widehat{y_i} - y_i\right|^2}{n}}$$

$$R^2 = \frac{SSR}{SST} = 1 - \frac{SSR}{SST} = 1 - \frac{\Sigma\left(y_i - \hat{y}_i\right)^2}{\Sigma\left(y_i - \bar{y}_i\right)^2}$$

where y_i, y_i, and y_i represent the actual value, the method's projected value and the average of $[y_1, y_2, \ldots, y_n]$, furthermore, n means the all-out number of tests, individually.

Among various features, through the analysis, it has been found that the most optimal feature is the combination of all 93 features. Features 9, 10, 11, 28, 35, and 51 are directly linked with REM, MoCA years 0 and 1, STAIA, and others. However, these features are generally selected by applying 10 predictor algorithms. From this, it can be seen that FSSAs and SA have reached their acceptable results. Among 10 different machine learning algorithms, three major predictor algorithms such as MLP-BP, LOLIMOT and RFA, were used to perform the task of prediction using the MoCA score at year zero. These results are shown in the below table.

Table 1: Performance table of results for prediction using machine learning algorithms.

	Mean Absolute Error (MoCA at year zero)	Mean Absolute Error (Six vital features)
RFA	2.97 ± 0.25	2.56 ± 0.37
MLP-BP	2.37 ± 0.20	2.22 ± 0.45
LOLIMOT	2.17 ± 0.20	1.88 ± 0.11

Vectors were then subjected to the clustering procedure using k-means, demonstrated in Figure below. Researchers retrieved five groups, of which two connect with control subjects or NPD subjects; the remaining three clusters correspond to the patients from the original study. These make up an extracted condensed representation C set, which is made up of five 128-dimensional matrices as a result (Alsenan *et al.*, 2021). The image below displays the DaTscans associated with the cluster centres that were retrieved. With the assistance of medical experts, we were able to validate that the three DaTscans associated with patient cases demonstrate various PD stages. The first (C3) shows an early incident that occurred between stages 1 and 2, the second (C4) depicts a neurotic instance that is in stage 2, and the third (C5) discusses a case that has advanced to phase 3 of Parkinson's. Due to controls, there are limitations between the first (C1), a straightforward NPD model, and the second (C2), a more nuanced example (Mandal *et al.*, 2020).

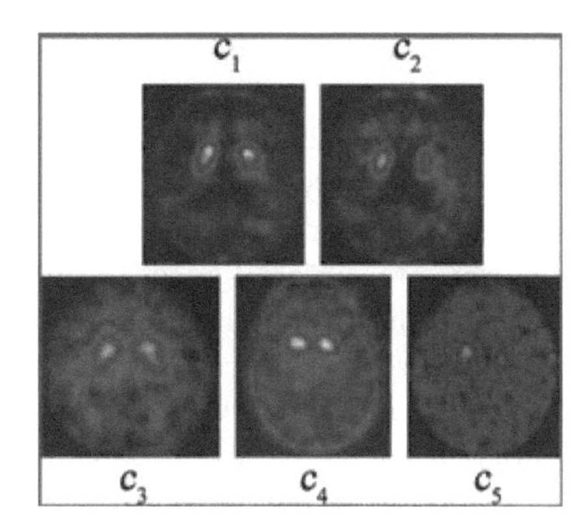

'The DaTscans of the 5 selected cluster centres: c_1 and c_2 correspond to NPD cases, whilst $c_3 - c_5$ to progressing stages of Parkinson's'

As per the previously mentioned comments, it very well may be reasoned that determined portrayals provide better insight into the status of the subjects than taught DNN outputs. When it comes time to analyse the data for new subjects, medical professionals can utilise this information to assess the predictions provided by the initial DNN (Shamrat *et al.*, 2019). The calculated V5 models in the latest instances can be quickly assigned to the cluster centre category of C, which will be used along with the group community's DaTscan, X-rays, and explanations to straightforwardly uphold the expectation made.

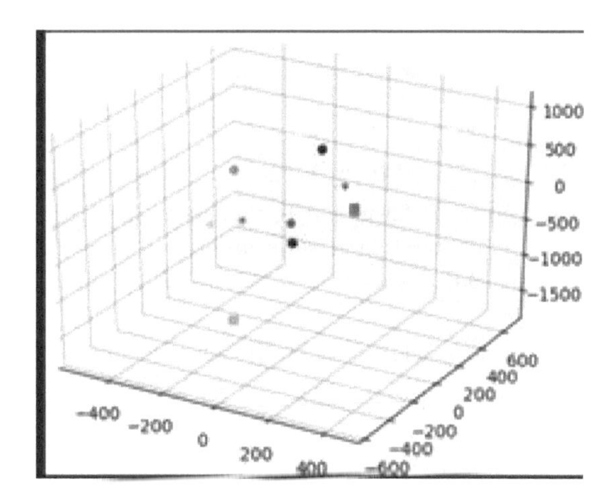

'The obtained ten cluster centres in 3-D: 5 of them (squares with red/rose colour, & plus (+) symbols with green colour) depict patients; five of them (stars with blue colour & circles with black/grey colour) depict non-patients'

Researchers display the proportion of training inputs that are present in each cluster category in Table below. Since many instances of Parkinson's sickness (PD) are in the beginning phases, having tools is critical, like the one that is being suggested, that can make extremely precise predictions in a variety of medical settings and data sets.

5. CONCLUSION

A serious neurodegenerative condition impacted by genetics as well as the environment is Parkinson's disease (PD). The brain's dopamine-producing cells degenerate in Parkinson's sickness (PD), is the second most common neurodegenerative condition, prompting both engine and non-engine side effects.

Neuroscience research significantly favours intervention during the first few therapeutic windows. Despite the accuracy of positron emission tomography and computed tomography, clinical characteristics and scores are currently the primary determinants of PD diagnosis. Neuroimaging has been involved all the more habitually as of late to aid the early discovery of PD. Various neuroimaging procedures, including positron emanation tomography, electroencephalography, and practical X-ray (fMRI), have tracked down boundless use. For unequivocally pinpointing the atypical unconstrained movement in neuropsychological illness, resting-state working MR imaging (rs-fMRI) is one of these techniques. An undertaking-free methodology is presented by various rs-fMRI-based procedures, like territorial homogenization (ReHo), the force with varieties, including working availability (FC), to look at unscheduled cerebrum movement and association among networks in different mind areas of PD patients. Hence, the research study about a machine learning approach to predict PD using big data has provided essential insights into the potential insights regarding capabilities of machine learning. However, the researchers have identified some areas of improvement and further research. It is essential to test and validate the machine learning models, refine and optimize the feature selection process, and incorporate additional data sources to provide a comprehensive understanding of the research study.

REFERENCES

Alsenan, S., I. Al-Turaiki, and A. Hafez (2021). "A deep learning approach to predict blood-brain barrier permeability," *PeerJ Computer Science*, 7, p.e515.

Bi, X. A., X. Hu, Y. Xie, and H. Wu (2021). "A novel CERNNE approach for predicting Parkinson's Disease-associated genes and brain regions based on multimodal imaging genetics data," *Medical Image Analysis*, 67, 101830.

Cao, X., X. Wang, C. Xue, S. Zhang, Q. Huang, and W. Liu (2020). "A radiomics approach to predicting Parkinson's disease by incorporating whole-brain functional activity and gray matter structure," *Frontiers in Neuroscience*, 14, 751.

Kleanthous, N., A. J. Hussain, W. Khan, and P. Liatsis (2020). "A new machine learning based approach to predict Freezing of Gait," *Pattern Recognition Letters*, 140, 119–126.

Mandal, S., A. Guzmán-Sáenz, N. Haiminen, S. Basu, and L. Parida (2020). "A topological data analysis approach on predicting phenotypes from gene expression data," *Algorithms for Computational Biology: 7th International Conference, AlCoB 2020, Missoula, MT, USA, April 13–15, 2020, Proceedings 7* (pp. 178–187). Springer International Publishing.

Salmanpour, M. R., M. Shamsaei, A. Saberi, S. Setayeshi, I. S. Klyuzhin, V. Sossi, and A. Rahmim (2019). "Optimized machine learning methods for prediction of cognitive outcome in Parkinson's disease," *Computers in Biology and Medicine*, 111, 103347.

Shahid, A. H., and M. P. Singh (2020). "A deep learning approach for prediction of Parkinson's disease progression," *Biomedical Engineering Letters*, 10, 227–239.

Shamrat, F. J. M., M. Asaduzzaman, A. S. Rahman, R. T. H. Tusher, and Z. Tasnim (2019). "A comparative analysis of parkinson disease prediction using machine learning approaches," *International Journal of Scientific & Technology Research*, 8(11), 2576–2580.

Singh, B., and H. Tawfik (2019). "A machine learning approach for predicting weight gain risks in young adults," *2019 10th International Conference on Dependable Systems, Services and Technologies (DESSERT)* (pp. 231–234). IEEE.

Song, Y. X., X. D. Yang, Y. G. Luo, C. L. Ouyang, Y. Yu, Y. L. Ma, H. Li, J. S. Lou, Y. H. Liu, Y. Q. Chen, and J. B. Cao (2023). "Comparison of logistic regression and machine learning methods for predicting postoperative delirium in elderly patients: a retrospective study," *CNS Neuroscience & Therapeutics*, 29(1), 158–167.

Wingate, J., I. Kollia, L. Bidaut, and S. Kollias (2020). "Unified deep learning approach for prediction of Parkinson's disease," *IET Image Processing*, 14(10), 1980–1989.

Zeroual, A., F. Harrou, A. Dairi, and Y. Sun (2020). "Deep learning methods for forecasting COVID-19 time-series data: a comparative study," *Chaos, Solitons & Fractals*, 140, 110121.

28. A Critical Evaluation Related to the Rise of Blockchain Technology in Supply Chain Management

Rakhi Mutha[1], Nelli Sreevidya[2], Neeraj[3], Attili Venkata Ramana[4], Dalia Younis[5], Richa Nangia[6], and Richa Arora[7]

[1]Associate Professor, Department of Information Technology, Amity University, Jaipur, Rajasthan, India

[2]Assistant Professor, Department of IT, Sreenidhi Institute of Science and Technology, Hyderabad, India

[3]Department of Computer Applications, Chitkara University School of Engineering & Technology, Chitkara University, Himachal Pradesh, India

[4]Associate Professor, Department of CSE (Data Science), Geetanjali College of Engineering and Technology, Hyderabad, India

[5]College of International Transport and Logistics, AASTMT University, Egypt

[6]Associate Professor, School of Management and Commerce, KR Mangalam University, Sohna Road, Gurgaon, Haryana, India

[7]Associate Professor, School of Management, Delhi Metropolitan Education, Noida, Uttar Pradesh, India

ABSTRACT: Supply chains in advanced times have formed into extraordinarily complex worth organisations that are currently a vital wellspring for acquiring an upper hand. Confirming the beginning of natural substances, notwithstanding, and keeping up with item and product permeability as they travel through the worth cycle network has developed increasingly troublesome. In this particular research study, only a secondary qualitative research method has been utilised by considering various literature, academic sources, journal articles and industry reports. Moreover, thematic analysis was used to identify the common themes among these data to provide essential information. While the study's findings made it abundantly clear that cutting-edge innovations in technology, like blockchain technology, have the potential to completely transform the way supply chains are managed by emphasising the streamlining of procedures, increasing transparency and reducing fraudulent activity. However, there are several difficulties with blockchain technology that call for cooperation and effective execution. In order to overcome possible issues and realise the full potential of blockchain technology, there is also a critical need for more study in this area. Future research must concentrate on pilot programmes, practical application, and empirical analysis to assess the influence of blockchain-based technology on management of supply chains.

KEYWORDS: Blockchain, supply chain management, technologies, IoT

1. INTRODUCTION

Because of developing interest needs between intra-hierarchical associations, which is made conceivable by improvements in contemporary innovation and firmly coupled corporate cycles, supply chains are turning out to be more heterogeneous and complex. Organisations are carrying out state-of-the-art advancements like the Internet, the Web of The conditions (IoT), registering by means of the cloud, organisation examination, AI, computer-based intelligence, and the utilisation of Blockchain, as well as weighty thoughts like the purported actual web to manage this quickly changing setting and the developing need to digitise supply chains and increment seriousness (Park and Li, 2021). The overflow of connected gadgets, additionally alluded to as 'savvy' articles or contraptions, as well as the range of innovations

DOI: 10.1201/9781003532026-28

that are habitually sent off simultaneously, empower esteem chain move (or exchanging) counterparties to accomplish more elevated levels of viability as well as adequacy (Rejeb *et al.*, 2019).Through superior information gathering, data trade and examination across coordinating supply chain accomplices, these developments vow to change the functional model of contemporary supply organisations (Philsoophian *et al.*, 2022). There is a review hole about what these innovations mean for supply chains that is significant for the two scholastics and professionals. IoT is the following phase of the alleged business 4.0's mass supply chain digitisation (Shen *et al.*, 2022).

2. LITERATURE REVIEW

Blockchain technology has lately generated a lot of interest since it has the potential to upend several industries, including supply chain management. This review of the literature examines how blockchain technology is being used to manage supply chains, highlighting the benefits, drawbacks, and potential uses of the technology (Park and Li, 2021).

2.1. Blockchain Technology's Advantages for Supply Chain Management Includes

The use of 'blockchain technology in supply chain management' has various benefits. First off, it improves traceability and transparency by establishing an immutable log of transactions, enabling visibility into the transportation of items, and confirming their provenance (Zhu *et al.*, 2022). Through its capacity to store and exchange data across the entire supply chain network, blockchain enables real-time tracking and effective inventory management. Finally, by creating a shared and accessible record of transactions, blockchain encourages cooperation and trust among supply chain players.

2.2. Limitations and Challenges

Blockchain technology has a lot of potential applications, but there are also many obstacles and restrictions. Issues with scalability and performance, particularly in public blockchains, can prevent supply chains with high transaction volumes from adopting these technologies widely (Kamilaris *et al.*, 2019).

2.3. Blockchain Applications in the Supply Chain Management

Supply chain management offers a number of uses for blockchain technology. Assuring authenticity and compliance, traceability and authenticity verification allow for the tracking of commodities from their point of origin to their destination (Rejeb *et al.*, 2021).

2.4. Enhanced Data Integrity and Security

Information may be shared and stored on a secure, unchangeable network thanks to blockchain technology. All parties engaged in the process of supply will have access to reliable and precise details as a result of this, making it nearly impossible to change or manipulate data.

2.5. Process Simplification and Fewer Intermediaries

Blockchain can lessen the need for intermediaries in the supply chain, such as banks and brokers, by enabling direct peer-to-peer transactions. This can simplify processes, increase efficiency, and save costs (Shen *et al.*, 2022).

2.6. Case Studies and Practical Applications

Numerous noteworthy case studies demonstrate how blockchain-based technologies are successfully used to manage the supply chain. IBM Food Confidence transformed food traceability by creating a blockchain-based infrastructure that encourages transparency and confidence in the nutritional supply chain (Kim, and Shin, 2019).

2.7. Research Directions and Gaps

Although there has been great advancement in the use of ledger technology in managing supply chains, there are still a number of research gaps and open questions.

3. RESEARCH METHODOLOGY

Sometimes, gathering secondary data is quicker than gathering main data. This is due to the fact that the data is already available and is simple to get. Because it has been collected and confirmed by other researchers or organisations, secondary data collecting frequently yields accurate results. Additionally, secondary data gathering might give access to people that are hard to reach or expensive to do so through primary data collecting. Data was gathered from a variety of sources, including academic databases, government publications, industry reports, journals, websites, industry reports, and other internet resources, since only secondary qualitative data were used in this study. A wide range of data has been collected relevant to the research topic using various search terms such as blockchain technology, supply chain management, IoT, and supply chain performance. In terms of data analysis, thematic analysis was employed to identify the recurring themes to draw patterns so the key findings can be gathered from the databases.

4. ANALYSIS AND DISCUSSION

IoT is basically an organization of data that joins sensors on or in genuine items (otherwise called 'things'), like shopper items, beds of wares, normal family things, machines, and modern stuff. Furthermore, cloud-based registering and the later thought of haze figuring (a decentralised processing structure that extends the reason of distributed computing through the neighbourhood viability of calculation, stockpiling, and cooperations through purported 'edge gadgets') offer registering assets and flexibility to interface, store, and investigate IoT information (frequently alluded to as large information) produced by associated gadgets and sources like WSNs, worldwide situating frameworks (GPS), GPRS, and other remote organisations.

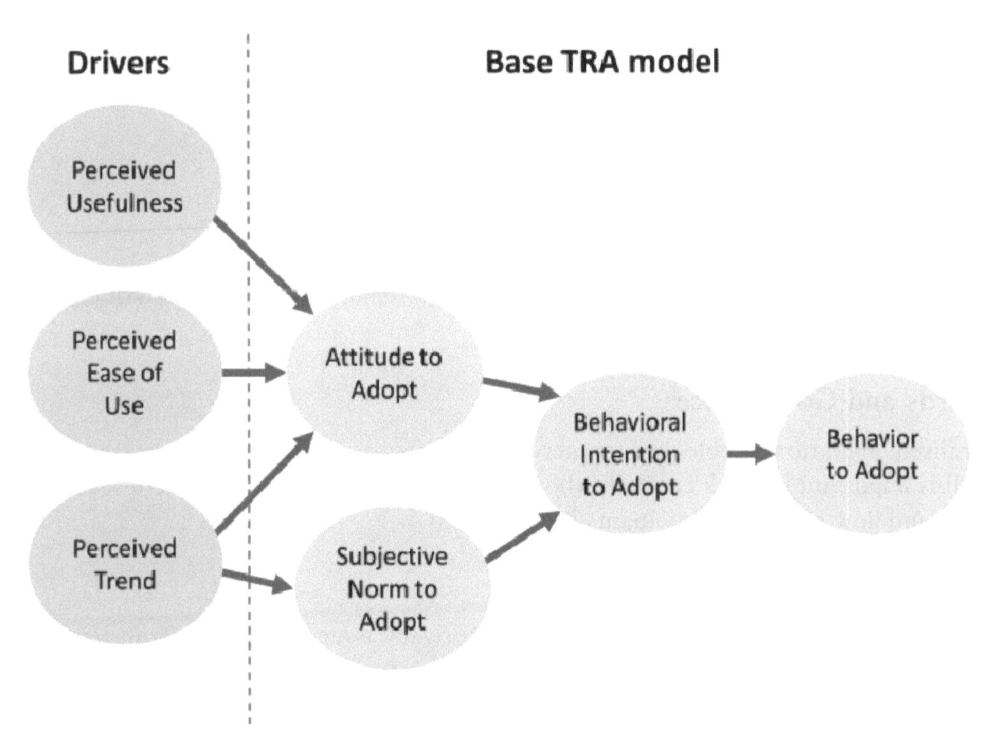

Figure 1. Blockchain adoption model.

The investigation of IoT information might bring about mechanising or creating esteem prescient examination abilities, which can assist organisations with perceiving occasions and answering continuously. Moreover, IoT smoothest out vital business processes by catching information like the discovery of human labourers and outside conditions (e.g., temperature, moistness, vibration and air flows) through the network of a variety of equipment parts (e.g., sensors).

4.1. Scalability and Effectiveness

Scalability is one of the main issues facing blockchain technology in supply chain management. The blockchain network's capacity may be stressed as the volume of transactions rises, which may cause transaction times to lag and costs to rise.

4.2. Integration and Interoperability

It is extremely difficult to integrate blockchain technology with legacy infrastructure and supply chain systems that are already in place. It is challenging to integrate blockchain solutions into supply chains that use many different software programmes, databases, and platforms (Madumidha *et al.*, 2019).

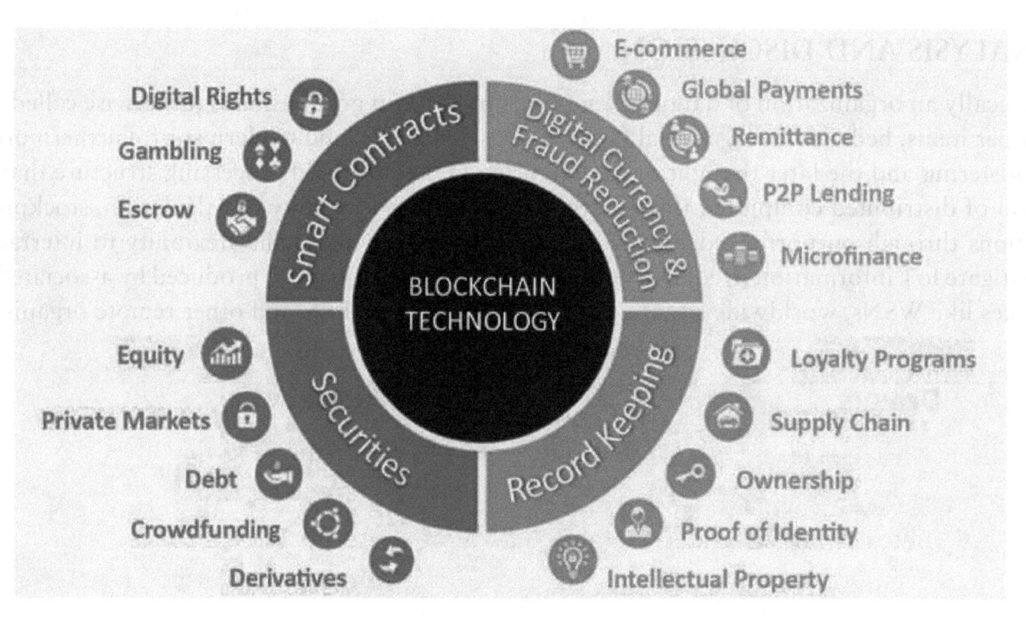

Figure 2. Types of blockchain modelling processes.

4.3. Standards and Governance

The decentralised operation of blockchain networks raises questions regarding standardisation and governance. It is important to think carefully about how selections are made, how consensus processes are put into place, and how network participants are governed.

4.4. Data Security and Privacy

Although blockchain offers a safe and open network, privacy issues still exist. Transparently recording transactions on public blockchains could expose private data to unauthorised parties.

4.5. Legal and Regulatory Considerations

Regulatory and legal factors must be taken into account when implementing the use of 'blockchain in supply chains'. To avoid legal issues, complying with laws pertaining to data protection, creativity rights, and sector-specific legislation is essential.

4.6. Return on Investment and Cost

'Blockchain technology implementation' calls for a large commitment in terms of system integration, infrastructure and training. While blockchain has the ability to improve supply chain efficiency, cut costs and streamline processes, it is crucial to calculate the financial return on investment (also known as ROI) before deciding whether the advantages outweigh the expenses related to its implementation (Madumidha *et al.*, 2019).

4.7. Management of Organisational Change

The supply chain's stakeholders' responsibilities and processes must alter as a result of the implementation of blockchain technology. In order to overcome resistance, inform participants and address any potential operational or cultural issues that may develop during the transition.

4.8. Effect on the Environment

Blockchain technology has environmental implications even though it has the potential to increase sustainability in the supply chain and lower carbon emissions through increased efficiency (Yousuf and Svetinovic, 2019).

"Productivity Gain:
Numerical condition: Productivity Gain = (Time Saved/Complete Time) * 100"

Assess the decrease in time and assets expected for different store network processes, like following, confirmation, and documentation, utilizing blockchain innovation. This condition estimates the rate of improvement in effectiveness accomplished.

"Cost Decrease:
Numerical condition: Cost Decrease = (Cost Saved/Absolute Expense) * 100"

Evaluate the decrease in costs related to go-betweens, administrative work, debates, misrepresentation, and stock administration utilising blockchain innovation. This condition estimates the rate decline in generally speaking expenses.

"Straightforwardness and Recognizability:
Numerical condition: Straightforwardness = (Number of Straightforward Exchanges/Absolute Exchanges) * 100"

Assess the number or level of exchanges that are straightforward and detectable inside the inventory network utilising blockchain innovation. This condition estimates the degree of straightforwardness accomplished.

"Security and Trust:
Numerical condition: Security and Trust = (Number of Checked Exchanges/Absolute Exchanges) * 100"

Survey the number or level of confirmed exchanges, brilliant agreements, and unchanging records in the store network utilising blockchain innovation. This condition estimates the degree of safety and trust laid out.

"Versatility and Hazard Moderation:
Numerical condition: Versatility = (Number of Interruption Occasions Forestalled/ Absolute Number of Disturbance Occasions) * 100"

Assess the capacity of blockchain innovation to forestall and moderate inventory network interruptions (Batwa and Norrman, 2021).

These highlights open up new business prospects and prepare for imaginative and inventive IoT use cases among trade accomplices in both straightforward and convoluted supply chains. IoT is expected to grow rapidly and become broadly coordinated in society because of the transition to 5G innovation, which offers faster information move rates. Cisco, a notable seller of organisation innovation, extends that by 2030, there will be 500 billion web-associated IoT gadgets.

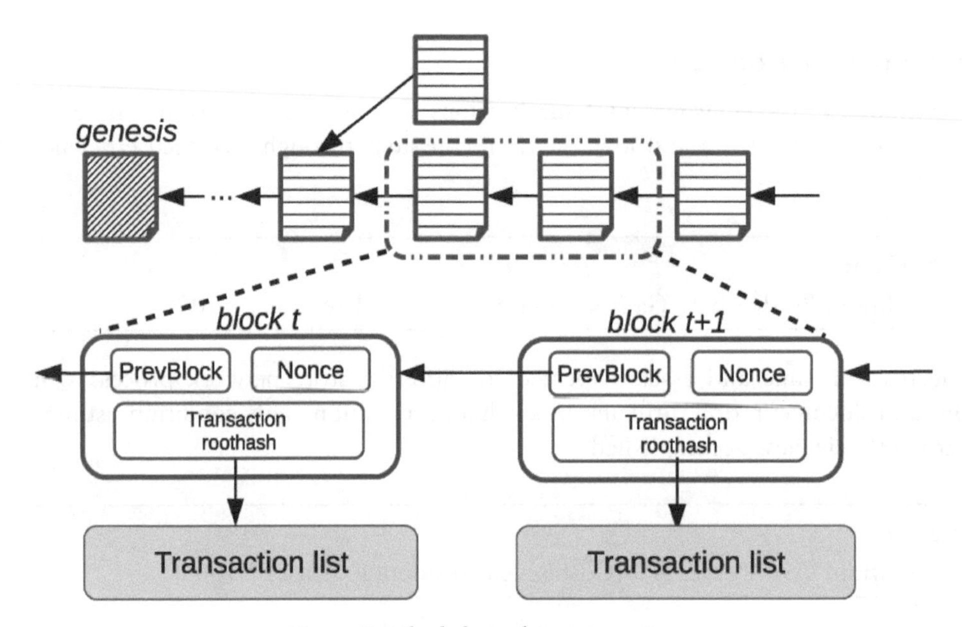

Figure 3. *Blockchain data processing.*

Overall, even if blockchain technology presents positive prospects for supply chain management, rigorous assessment is required to solve issues with scalability, interoperability, privacy, legal adherence, cost-effectiveness and change management (Philsoophian *et al.*, 2022).

5. CONCLUSION

Even if peer-to-peer communication between logistics exchange partners could be improved by blockchain technology, the mix of IoT and blockchain innovation questions a few institutional assumptions that are commonplace in worldwide exchange. How different Blockchain frameworks and innovations will cooperate and interface with other specialised headways is as yet a question of discussion. This is aggravated by the utilisation of incapable interchanges conventions and guidelines that block the trade gatherings' capacity to share data.

Besides, an IoT climate is dynamic, capricious and influenced by continually developing legitimate necessities for security and other interoperability norms. There is an earnest requirement for new standards and guidelines because of this startling change and turbulent nature. By utilising permissioned Blockchains, or conveyed which are less asset serious, security and adaptability concerns can be decreased. Furthermore, the possibility of 'Blockchain pruning' (i.e., disposing of unnecessary information or old exchanges) has been proposed as an expected cure.

REFERENCES

Batwa, A., and A. Norrman (2021). "Blockchain technology and trust in supply chain management: a literature review and research agenda," *Operations and Supply Chain Management: an International Journal*, 14(2), 203–220.

Gurtu, A., and J. Johny (2019). "Potential of blockchain technology in supply chain management: a literature review," *International Journal of Physical Distribution & Logistics Management.*

Kamilaris, A., A. Fonts, and F. X. Prenafeta-Boldv̌ (2019). "The rise of blockchain technology in agriculture and food supply chains," *Trends in Food Science & Technology*, 91, 640–652.

Kim, J. S. and N. Shin (2019). "The impact of blockchain technology application on supply chain partnership and performance," *Sustainability*, 11(21), 6181.

Madumidha, S., P. S. Ranjani, U. Vandhana, and B. Venmuhilan (2019). "A theoretical implementation: Agriculture-food supply chain management using blockchain technology," *2019, TEQIP III Sponsored International Conference on Microwave Integrated Circuits, Photonics and Wireless Networks (IMICPW)* (pp. 174–178). IEEE.

Park, A., and H. Li (2021). "The effect of blockchain technology on supply chain sustainability performances," *Sustainability*, 13(4), 1726.

Philsoophian, M., P. Akhavan, and M. Namvar (2022). "The mediating role of blockchain technology in improvement of knowledge sharing for supply chain management," *Management Decision*, 60(3), 784–805.

Rejeb, A., J. G. Keogh, and H. Treiblmaier (2019). "Leveraging the internet of things and blockchain technology in supply chain management," *Future Internet*, 11(7), 161.

Rejeb, A., K. Rejeb, S. Simske, and H. Treiblmaier (2021). "Blockchain technologies in logistics and supply chain management: a bibliometric review," *Logistics*, 5(4), 72.

Shen, B., C. Dong, and S. Minner (2022). "Combating copycats in the supply chain with permissioned blockchain technology," *Production and Operations Management*, 31(1), 138–154.

Yousuf, S., and D. Svetinovic (2019). "Blockchain technology in supply chain management: preliminary study," *2019 Sixth International Conference on Internet of Things: Systems, Management and Security (IOTSMS)* (pp. 537–538). IEEE.

Zhu, Q., C. Bai, and J. Sarkis (2022). "Blockchain technology and supply chains: The paradox of the atheoretical research discourse," *Transportation Research Part E: Logistics and Transportation Review*, 164, 102824.

29. Analysing Challenges and Solutions of Implementing Enterprise Resource Planning and Development

G. Venkatakotireddy[1], Ansuman Samal[2], PellakuriVidyullatha[3], Y. Venkata Rangaiah[4], Ivanenko Liudmyla[5], Shikha Dutt Sharma[6], and Manasvi Maheshwari[7]

[1]Associate Professor, Department of Computer Science and Engineering, Holy Mary Institute of Technology & Science, Telangana, India

[2]Professor, Faculty of Hospitality and Tourism Management, Siksha O Anusandhan Deemed to be University, Bhubaneswar, Odisha, India

[3]Associate Professor, Department of Computer Science and Engineering, Koneru Lakshmaiah Education Foundation, Vaddeswaram, Guntur, Andhra Pradesh, India

[4]Associate Professor, School of Management, Presidency University, Bengaluru, India

[5]Chernigiv Post Graduate Pedagogic Institute of K.D. Ushinsky, Ukraine

[6]Assistant Professor, School of Humanities, K R Mangalam University, Gurugram, India

[7]Associate Professor, Media School, Delhi Metropolitan Education, Noida, India

ABSTRACT: Any enterprise resource planning (ERP) framework's essential parts are often made as brought-together frameworks. The present blockchain innovation, or all the more comprehensively disseminated record innovation (DLT), applications can change the engineering, get around, and ease a portion of the disadvantages of concentrated frameworks, most strikingly security and protection. The objective of this research study is to critically analyse the challenges associated with the ERP integration process into organisations. Moreover, it will also focus on examining the impacts of these challenges and employing the strategies and solutions that can be employed to overcome these challenges. In terms of research methodology, the researchers have adopted a secondary qualitative research approach to get a comprehensive overview of the research topic from various secondary sources of data. Thematic analysis was selected for analysing and interpreting the collected data. From the research findings, it can be said that ERP implementation is indeed a complex process. The key challenges associated with this process include data integration, lack of integration process, cost overruns and many more. Moreover, due to the rise of technological advancements, further research is required to stay updated about the emerging challenges and solutions of ERP integration into organisational functions. The role of leadership, organisational culture, and training and development need to be examined to provide a comprehensive understanding of the research topic.

KEYWORDS: Enterprise resource planning (ERP), bookkeeping, security, blockchain

1. INTRODUCTION

Bookkeeping, buying, projecting the executives, hazard and consistency the board, and store network tasks are only a couple of the business exercises that are overseen by ERP, a type of programming utilised by firms (Namugenyi *et al.*, 2019). Enterprise execution of the board, programming that guides in planning, planning, gauging and revealing an association's monetary outcomes, is likewise a part of a full ERP suite. Bookkeeping data frameworks (AISs) gather, store, and cycle bookkeeping and monetary information that is then utilised by inward clients to give information to financial backers, lenders, and duty specialists. AISs are, much of the time PC, based methods that utilise data innovation resources to screen bookkeeping action. An AIS mixes laid-out bookkeeping systems, like commonly perceived bookkeeping norms, with state-of-the-art IT apparatuses (Stahl *et al.*, 2020). Planning and it is hard to investigate an AIS. Since

DOI: 10.1201/9781003532026-29

numerous business applications should be coordinated into ERPs to address different issues, they are considerably more muddled (Faccia and Petratos, 2021). To give security models like classification, controllability (direct access and liberated hierarchy of leadership), and cost-adequacy, AISs (in any event, when cloud-based) are at present utilised by all organisations (Yong *et al.*, 2020).

2. LITERATURE REVIEW

ERP solutions are commonly used in businesses to enhance operations and decision-making. It can be difficult to implement ERP systems, and research has found a number of elements that might make an adoption successful or unsuccessful (Lu *et al.*, 2020).

2.1. ERP System Implementation Challenges

Reluctance to Change: One of the biggest obstacles to putting an ERP system into place is reluctance to change. Employees could be hesitant to adopt new technology, particularly if it necessitates them picking up new skills or altering how they conduct their jobs. As a result, productivity may suffer and frustration levels may rise.

2.1.1. Cost

For small and medium-sized organisations in particular, ERP systems can be pricey. Hardware, software, consultancy fees and employee training are all possible implementation expenses.

2.1.2. Customisation

Businesses may have certain procedures or needs that are incompatible with ERP software. Customisation can be time- and money-consuming, and it may call for specialised expertise that isn't in-house.

2.1.3. Integration

ERP systems are made to integrate various organisational activities and functions. It can be difficult to integrate the system with current systems and data, though. Data discrepancies, mistakes, and delays may result from this (Koval *et al.*, 2021).

2.1.4. Undertaking Management

Setting up an ERP system is a challenging undertaking that need for meticulous preparation and administration. Organisations might not have the knowledge or resources necessary to manage the project successfully. Delays, cost overruns, and a failure to meet project objectives may result from this.

2.2. Solutions for Putting ERP Systems in Place

Change management: By including staff members in the implementation procedure, organisations can overcome resistance to change. This may entail offering instruction, requesting criticism, and promoting involvement.

Enhanced Decision-Making: ERP systems give businesses real-time data access, which can assist decision-makers in making wise choices. Organisations can provide reports and insights through the use of analytics technologies that can assist in trend identification, outcome forecasting, and process optimisation.

Cost Control: By carefully planning the project and finding areas where cost reductions can be made, organisations can control the costs associated with deploying an ERP system (Namugenyi *et al.*, 2019).

Integration: By carefully planning the integration process and recognising potential concerns, organisations may guarantee that the ERP system integrates with no hitches. To facilitate data flow between the ERP system and current systems, they may also take into account employing middleware or integration tools.

Project Management: By investing in project management skills or collaborating with a vendor or consulting firm that specialises in ERP deployment, organisations can assure the effective implementation of an ERP system. This can assist guarantee that the project is well-planned, properly carried out, and achieves its objectives (Leng *et al.*, 2020).

3. RESEARCH METHODOLOGY

Researchers can use secondary data to compare and validate their findings against existing studies or data sets. This helps to establish the reliability and generalizability of research results by corroborating them with previous research or data from different sources. Secondary data is often publicly available or accessible through various institutions, databases, or repositories. Among various research methods, the reason behind choosing a secondary qualitative research method includes its reliability, validity and availability. The researchers reviewed all the essential resources relevant to the research to gather valuable insights. Concerning data collection, the researchers first conducted a thorough literature review can collect data from various sources such as business publications, academic databases, case studies and industry reports. Furthermore, thematic analysis was employed to identify themes, draw patterns and develop the key findings. By using this qualitative analysis approach, the researcher was able to identify effective information about the challenges and solutions of ERP implementation.

4. ANALYSIS AND DISCUSSION

A crucial component of any firm is the accounting process (see Figure below), mostly because it is required by law in every country. The books and bookkeeping records required by common, monetary, corporate, and work prerequisites should be ready and kept up with by all organisations. Contingent to the nature and size of the organisation, different bookkeeping records are required.

Systems for ERP are created to streamline and combine several corporate processes like accounting, inventory control, production, and human resources (Yong *et al.*, 2020). The creation and application of ERP systems involve the use of numerous mathematical equations. Here are a few illustrations: Equation for Economic Order Quantity (EOQ): In order to reduce the overall cost of acquiring and maintaining inventory, this equation is used to find the ideal quantity of inventory to order. It goes like this:

'Sqrt((2DS)/H) = EOQ', where:

'D is the product's annual demand, and S is the order placement fee'.

'H is the annual cost of keeping one unit of inventory'.

COGS, or cost of goods sold Equation: This formula is used to figure out how much it costs to make and sell a product. It goes like this:

Beginning inventory is the value of the inventory at the beginning of the period, and Gear teeth are equivalent to starting stock in addition to buying short consummation stock.

Inventory is the worth of the inventory at the conclusion of the period, and purchases are the cost of all purchases made during that period (Ivanov *et al.*, 2021).

Equation for Return on Investment (ROI): This formula is used to calculate the return on an investment. It goes like this:

'ROI is calculated as (gain from investment – investment cost) / investment cost'.

where:

gain from investment is the sum of the investment's earnings.

Investment cost equals the sum of all investment costs

Equation for capacity planning: This equation is used to calculate a manufacturing system's capacity. It goes like this:

Various machines allude to the complete number of machines in the creation framework; working time alludes to how much time each machine is available for creation; and use alludes to how well each machine is used.

Utilisation is the proportion of time that a machine is in use, while processing time is the amount of time needed to create one unit of a product (Davidescu *et al.*, 2020).

These are only a few of the numerous mathematical equations that are employed in the creation and use of ERP systems.

Bookkeeping, book arrangement and safeguarding are fundamental parts of any business. In this sense, all organisations are expected to keep two books: (a) the diary, which is the book where all administration exercises—like buys, deals, assortments, and instalments—are recorded sequentially; and (b) the record, which relates to all the record accounts that are utilised during the bookkeeping section processes.

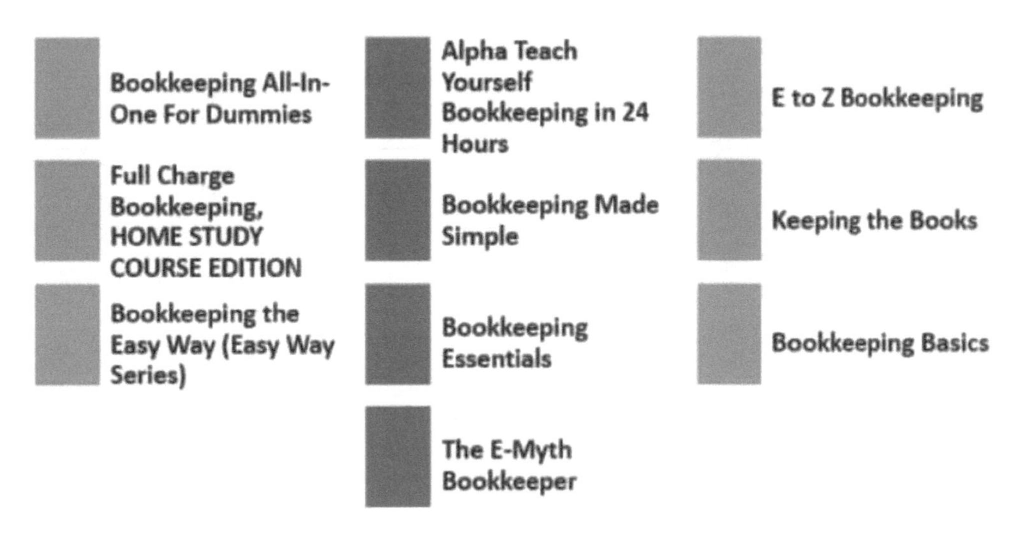

Figure 1. Bookkeeping book arrangement. Source: Leng et al. (2020).

4.1. The Accounts Flow

New blocks can be added to the 'chain' utilising a common convention to ensure consistency among the different duplicates. Each time another block gets agreement and is endorsed fastened, all hubs update their confidential duplicates. This design ensures that information can't be adjusted, erased or changed. Subsequently, the blockchain is a decentralised framework.

4.2. Centralised Vs Decentralised Vs Distributed

The blockchain fills in as a public, unchangeable data set. It is portrayed as a computerised register whose trustworthiness is safeguarded by cryptography and whose exchanges are coordinated into blocks and associated with sequential requests (Mousa and Othman, 2020).

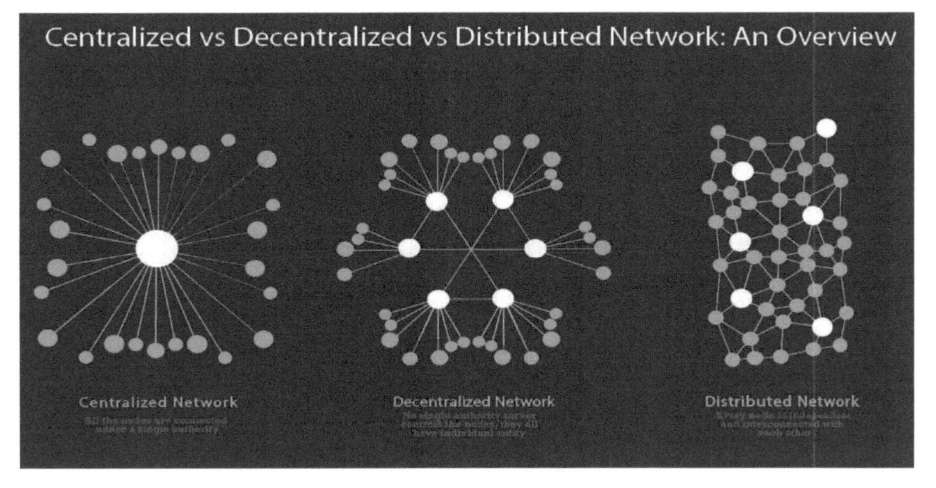

Figure 2. Centralised vs decentralised vs distributed. Source: Lu et al. (2020).

Each time another block gets agreement and is supported bound, all hubs update their confidential duplicates. This design ensures that information can't be adjusted, erased, or changed (Fenech *et al.*, 2019).

4.3. Decentralised Business System Network

The two modules that are supposed to be introduced by SAP, the pristine e-acquisition module and bookkeeping data frameworks (AISs), will be the main ones in this joining system. Hash hub and shared records ought to be remembered for bookkeeping data frameworks as extra accounting systems.

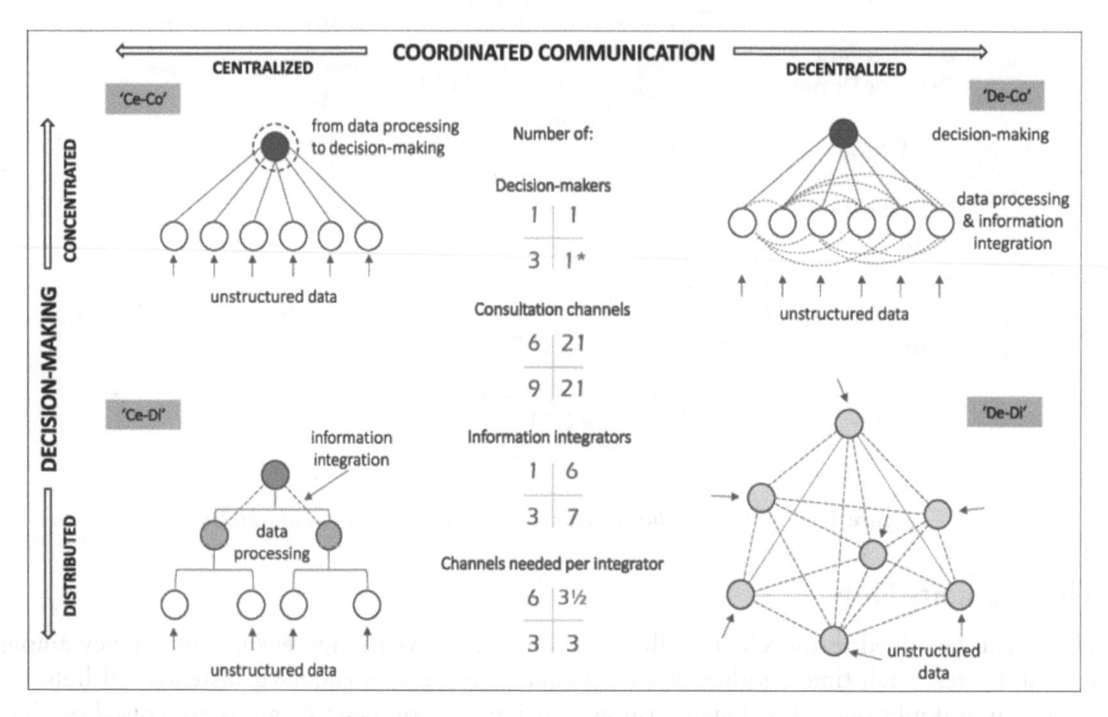

Figure 3. Decentralised business system network. Source: Mousa and Othman (2020).

4.4. AIS and e-Procurement Processes Combination

Since they might protect, work with, and robotise the impacts settled upon by organisations and people that marked the agreements, brilliant agreements are habitually viewed as an urgent part of cutting-edge blockchain frameworks. At the point when certain standards are met, savvy get that have been made and put away on a blockchain consequently works without the requirement for a middle person or extra affirmation (Stahl *et al.*, 2020).

In order to implement enterprise resource planning, there include many factors that directly or indirectly affect small and medium-scale businesses. (Stahl *et al.*, 2020).

Table 1: Profile of different categories.

	Frequency	Percent	Valid Percent	Cumulative Percent
Top executives	21	33.33	33.3	33.3
Employees	42	33.33	33.3	66.7
Customers	15	33.33	33.3	100.00
Total	78	100	100	

Source: Self-made

4.5. The ERP System Pivotal Role

Many authors are in favour of AIS and blockchain existing side by side. Blockchain applications in accounting are fast evolving and developing as a topic of study. This has an impact on AIS and ERP as well. Researchers, please offer a review of the relevant literature that, in addition to making a contribution, will help with the rest of the analysis. This study's practical scientific implications apply to IT systems as well as management theory and practise. DLT and blockchain applications transform technological and scientific fields via innovation and can benefit both businesses and people.

5. CONCLUSION

'Blockchain applications on ERP, AIS', and, generally, the wider use of FinTech, result in significant value creation and benefits. We find proof that there are advantages to faster and more efficient procurement processes, IT automation and productivity, cost-cutting, 'security, dependability, and safety', as well as flexibility and quality. Since they further develop as of now existing logical applications and, all the more urgently, lead to new disclosures and empower inventive applications, this large number of benefits can prompt abundance creation. The coordination of DLT with AIS and ERP frameworks may likewise enjoy benefits, generally as different effectiveness gains. Justifying and, more importantly, determining if a certain application may make it simpler to combine ERP and AIS systems are complex tasks.

REFERENCES

Chaudhary, R. (2020). "Green human resource management and employee green behavior: an empirical analysis," *Corporate Social Responsibility and Environmental Management*, 27(2), 630–641.

Davidescu, A. A., S. A. Apostu, A. Paul, and I. Casuneanu (2020). "Work flexibility, job satisfaction, and job performance among Romanian employees—implications for sustainable human resource management," *Sustainability*, 12(15), 6086.

Faccia, A., and P. Petratos (2021). "Blockchain, enterprise resource planning (ERP) and accounting information systems (AIS): research on e-procurement and system integration," *Applied Sciences*, 11(15), 6792.

Fenech, R., P. Baguant, and D. Ivanov (2019). "The changing role of human resource management in an era of digital transformation," *Journal of Management Information & Decision Sciences*, 22(2).

Hamouche, S. (2021). "Human resource management and the COVID-19 crisis: Implications, challenges, opportunities, and future organizational directions," *Journal of Management & Organization*, 1–16.

Ivanov, D., C. S. Tang, A. Dolgui, D. Battini, and A. Das (2021). "Researchers' perspectives on Industry 4.0: multi-disciplinary analysis and opportunities for operations management," *International Journal of Production Research*, 59(7), 2055–2078.

Koval, V., I. Mikhno, I. Udovychenko, Y. Gordiichuk, and I. Kalina (2021). "Sustainable natural resource management to ensure strategic environmental development," *TEM Journal*, 10(3), 1022.

Leng, J., G. Ruan, P. Jiang, K. Xu, Q. Liu, X. Zhou, and C. Liu (2020). "Blockchain-empowered sustainable manufacturing and product lifecycle management in industry 4.0: a survey," *Renewable and Sustainable Energy Reviews*, 132, 110112.

Lu, H., Y. Zhang, Y. Li, C. Jiang, and H. Abbas (2020). "User-oriented virtual mobile network resource management for vehicle communications," *IEEE Transactions on Intelligent Transportation Systems*, 22(6), 3521–3532.

Mousa, S. K., and M. Othman (2020). "The impact of green human resource management practices on sustainable performance in healthcare organisations: a conceptual framework," *Journal of Cleaner Production*, 243, 118595.

Namugenyi, C., S. L. Nimmagadda, and T. Reiners (2019). "Design of a SWOT analysis model and its evaluation in diverse digital business ecosystem contexts," *Procedia Computer Science*, 159, 1145–1154.

Stahl, G. K., C. J. Brewster, D. G. Collings, and A. Hajro (2020). "Enhancing the role of human resource management in corporate sustainability and social responsibility: a multi-stakeholder, multidimensional approach to HRM," *Human Resource Management Review*, 30(3), 100708.

Yong, J. Y., M. Y. Yusliza, T. Ramayah, C. J. Chiappetta Jabbour, S. Sehnem, and V. Mani (2020). "Pathways towards sustainability in manufacturing organizations: empirical evidence on the role of green human resource management," *Business Strategy and the Environment*, 29(1), 212–228.

30. The Impacts of Artificial Intelligence on Management Decision-Making

G. Nagaraj[1], Regonda Nagaraju[2], M. Siva Sangari[3], Satish Kumar Kalhotra[4], Tomasz Turek[5], N.V.S. Suryanarayana[6], and M. Harikumar[7]

[1]Associate Professor, Department of Mechanical Engineering, Sethu Institute of Technology, Tamil Nadu, India

[2]Professor and HoD, Department of IT, St. Martin's Engineering College, Dhulapally, Secunderabad, Telangana, India

[3]Associate Professor, Department of CSE, KPR Institute of Engineering and Technology, Coimbatore, India

[4]Department of Education, Rajiv Gandhi University, Rono Hills, Doimukh, Arunachal Pradesh, India

[5]Faculty of Management, Czstochowa University of Technology, Poland

[6]Administrative Officer, Central Tribal University of Andhra Pradesh, Vizianagaram, Andhra Pradesh, India

[7]Assistant Professor, Department of Information Technology, St. Martin Engineering College, Secunderabad, Telangana, India

ABSTRACT: This paper investigates the effect of AI (artificial intelligence) on management decision-making. AI alludes to the capacity of PCs to execute exercises that customarily include human knowledge, similar to decision-making, thinking, critical thinking, and learning. This paper looks at how AI can be utilised to improve decision-making, by providing a better understanding of the underlying data, automating certain tasks, and enhancing the accuracy of decision-making. It examines the potential benefits and challenges associated with implementing AI in management decision-making and provides a discussion of the current research in this area. The paper presumes that artificial intelligence has the capacity to change the manner in which organisations decisions and that further examination is expected to comprehend its effect completely.

KEYWORDS: Decision making, AI, management, data

1. INTRODUCTION

Around the globe, scholars frequently disagree about AI. This discussion includes issues related to accounting, management, and business. Strategic decision-making and technology have recently become hybridised. A growing number of users may have access to fresh decision variables and procedures owing to advances in computing, the accessibility of fresh data sources, and the declining cost of technical instruments. Artificial intelligence is perhaps of the most fascinating development in this regard Secinaro *et al.* (2021). In understanding to Shrestha *et al.* (2019), artificial intelligence achieves the required adjustments in the decision-making by helping novel techniques for perceiving the critical boundaries of the choice space, the deciphering the dynamic technique and result, the assessment of substitutes, the speed of the choice, and the repeat of the system, which arises captivating roads for studies and practice conditions.

Developments in AI help managers make better decisions, but they may also be used to introduce brand-new products and solutions to consumers (Mehbodniya *et al.*, 2022). AI might assist client decision-making, for instance, in the case of personalised financial budgeting and banking services. It is

DOI: 10.1201/9781003532026-30

significant that these new possibilities constitute an untapped study setting for advancing the significance of expertise.

Companies' data collection and analysis efforts are contributing to the value they provide to customers Bagnoli *et al.*, (2019). In an analogous way, AI creates new potential for customer relationship research. For instance, in the retail sector, personalised marketing and advertising are revolutionising both the online and in-person shopping experience. AI expectation calculations work with organisations to exactly project interest and achieve the engaged purchasers through remarkable sorts of dissemination and correspondence channels Talwar and Koury (2017). In expansion, AI gives information-driven information across the inventory network organisation to make wise information hotspots for bunch critical thinking, particularly for orchestrating issues, site association, wares reconciliation, and stock administration.

1.1. Research Gap

The usage of AIin decision-making is a relatively new area of research, with few studies exploring the potential societal impacts of AI on management decision-making. While AI has been used in numerous areas of research, including healthcare, finance and marketing, its application in the context of management remains largely unexplored. Most existing research focuses on the technical and economic aspects of AI, rather than its potential implications for the management process.

2. LITERATURE REVIEW

El Khatib and Al Falasi (2021), this paper intends to look at how AI impacts project the board direction. All the more especially, the article explores the amount and realness of the information that AI frameworks secure and propose to the undertaking administrator for direction. In the wake of executing an essential review, in which 25 administrators were examined, and an optional examination, in which a few logical papers were considered, and information was gained from different areas, it was concluded that AI applications improve information quality and believability, which speeds up and execution in both single-project and multi-project conditions (Jurczyk-Bunkowska and Pawełoszek, 2015).

Lawrence (1991) of every one of the new developments that showed up in the later 50% of the twentieth 100 years, the progression of AI might highestly affect the authoritative direction. Since the advancement of AI methods and models was, to a great extent, founded on mental speculations of human cognisance, the results of their utilisation in complex social circumstances have not been very much examined. This essay aims to stimulate research that will advance our knowledge of AI's effects and place in complex organisations. This article puts forth the defence that executing master frameworks will bring about less convoluted and political decision-making, however, carrying out natural language frameworks will bring about a more troublesome and political independent direction.

Bokhari and Myeong (2022) this exploration plans to investigate the immediate and aberrant associations between AI, social development, and smart decision-making (SDM). Cross-sectional information from South Korea and Pakistan were assembled for this study using a review plan and overview polls. Utilising SPSS numerous relapse, 460 respondents from people in general and confidential areas were accumulated and observationally broke down. As expected, the study found that SI had a significant and beneficial mediating influence on the connection between AI and SDM. Some of the decision-making-influencing elements have been the subject of prior research. By investigating the influence of a mediating factor on decision-making, this study provides the body of social scientific studies.

Dietzmann and Duan (2022) managers should dissect a developing amount of organized and unstructured information consistently in the big data period to simply decide. In this present circumstance, administrative information processing (IP), which fills in as the establishment for managers decision-making, can be upheld by artificial intelligence functionalities. Little is presently perceived about the issues supervisors experience while integrating AI into their IP and direction. In the ongoing article, these are recognised through three centre gatherings with chiefs from the monetary business, approved through an overview, and hierarchical outcomes are drawn from them.

3. RESEARCH METHODOLOGY

The Adaptive approach was chosen for this study since it offers a more comprehending of study phases and is straightforward and simple to make and are measuring. It is a sequential advancement process that progresses progressively until the project's conclusion, simulating an adaptive system's development. Case studies, questionnaires, and interviews with subject-matter experts are the main data sources. The secondary information is compiled from a wide range of online journals, papers, and e-books that are accessible on different academic websites.

For this study, a descriptive research design which seeks to paint a depiction of a circumstance as it occurs is used. Additionally, the deductive method, sometimes known as the 'top-down' method, produces hypotheses relying on accepted principles and concentrates on confirming accepted beliefs. To choose the study subjects for this investigation, a straightforward random selection procedure was used. To address the issues raised by the study and evaluate its presumptions, the data was analysed using IBM SPSS software. To assess subjective coherence and the connection between control and response parameters, Pearson's correlation coefficient was utilised.

4. RESULT AND DISCUSSION

220 workers of commercial companies in South Africa that used AI in DM for their operations participated in this study. Males made up the majority (68.18%), while women made up the remaining (31.82%). The highest percentage of respondents (43.18%) belonged to the 41–50 age groups, while the lowest percentage of participants (22.73%) came from the 21–30 age groups. The postgraduate degree represented the highest percentage in terms of education level (40.91%), while the bachelor's degree represented the lowest percentage (22.73%). The greatest percentage of respondents who use more than 50% of AI methods at work is 45.46%, while the lowest percentage of respondents who use less than 90% of artificial intelligence (AI) methods at work is 22.73%, based on the variable of dependence on AI methods at work. (Table 1)

Table 1: Demographic characteristics.

Characteristics	Frequency	Percent
Gender		
Male	150	68.18
Female	70	31.82
Age		
Less than 30 years	75	34.09
21–30	50	22.73
41–50	95	43.18
Education		
High school	10	4.54
Bachelor's degree	50	22.73
Post graduate degree	90	40.91
Others	70	31.82
Reliance on AI		
Less than 10%	50	22.73
30–40%	70	31.82
More than 50%	100	45.46

Source: Author's compilation.

AI Applications as control parameter have three aspects: learning and development, Conformity, and efficiency. Table 2 shows the rank of the independent variables. The overall mean of training & development, conformity and efficiency is 4.71 ± 0.11/agree, 4.52 ± 1.23/disagree, 4.39 ± 0.78/agree.

Table 2: AI applications (Control parameter).

Control Parameter	Mean	SD	Scale
Learning & development	4.71	0.11	Agree
Conformity	4.52	1.23	Disagree
Efficiency	4.39	0.78	Agree

Source: Author's compilation.

The decision-making excellence responsiveness factor also includes the three factors of decision acceptance, decision acceptance speed and decision quality. The order of the independent variables is shown in Table 3. The total mean of decision standard, decision speed, and decision endorsement is 3.79 ± 1.02 (disagree), 2.32 ± 0.43 (agree), and 2.91 ± 0.57 (agree).

Table 3: Decision quality (Response parameters).

Response Parameters	Mean	SD	Mean
Speed of decision	3.79	1.02	Disagree
Standard of decision	2.32	0.43	Agree
Endorsing decision	2.91	0.57	Agree

Source: Author's compilation.

The connection among the control parameter 'AI applications' and its factors and the response parameter 'DM quality' and its components (the swiftness of decision making, the standard of decision, and endorsing decision) was examined using Pearson correlation analysis. All the control and response parameters have statistically positive correlations, as shown in Table 4. The range of Pearson correlation coefficients (r) is 0.31–0.81. Positive and strong correlations exist between the control and response parameters for each of these coefficients.

Table 4: Pearson correlations.

	Speed of Decision	Standard of Decision	Endorsing Decision	Decision Making Quality
Training & Development	0.75^{**}	0.56^{**}	0.49^{**}	0.81^{**}
Conformity	0.58^{**}	0.31^{**}	0.68^{**}	0.71^{**}
Efficiency	0.38^{**}	0.29^{**}	0.38^{**}	0.33^{**}
AI applications	0.64^{**}	0.48^{**}	0.56^{**}	0.82^{**}

Source: Author's compilation.

5. CONCLUSION

In conclusion, AI has had a tremendous impact on management decision-making. AI can help managers make better decisions that are more informed, and informed decisions lead to better outcomes. AI can help managers to identify opportunities and make decisions quickly, accurately, and cost-effectively. Additionally, AI may increase decision-making speed while lowering decision-making costs. Even managers may benefit from using AI to help them comprehend their markets and consumers more effectively and create long-term company choices. Artificial intelligence (AI) has the power to completely change how managers makes choices. The opportunities for using AI to enhance decision-making in management situations are almost limitless as the technology develops.

REFERENCES

Bagnoli, C., F. Dal Mas, and M. Massaro (2019). "The 4th industrial revolution: Business models and evidence from the field," *International Journal of E-Services and Mobile Applications (IJESMA)*, 11(3), 34–47.

Bokhari, S. A. A., and S. Myeong (2022). "Use of artificial intelligence in smart cities for smart decision-making: a social innovation perspective," *Sustainability*, 14(2), 620.

Dietzmann, C., and Y. Duan (2022). "Artificial intelligence for managerial information processing and decision-making in the era of information overload," *Proceedings of the 55th Hawaii International Conference on System Sciences.*

El Khatib, M., and A. Al Falasi (2021). "Effects of artificial intelligence on decision making in project management," *American Journal of Industrial and Business Management*, 11(3), 251–260.

Jurczyk-Bunkowska, M., and I. Pawełoszek, "The concept of semantic system for supporting planning of innovation processes," *Polish Journal of Management Studies*, 11(1), 2015.

Lawrence, T. (1991). "Impacts of artificial intelligence on organisational decision making," *Journal of Behavioral Decision Making*, 4(3), 195–214.

Secinaro, S., D. Calandra, A. Secinaro, V. Muthurangu, and P. Biancone (2021). "The role of artificial intelligence in healthcare: a structured literature review," *BMC Medical Informatics and Decision Making*, 21, 1–23.

Shrestha, Y. R., S. M. Ben-Menahem, and G. Von Krogh (2019). "Organizational decision-making structures in the age of artificial intelligence," *California Management Review*, 61(4), 66–83.

Talwar, R., and A. Koury (2017). "Artificial intelligence–the next frontier in IT security?" *Network Security*, 2017(4), 14–17.

31. Development and Deployment of AI in Marketing Financial Services – Understanding the Managers and Study Insights

Nishikant Jha[1], Sagar Balu Gaikwad[2], Shilpi Kulshrestha[3], Ravindra Pathak[4], Tomasz Turek[5], Supriya Pathak[6], and G. Nagaraj[7]

[1]Vice Principal (Commerce), HoD Accounting and Finance, Thakur College of Science & Commerce (Affiliated to University of Mumbai), Mumbai, Maharashtra, India

[2]Assistant Professor, Department of Management, MET Institute of Management

[3]Associate Professor, Department of Management Studies, Global Academy of Technology, Bangalore, Karnataka, India

[4]Associate Professor, Department of Mechanical Engineering, Medi-Caps University, Indore, Madhya Pradesh, India

[5]Czestochowa University of Technology, Poland

[6]Assistant Professor, Faculty of Management and Commerce, Oriental University, Indore, Madhya Pradesh, India

[7]Associate Professor, Department of Mechanical Engineering, Sethu Institute of Technology, Tamil Nadu, India

ABSTRACT: The financial sector is becoming increasingly interested in Artificial intelligence (AI) chatbots as an emerging platform for the electronic shift. The objective of this research is to better comprehend managers' understanding of AI and its difficulties, in addition, to emphasising important players and their joint attempts to offer financial services, provided that managers serve an essential part in the deployment and development of AI for marketing financial services. This investigation used a core-periphery evaluation of societal illustrations along with managers amid AI chatbot services in financial companies. The findings indicate that managers were all aware that the source of financial aid has undergone important electronic alterations as a result of AI. Although the role of AI chatbots in this sector has been widely recognised, the managers noted that they lacked the precise kind of AI platform required and stressed a range of potential approaches for integrating the marketing of financial services and AI.

KEYWORDS: Artificial intelligence, marketing, financial services, banks, and managers

1. INTRODUCTION

AI innovations are being utilised in marketing, in which big data are employed to create highly customised customer accounts, forecast consumer needs, and produce intended commercials (Manser Payne *et al.*, 2021). Given that data, the increasingly particular customer, as well as transaction-related information, is the primary resource that insurance firms and banks constantly gather, sort, manage, and connect to, the financial services (FS) sector exhibits significant prospective for AI (Groot, 2017). Several FS firms are beginning to roll out chatbots or robot managers, frequently in their mobile applications utilising social networking sites. To enhance customer service, the German insurance business VHV Versicherungen, for example, recently unveiled its novel chatbot 'Mia' (Kruse *et al.*, 2019). For example, Deutsche Bank has emerged as the initially associated bank in Germany to employ 'Robin', referred to as an electronic resource manager (Raff, 2020). Accenture projects $4.6 trillion in the starting point development for financial

DOI: 10.1201/9781003532026-31

services using AI innovations through 2035 (Raff, 2020). Financial assistance companies are reacting by executing AI to raise their company functions. The rapid acceptance is occurring at an unparalleled rate, posing new conceptual and managerial challenges (Bussmann *et al.*, 2020).

Managers of financial companies anticipate that AI chatbots would increase client ease and reduce expenses for operations, provided the widespread acceptance of financial chatbots in comparison to other service industries (Azemi *et al.*, 2019). Few investigations have examined the viewpoint of managers, who drive the acceptance of and manage AI chatbots, as opposed to the majority of research, which has examined AI chatbots from the perspectives of clients and programmers. Nevertheless, managers' perspectives are crucial since they have an impact on the way novel information technology (IT) assistance is adopted, as well as the way those services execute and the way customers – particularly tech-savvy ones – react (Leclercq-Vandelannoitte, 2015). The same holds for AI chatbots. When it comes to effectively developing and running AI chatbots, managers are an essential link between developers and clients. In addition, it's important to comprehend what managers believe and encounter in the AI chatbot areas to prevent the so-called IT paradox condition in AI chatbots (Barlette *et al.*, 2021). To ensure that AI chatbots are more than just passing trends, the possibilities and difficulties identified by managers at an initial phase would be essential resources. In particular, the investigation offers an analytical structure for AI in connection with marketing financial service investigation, emphasising important players and their collaborative efforts in offering clients improved financial services.

2. LITERATURE REVIEW

In the finance sector, insurance firms and banks function in a complex, cutthroat sector that is under severe strain to change to survive. Maintaining customer commitment in this setting, which entails contentment, confidence, dedication, and perceived worth, is crucial (Ansari and Riasi, 2016). Financial fragility, as stated (Mogaji, 2020), might be established from both an individual's viewpoint as well as a market structure's viewpoint. On the other hand, the individual viewpoint happens when an individual lacks access to financial services and is unable to efficiently settle their financial obligations as an outcome. Sheth *et al.* (2022) investigated, AI-powered financial assistance for customised expertise in a contemporary world, emphasising the importance of AI mediation in emerging markets. Furthermore, they emphasise the significance of human assistance powered by AI financial services by implementing customised assistance encounter components and emphasising the function of client encounters powered by AI financial service. Based on a cross-cultural standpoint, Mogaji *et al.* (2020) note that finance managers possess an understanding of the potential of AI and are working to tackle AI as a commercial requirement, yet they also note that there are some difficulties in speeding up AI adoption. Furthermore, they particularly challenge some assumptions about AI and its function in financial assistance, the AI applications used for this purpose, and the part played by advertising professionals in the development of AI (Ahamad *et al.*, 2022; Isabona *et al.*, 2023).

3. RESEARCH METHODOLOGY

The core-periphery investigation of the societal representation concept has been utilised to understand managers' insights into AI chatbot assistance for marketing financial services. This investigation used an exploratory, inductive methodology to gain an understanding of managers' insights into the uses of AI in advertising in the financial sector. Secondary data were collected from the managers of every banking, financial services, and insurance (BFSI) firm in India that oversees chatbots , with 14 managers of Indian BFSI firms who had participated in the implementation of chatbot assistance. Data were collected with managers responsible for chatbot assistance. The retrieved pieces from the content evaluation of the interviews were then examined to create a representation framework using the core-periphery research. People in managerial positions in the financial services sector who have accountability for developing, deploying, sourcing, and implementing marketing and electronic evolution methods, including AI have been the intended participants(Mohanta *et al.*, 2022). Participants have been looked for from multiple nations and financial assistance providers with varying degrees of expertise to guarantee an array of perspectives and encounters to enhance the data and the following results. Table 1 contains in-depth data about the interviewees.

Table 1: Demographic data.

Demographic Details	Interview Number's
40s	7
30s	7
Female	4
Male	10
AI chatbot-based work expertise	
>3 years	5
>2 years	4
>1 years	2
<1 year	3
Stock market	1
Credit cards	2
Banking sector	5
Insurance	6

Source: Author's compilation.

4. RESULT AND DISCUSSION

According to the findings of the research investigation, there are 15 elements of AI chatbot assistance that managers understand in marketing financial services. According to the core-periphery investigation, 7 areas have been identified as core, while the rest of the eight have been categorised as periphery, as shown in Table 2. Figure 1 depicts managers' thorough understanding of AI chatbot assistance in marketing financial services.

Table 2: Managers' thorough understanding of AI chatbot assistance in marketing financial services.

No.	Topics	Coreness
T05	Advancing digital shifts	0.062
T06	Adapting to customers who prefer non-public channels of communication	0.131
T07	Government regulation	0.145
T08	Creating a fresh service paradigm	0.152
T01	AI	0.182
T09	Substituted position	0.184

No.	Topics	Coreness
T04	Investment	0.19
T11	Collaborating with the tech sector	0.219
T12	Experimentation	0.264
T03	Enhancing operational effectiveness	0.271
T14	Organizational opposition	0.28
T13	Absence of firm capacity	0.281
T10	Supportive function	0.304
T02	Enhancing customer experience	0.367
T15	Technological immaturity	0.505

Source: Author's compilation.

Technological immaturity (T15), Enhancing customer experience (T02), Supportive function (T10), Absence of firm capacity (T13), Organisational opposition (T14), Enhancing operational effectiveness (T03), and Experimentation (T12) tend to be the core elements. As per this empirical study, bank managers in both developing and developed nations have been understanding the potential of AI in financial assistance for marketing. They were all aware that AI had caused vital electronic shifts in the supply of financial assistance. While the function of AI chatbots has been widely accepted in this supply, the managers acknowledged that they do have not the kind of AI platform required and emphasized an array of alternative methods for incorporating AI into marketing financial services.

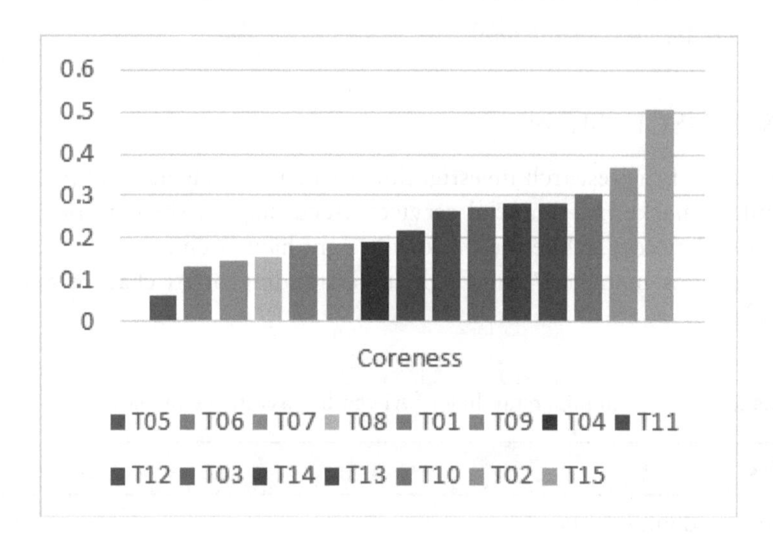

Figure 1. *Managers' thorough understanding of AI chatbot assistance in marketing financial services.*
Source: Author's compilation.

4.1. Opportunities

The majority of managers stated that AI chatbot services have been placed in place to improve customer experience and offer client-centered services (T02). The way that businesses communicate with their customers has changed as a result of chatbots (T06). They lacked faith, though, that the AI chatbot service might boost the clientele's experience. The relationship between Enhancing customer experience (T02) and

Addressing clients who opt for non-facing channels (T06) suggests that the way various user types react to AI chatbot services could vary. Many people started using AI chatbot services as a test (T12). Some people used the implementation of AI chatbot services as a test for working with tech companies (T11), or as a team to identify a novel market or service structure (T08). Cooperation with a tech company (T11) is related to the term experiment (T12).

4.2. Challenges

Additionally, according to certain managers, AI chatbot services were developed for operational effectiveness (T03), which possessed an effect on employees due to the service's replacement or back functions (T09, T10). One type of workforce change involves substituting the work of current employees. Existing employees were worried about this, which raised the possibility of organisational resistance (T14). Alternatively, to these worries, most managers claimed that AI chatbot assistance increased employment efficiency by assisting the current workforce instead of displacing current workers (Sharma *et al.*, 2022).

5. CONCLUSION

The purpose of this investigation was to investigate managers' perceptions of AI deployment in the field of marketing financial assistance. The current investigation has limitations, just like every other investigation, and these limitations must be taken into account when interpreting the results. Since the managers self-reported for the research, which used an empirical approach, the results might not be generalisable. People who live in distinct nations and possess distinct functions and degrees of AI knowledge might possess distinct experiences integrating AI into their company's activities. To completely measure the degree of understanding among bank managers, subsequent studies may try to take a quantitative approach. To determine if the concepts are reliable and relevant to various commercial activities, the conceptual structure developed is accessible for additional study and verification.

REFERENCES

Ahamad, S., N. Christian, A. K. Lodhi, U. Mamodiya, and I. R. Khan (2022). "Evaluating AI system performance by recognition of voice during social conversation," *2022 5th International Conference on Contemporary Computing and Informatics (IC3I)* (pp. 149–154). IEEE.

Ansari, A., and A. Riasi (2016). "Modeling and evaluating customer loyalty using neural networks: evidence from startup insurance companies," *Future Business Journal*, 2(1), 15–30.

Azemi, Y., W. Ozuem, K. E. Howell, and G. Lancaster (2019). "An exploration into the practice of online service failure and recovery strategies in the Balkans," *Journal of Business Research*, 94, 420–431.

Barlette, Y., A. Jaouen, and P. Baillette (2021). "Bring Your Device (BYOD) as reversed IT adoption: insights into Managers' coping strategies," *International Journal of Information Management*, 56, 102212.

Bussmann, N., P. Giudici, D. Marinelli, and J. Papenbrock (2020). "Explainable AI in fintech risk management," *Frontiers in Artificial Intelligence*, 3, 26.

Groot, M. (2017). *A Primer in Financial Data Management*. Academic Press.

Isabona, J., L. L. Ibitome, A. L. Imoize, U. Mamodiya, A. Kumar, M. M. Hassan, and I. K. Boakye (2023). "Statistical characterization and modeling of radio frequency signal propagation in mobile broadband cellular next generation wireless networks," *Computational Intelligence and Neuroscience*, 2023, 5236566, 9 pages.

Kruse, L., N. Wunderlich, and R. Beck (2019). "Artificial intelligence for the financial services industry: what challenges organizations to succeed."

Leclercq-Vandelannoitte, A. (2015). "Managing BYOD: How do organizations incorporate user-driven IT innovations?" *Information Technology & People*, 28(1), 2–33.

Manser Payne, E. H., J. Peltier, and V. A. Barger (2021). "Enhancing the value co-creation process: artificial intelligence and mobile banking service platforms," *Journal of Research in Interactive Marketing*, 15(1), 68–85.

Mogaji, E. (2020). "Consumers' financial vulnerability when accessing financial services," *Research Agenda Working Papers*, 2020(3), 27–39.

Mogaji, E., T. O. Soetan, and T. A. Kieu (2020). "The implications of artificial intelligence on the digital marketing of financial services to vulnerable customers," *Australasian Marketing Journal*, j-autism.

Mohanta, H. C., Geetha, B. T., Alzaidi, M. S., Dhanoa, I. S., Bhambri, P., Mamodiya, U., and R. Akwafo (2022). "An optimized PI controller-based SEPIC converter for microgrid-interactive hybrid renewable power sources," *Wireless Communications and Mobile Computing*.

Raff, S. (2020). *Products and Services in the Digital Age: Findings of a Multi-Perspective Approach* (Doctoral dissertation, Dissertation, Rheinisch-Westfälische Technische Hochschule Aachen, 2019).

Sharma, S., U. Mamodiya, T. Saini, K. A. Kumari, P. C. S. Reddy, and A. W. Hashmi (2022). "Investigating the role of image processing to identify patterns for industrial automation," *2022 2nd International Conference on Advance Computing and Innovative Technologies in Engineering (ICACITE)* (pp. 250–254). IEEE.

Sheth, J. N., V. Jain, G. Roy, and A. Chakraborty (2022). "AI-driven banking services: the next frontier for a personalized experience in the emerging market," *International Journal of Bank Marketing*, (ahead-of-print).

32. AI-Driven Dynamic Resource Management in Cloud Computing Environments

M. Siva Sangari[1], Nupur Mistry[2], Gouri Desai[3], Pugalendhi G. S.[4], Astha Bhanot[5], S. Durga[6], and Binod Kumar[7]

[1]Associate Professor, Department of CSE, KPR Institute of Engineering and Technology, Coimbatore, India

[2]Associate Professor, Department of Architecture, Anantrao Pawar college of Architecture, Pune, India

[3]Associate Professor, Department of Architecture, Goa College of Architecture, Goa, India

[4]Assistant Professor, Department of Artificial Intelligence and Data Science, Sri Krishna College of Engineering and Technology, Coimbatore, India

[5]PNU University, Riyadh, KSA

[6]Assistant Professor, KL Business School, Koneru Lakshmaiah Education Foundation, Vaddeswaram, Andhra Pradesh, India

[7]Professor, JSPM'S Rajarshi Shahu College of Engineering, Pune, Maharashtra, India

ABSTRACT: This article explores the potential of artificial intelligence (AI) to improve resource management in cloud computing environments. It examines the development of AI-driven approaches to dynamic resource allocation and investigates the implications for cloud computing performance. The article will provide an overview of the current state of AI-driven resource management and discuss its potential to automate resource allocation decisions, including the trade-offs between cost, performance and efficiency. The article will also explore the challenges of implementing AI-driven dynamic resource management and the potential implications for cloud computing providers and users. Finally, it will consider the implications of AI-driven dynamic resource management for the future of cloud computing.

KEYWORDS: Artificial intelligence, cloud computing, resource management

1. INTRODUCTION

Cloud computing is a rapidly evolving technology that has revolutionised the way businesses operate. As cloud computing has gained in popularity, its use in various industries has grown exponentially. This has resulted in an increased demand for resources, such as servers and storage, from cloud providers. As such, resource management is a critical component of cloud computing, as it ensures the effective utilisation of resources to meet the needs of customers. To address this challenge, AI-driven dynamic resource management has emerged as a viable solution Al-Fuqaha *et al.* (2015). This article provides an overview of AI-driven dynamic resource management in cloud computing environments, including its key features and benefits. AI-driven dynamic resource management (DRM) is a technology that enables cloud providers to dynamically adjust the resources allocated to customers based on their current and future needs. It uses artificial intelligence (AI) algorithms to analyse the customer's usage patterns, predict their future needs and adjust resources accordingly (Kulkarni and Hedman, 2020). This approach allows cloud providers to optimise resource utilisation and improve customer satisfaction.

The key features of AI-driven DRM in cloud computing environments include the following: automated resource allocation: AI-driven DRM automates the process of allocating resources to customers, ensuring that resources are used efficiently and that customer needs are met. Real-time adjustments: AI-driven DRM allows cloud providers to make real-time adjustments to resource allocations, ensuring that resources are optimised in response to changing customer needs (Kumar and Jayaraman, 2020). Predictive analytics: AI-driven

DOI: 10.1201/9781003532026-32

DRM uses predictive analytics to anticipate customer needs and adjust resources accordingly. Cost optimisation: by optimising resource utilisation, AI-driven DRM can help cloud providers reduce costs and increase efficiency. Improved customer satisfaction: AI-driven DRM can help cloud providers improve customer satisfaction by ensuring that resources are allocated in an optimal manner (Paweloszek, 2015).

The benefits of AI-driven DRM in cloud computing environments include improved resource utilisation, cost optimisation and improved customer satisfaction. By leveraging AI to optimise resource utilisation, cloud providers can reduce costs, improve customer satisfaction and ensure that resources are used efficiently (Liu and Zhang, 2020). In conclusion, AI-driven dynamic resource management is a powerful technology that allows cloud providers to optimise resource utilisation, reduce costs and improve customer satisfaction. As such, it is an important part of any cloud computing strategy (Li and Zhang, 2020).

1.1. Objective

The objective of this study is to develop and evaluate an AI-driven dynamic resource management framework for cloud computing environments. The framework aims to optimise resource allocation, scalability and energy efficiency by leveraging AI techniques to intelligently allocate and reallocate resources based on workload patterns and system conditions, leading to improved performance and cost-effectiveness.

1.2. Hypothesis

- AI-driven dynamic resource management in cloud computing environments improves resource utilisation and optimises cost-efficiency.
- AI-driven dynamic resource management in cloud computing environments enhances system scalability and responsiveness by effectively allocating resources based on real-time demand and workload patterns.

2. LITERATURE REVIEW

The increasing need for services from users and the growth of cloud computing technology has made it unfeasible for human operators to oversee cloud resources individually (Nzanywayingoma and Yang, 2019). In this study, we reviewed the current level of architecture as a service cloud resource management. We focused on resource management approaches like resource provisioning, discovery, tracking, mapping, allocation, consolidation, modelling, and scheduling while providing a summary of current research findings and technological advances.

Many new applications and services are targeted by the 5G system and beyond, which will increase network traffic (Boudi *et al.*, 2021). The network faces a challenging mission to accomplish the intended objectives because of the aggressive, contentious, and competing expectations of these industrial verticals. It is anticipated to receive demands for nearly instantaneous latency, high data rates and network dependability. Due to this, industry and academics are under enormous pressure to utilise these methodologies by developing a new idea known as a CCN environment that can coexist and adapt depending on the network and resource condition as well as observed KPIs.

For use in upcoming cellular systems, this study provides a distributed core network design (Mukherjee *et al.*, 2018). The centralised gateways utilised in current mobile core systems cause latency and efficiency limitations that are addressed by the suggested structure. Utilising identifier-based protocol extensions to IP that operate on base stations and routers without the requirement for centralised gateways, a completely distributed framework for the mobile core is realised. In comparison to existing systems, the resultant 'flat' mobile core system may serve an assortment of IoT services with much minimal latency and increased throughput.

3. RESEARCH METHODOLOGY

Three main parts make up the dynamic resource management algorithm (DRMA): an allocation, a migration and a combined algorithm for job migration and allocation onto the server. Before getting into the specifics, we provide the reader a broad overview of DRMA in this part by briefly describing each of these elements.

The method uses the best-fit theory to reduce server costs. The algorithm cycles over a list of jobs that have been arranged in descending order by CPU need. As a result, the algorithm is given the activity with the largest CPU needs during the first iteration. Once the current work has been selected, the algorithm searches for a server that can perfectly accommodate it, that is, a server that has the same amount of free CPU accessible as the current task's CPU requirements. The algorithm will choose a server from the sorted server list that has the closest free CPU accessible in relation to the CPU need of the present task if it is unable to locate an exact match. If this occurs, the method will search for the closest match. The method assigns the current work to the server once the closest match has been identified. The next task is selected by the algorithm from the list of sorted tasks. The algorithm starts over by determining the exact match and then the closest match.

4. RESULT AND DISCUSSION

The prediction method uses past data, and Table 1 shows the RMSE values computed over 24 hours while tossing the first 30 results. As can be observed, seasonality is not a good fit for our situation because the RMSE value of the DES algorithm is lower for the three workload samples. As a result, the RMP approach's considerable phase of CPU utilisation in overload detection strategies was predicted using the DES algorithm (Table 1).

Table 1: Root mean square error for DES and SHW algorithms.

	DES	SHW
Workload #	9.11	11.23
Workload #	10.67	10.41
Workload #	12.45	11.25

Source: Author's compilation.

In this experiment, we consider four servers where jobs could be distributed or relocated. The highest limit of our resources' utilisation threshold must be kept to a minimum. Therefore, we choose the range of 10% to 60% before migration and between 70% and 100% after migration for our experiment. The total utilisation percentages of 70% and 40%, for instance, were chosen at random. Figure 1 provides a detailed explanation of this.

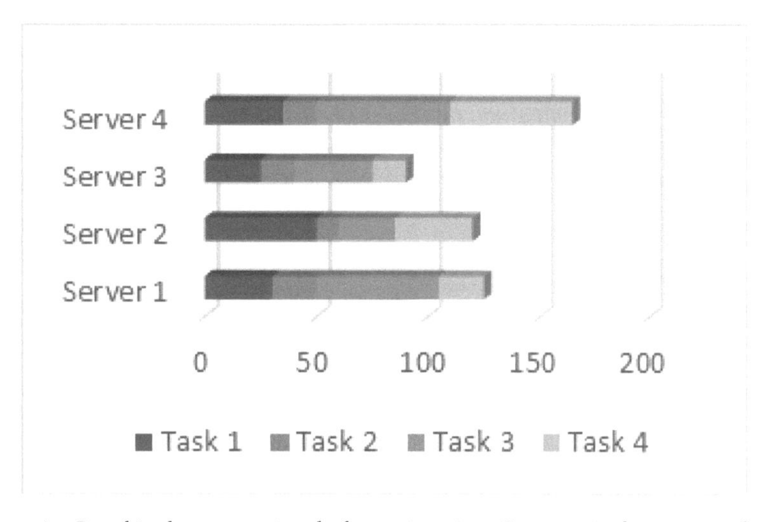

Figure 1. *Graphical presentation before migration. Source: Author's compilation.*

The outcome of the table depicts allocation upon migration, as it is represented in Table 2. It appears that server 2 is not being used and that we can get by with just three servers. Since we can fit the duties on three servers without using the fourth during the migration, our goal is to employ fewer servers while still doing the move efficiently. When servers are full, we move tasks in order to optimally distribute resources without turning on more servers.

Table 2: Allocation upon migration.

List of servers	Task 1	Task 2	Task 3	Task 4
Server 1	65	85	90	85
Server 2	0	0	0	0
Server 3	70	75	80	65
Server 4	75	85	65	75

Source: Author's compilation.

5. CONCLUSION

AI-driven dynamic resource management in cloud computing environments offers significant potential for improved performance and cost savings. It allows companies to better predict and respond to changing demand and workloads and to dynamically allocate resources as needed. This helps to ensure that resources are used efficiently and cost-effectively and that applications are running optimally. AI-driven dynamic resource management techniques also have the potential to improve the scalability of cloud computing environments, allowing for faster response times and improved utilisation of resources.

REFERENCES

Boudi, A., M. Bagaa, P. Pöyhönen, T. Taleb, and H. Flinck (2021). "AI-based resource management in beyond 5G cloud native environment," *IEEE Network*, 35(2), 128–135.

Kulkarni, V., and T. Hedman (2020). "AI-driven dynamic resource management for cloud computing," *IEEE International Conference on Cloud Computing*, pp. 98–105.

Kumar, V., and S. Jayaraman (2020). "Optimization of resource utilization in cloud computing using AI-driven dynamic resource management," *IEEE International Conference on Cloud Computing*, pp. 483–486.

Li, Y., and K. Zhang (2020). "AI-driven dynamic resource management for cloud computing: a survey," *IEEE International Conference on Cloud Computing*, pp. 86–91.

Liu, S., and Y. Zhang (2020). "An AI-driven dynamic resource management system in cloud computing," *IEEE International Conference on Cloud Computing*, pp. 211–214.

Nzanywayingoma, F., and Y. Yang (2019). "Efficient resource management techniques in cloud computing environment: a review and discussion," *International Journal of Computers and Applications*, 41(3), 165–182.

33. Entrepreneurship and Innovation in the Age of Artificial Intelligence: A Machine Learning Approach

T. Arul Raj[1], A. Parvathavarthine[1], Punit Kumar Dwivedi[2], Tripti Tiwari[3], Rania Mohy El Din Nafea[4], Ravindra Pathak[5], and G. Nagaraj[6]

[1]Assistant Professor, Department of Computer Science, Sri Paramakalyani College, Tamil Nadu, India

[2]Professor and Group Director, Modern Institute of Professional Studies, Modern Group of Institutions, Indore, India

[3]Assistant Professor, Department of Management Studies, Bharati Vidyapeeth (Deemed to be University) Institute of Management and Research, Delhi, India

[4]College of Administrative and Financial Sciences, University of Technology Bahrain, Bahrain

[5]Associate Professor, Department of Mechanical Engineering, Medi-Caps University, Indore, Madhya Pradesh, India

[6]Associate Professor, Department of Mechanical Engineering, Sethu Institute of Technology, Tamil Nadu, India

ABSTRACT: In the current industrial age, innovations in the field of artificial intelligence (AI) are drastically altering the commercial and entrepreneurial environment. AI has already infiltrated numerous facets of the company, professional, personal and perhaps daily lives. AI stimulated the expansion of entrepreneurship, where information is the primary means of production for a better understanding of the overall scenario, and it is required to evaluate the entrepreneurship procedure from a fresh perspective. From the standpoint of an entrepreneur, this research investigates the influence of entrepreneurship utilising traditional decision-making procedures and AI innovation using a machine learning approach. The research employed a sample of 50 replies from across India. Entrepreneurs have been invited to carry out structured surveys and personal interviews to analyse their perspectives on conventional vs. AI, in addition to the relationship between both of them and their influence on entrepreneurial decisions. According to the findings, female and male entrepreneurs evaluate personality and compatibility assessments distinctly.

KEYWORDS: Artificial intelligence, machine learning, innovation, and entrepreneurship

1. INTRODUCTION

In the current era of profound advancements in scientific technology, businesses are not just developing innovative products and assistance yet are additionally altering entire sectors, erasing geographical limits and posing challenges to the pre-existing regulatory structures (Xu *et al.*, 2018a). Business investigators and experts might find it trying to stay aware of the ongoing areas' quick improvement because of state of the art innovation like AI, blockchain, robotics and IoT. The arising developments of fourth modern transformation are essentially affected by man-made intelligence (Xu *et al.*, 2018b). Information, thoughts, advancement and specialised propels are key parts of business and financial advancement. Most associations try to utilise AI and robotics developments to create countless advancement, delivering novel pioneering prospects and empowering monetary and social change in the improvement of more noteworthy degrees

DOI: 10.1201/9781003532026-33

of living (Upadhyay *et al.*, 2022). Furthermore, AI has a large impact on administrations, organisations and communities. At the moment, AI is developing into a facilitator and instigator to achieve an elevated degree of entrepreneurial awareness (Zhiyang and Zenim, 2020).

Entrepreneurship has been practiced for decades, yet history identifies it as an area for research instead of an event. Entrepreneurship is a practice that includes three basic parts: taking risks, innovation and commercialising goods and services for revenue (Keisner *et al.*, 2016). Behind every successful entrepreneur is a combination of innovation and entrepreneurship. Entrepreneurs are required to constantly look for chances for effective innovation, as well as the changes and signs that signal these possibilities (Olanrewaju *et al.*, 2018). Therefore, the purpose of this research is to investigate how businessmen see the machine learning (ML) method to using AI in business.

2. LITERATURE REVIEW

The study (Olanrewaju *et al.*, 2020) conducted an investigation on social media entrepreneurs between 2018 and 2020. This research explores entrepreneurial investigation in the most rigorous way possible. Furthermore, it has been revealed that entrepreneur incorporation and involvement with social media extends beyond marketing activities, allowing network advertising to help them in collecting money from a variety of resources. As per Audretsch and Moog (2022), entrepreneurship is inextricably linked to a core democratic ideal shared by Western industrialised nations, and entrepreneurship will perform an essential part of the framework. Three significant characteristics were uncovered during the research. The initial category is National Socialism, which evolved in Germany as a result of the prolonged oppression of both enterprise and democracy.

Based on the research, significant policy choices on entrepreneurship enable autonomous, decentralised and independent decision-making, which serves as an essential component of democracy. Lévesque *et al.*, (2022) investigated significant entrepreneurial investigation using AI through theory-driven investigations. It was determined that there have been actual methodologies available to entrepreneurial scientists to extract the true capacity of AI with rigor, allowing for potential expansion in entrepreneurial influence. It has been indicated that leveraging the possibilities of AI in research on entrepreneurship and minimising the risks associated with it provides a new 'grand issue' for the discipline, and experts have determined that AI has an important influence on entrepreneurs. Saini *et al.*, (2018) research sought to predict the future proportion and variances of human cognition and AI utilisation in industry 4.0 societal entrepreneurship. The study investigated future AI applications in social entrepreneurship and examined participants' enthusiasm and drive to carry out the tasks till 2030. As per the study, it will not entirely automate but will use human resources selectively and increase the influence of AI.

3. RESEARCH METHODOLOGY

The present research employs an empirical sampling method with a sample scale of 50 entrepreneurs from throughout India. The present data were acquired using a standardised survey that was individually given to business owners and graded on a Likert scale of 1 to 5, with 1 denoting highly agreement, 2 denoting agreement, 3 denoting neutrality, 4 denoting disagreement and 5 denoting severely disagreement (Girraj and Heena, 2018). Secondary sources of information include books, investigations, studies and web pages, in addition to journal and periodical articles (Somwanshi *et al.*, 2016).

3.1. Hypothesis

H1: The opinions of male and female entrepreneurs on the use of artificial intelligence in innovation differ significantly statistically.

H2: The opinions of female and male entrepreneurs on the use of AI in decision-making differ significantly statistically.

4. RESULT AND DISCUSSION

The opinions of male and female business owners on the development of artificial intelligence and its application to decision-making are contrasted in Table 1. The study's entrepreneurs' mean and standard deviation, as well as how they used AI for decision-making, are shown in Table 2 and Figure 1.The findings of a one-way ANOVA study on how male and female entrepreneurs leverage AI innovation are shown in Table 3.

Table 1: Views of men and women entrepreneurs about AI innovation and its use in decision-making.

		Frequency	Percentage (%)
Validity	Female	23	46
	Male	27	54
	Total	50	100

Source: Author's compilation.

Table 2: The mean and standard deviation of entrepreneurs, in addition to their utilisation of AI for decision-making.

Constructs	Mean	Standard Deviation
Research and development	2.88	1.34
Personality and compatibility	2.78	1.233
Financial decisions	2.58	1.40
Marketing	2.64	1.15
Investment decisions	2.38	0.92
Refund and return filing	1.60	0.39

Source: Author's compilation.

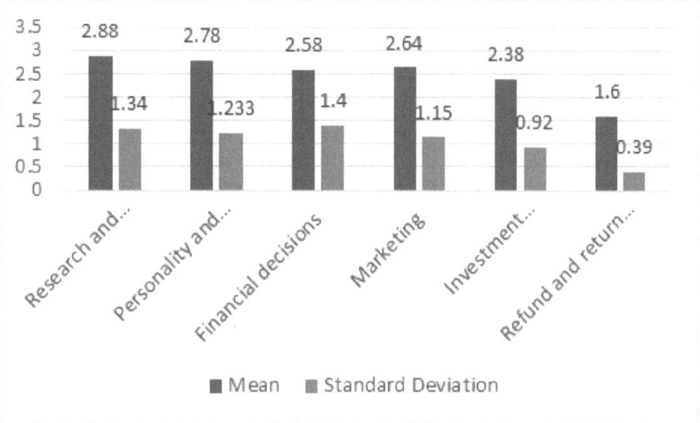

***Figure 1.** Graph showing entrepreneurs' means and standard deviations together with their use of AI in decision-making. Source: Author's compilation.*

Table 3: One-way ANOVA analysis on the use of AI innovation by male and female entrepreneurs.

	Groups	Sum of squares	Df	Mean square	F value	Sig.
Research and development	Between	1.824	1	2.324	0.021	0.322
	Within	87.456	48	3.922		
Personality and compatibility	Between	5.076	1	4.076	3.506	0.005*
	Within	69.504	48	2.448		
Financial decisions	Between	4.724	1	4.724	2.480	0.001*
	Within	91.456	48	1.905		
Marketing	Between	0.132	1	0.132	0.097	0.757
	Within	65.388	48	1.362		
Investment decisions	Between	0.321	1	0.211	0.212	0.002*
	Within	34.354	48	0.761		
Refund and return filing	Between	0.712	1	0.134	0.898	0.001*
	Within	84.189	48	2.977		

Source: Author's compilation.

Tables 1, 2 and 3 reveal a statistically significant disparity in female and male entrepreneurs' views when it arrives to the usage of AI innovation. Personality and compatibility testing (0.001), financial decisions (0.001), investment decisions (0.002) and refund and return filling (0.004) are statistically significant characteristics of entrepreneurship with a disparity in viewpoint towards female and male entrepreneurs (Koli and Mamodia, 2018). Table 3 demonstrates a statistically significant disparity in manufacturing and service entrepreneurs' attitudes towards AI for decision-making.

In the setting of ML, a precision value is frequently employed to assess a model's performance as shown in Table 4.

Table 4: Performance matrices of SVM.

Evaluation	Precision	Recall	F1 value	Support
1	0.65	0.71	0.68	0.18
2	0.72	0.68	0.70	0.32
3	0.63	0.72	0.68	0.25
4	0.81	0.68	0.74	0.21

Source: Author's compilation.

This work undergoes performance measurement for the AI innovation of entrepreneurship depending on the experimental investigation and ML algorithm using SVM to further elucidate the relationship between characteristics and more accurately predict the influence of AI innovation of entrepreneurship.

5. CONCLUSION

This study examines entrepreneurs' perspectives on the application of AI in business using the machine learning (ML) approach. The research employed a sample of 50 replies from across India. Entrepreneurs have been invited to carry out structured surveys and personal interviews to analyse their perspectives on conventional vs. AI, in addition to the relationship between both of them and their influence on entrepreneurial decisions. According to the findings, female and male entrepreneurs evaluate personality and compatibility assessments distinctly. AI may be used to solve accounting database problems. It has been discovered the present accounting database structures possess several problems. Not all accounting information is required. Business constitutes among the most common applications of computers. Text-based simulations have evolved into 3-D graphic simulations with enormous multiverses.

REFERENCES

Audretsch, D. B., and P. Moog (2022). "Democracy and entrepreneurship," *Entrepreneurship Theory and Practice*, 46(2), 368–392.

Lavanya, B. L., M. H. Mattaparti, S. R. Babu, and M. A. Aisha (2023). "Implementation of artificial intelligence in entrepreneurship: an empirical study," *Journal of Pharmaceutical Negative Results*, 231–242.

Lévesque, M., M. Obschonka, and S. Nambisan (2022). "Pursuing impactful entrepreneurship research using artificial intelligence," *Entrepreneurship Theory and Practice*, 46(4), 803–832.

Olanrewaju, A. S. T., M. A. Hossain, N. Whiteside, and P. Mercieca (2020). "Social media and entrepreneurship research: a literature review," *International Journal of Information Management*, 50, 90–110.

Olanrewaju, A. S. T., M. A. Hossain, P. Mercieca, and N. Whiteside (2018). "Identifying social media's capability for recognizing entrepreneurial opportunity: an exploratory study," *Challenges and Opportunities in the Digital Era: 17th IFIP WG 6.11 Conference on e-Business, e-Services, and e-Society, I3E 2018, Kuwait City, Kuwait, October 30–November 1, 2018, Proceedings 17* (pp. 344–354). Springer International Publishing.

Upadhyay, N., S. Upadhyay, and Y. K. Dwivedi (2022). "Theorizing artificial intelligence acceptance and digital entrepreneurship model," *International Journal of Entrepreneurial Behavior & Research*, 28(5), 1138–1166.

Xu, G., Y. Wu, T. Minshall, and Y. Zhou (2018a). "Exploring innovation ecosystems across science, technology, and business: a case of 3D printing in China," *Technological Forecasting and Social Change*, 136, 208–221.

Xu, M., J. M. David, and S. H. Kim (2018b). "The fourth industrial revolution: opportunities and challenges," *International Journal of Financial Research*, 9(2), 90–95.

Zhiyang, L., and W. Zemin (2020). "AI enabling entrepreneurship: comparison of theoretical frameworks," *Foreign Economics & Management*, 42(12), 3–16.

34. Evaluating Best Practices and Strategies to Protect Critical Assets Through Cybersecurity Management

Suresh Kumar M.V[1] and G. Vinodini Devi[2]

[1]Research Scholar, Department of Law, Koneru Lakshmaiah Education Foundation, Green Fields, Vaddeswaram, Guntur, Andhra Pradesh, 522302

[2]Assistant Professor, Department of Law, Koneru Lakshmaiah Education Foundation, Green Fields, Vaddeswaram, Guntur, Andhra Pradesh

ABSTRACT: The frequency, severity and sophistication of cybersecurity incidents are increasing, demonstrating that the next generation of cybersecurity risks cannot be fully mitigated by technological security controls alone. To increase cybersecurity in this age, end users must practice excellent cyber hygiene. According to the research, social culture has an impact on reducing cybersecurity risk. Since the majority of research on the topic is conducted in other nations, such findings could not apply to India because of its distinct societal culture. There is little research on how users of important assets behave in terms of cybersecurity. This study made an effort to fill the aforementioned research gap, and it was effective in creating a framework that enhances the cybersecurity behaviour of smart device users.

KEYWORDS: Cybersecurity, framework, behaviour, best practices

1. INTRODUCTION

Cyber threat risk is reduced by using security procedures. Technical, operational and physical controls are several types of security measures. Security software and firewalls are two examples of technical controls. An illustration of an administrative control is an information security policy, and an illustration of a physical control is a laptop lock. However, the increasing number of cybersecurity incidents shows that basic security measures are insufficient. The human user is regarded by cybersecurity as an asset that has to be safeguarded against dangers. Vulnerabilities can also be attributed to human users. Traditionally, the field of computer science has been used to view computer security. The interaction between human end users and security, however, can be seen as a system of technology and society in the context of cybersecurity (Neigel *et al.*, 2020).

The concerns about cybersecurity threats are growing as the digital era develops. Based on reports, RSA, 2020 after Canada, India is the nation that phishing assaults are most likely to target. Rogue mobile applications were the source of the bulk of attacks. Mobile virus targets Indian mobile consumers (Stitilis *et al.*, 2020). 3.17 lakh cybercrimes were recorded in India over the previous 18 months, providing an estimate of the size. Threats to mobile devices have increased by 33% in the past year. The smartphone is being used by attackers for a variety of activities, including cryptocurrency mining. For financial advantage, hackers use ransom ware to encrypt personal data. Users of smartphones leave themselves open to attack by not updating to the most recent version of their gadgets (Attaran *et al.*, 2019).

1.1. Problem Statement

Devices now have significantly more processing and storage space. Users are keeping and processing a lot of private information as a result of this development. In India, there is an increase in cybersecurity

DOI: 10.1201/9781003532026-34

incidents. End-user security practices are essential for reducing cyber dangers (Corallo *et al.*, 2020). It suggests that there is a need for enhanced organisational and individual cybersecurity awareness.

1.2. Research Objectives

- To comprehend how Indian smart device users now behave in terms of cybersecurity.
- To create a theory-driven structure for enhancing critical asset users' cybersecurity behaviour.

2. LITERATURE REVIEW

2.1. Aspects of Cybersecurity

It is obvious that technological cybersecurity safeguards cannot function in isolation and must collaborate with end users. One perspective on cybersecurity is as a socio-technical system. The viewpoint and behaviour of numerous stakeholders, such as decision-makers, designers of security systems and end users, are included in the sociotechnical perspective. However, security professionals fail to take into account the fact that for security to be effective, it must be useful and acceptable to the end user while creating such regulations. According to the data from numerous industry surveys (Attaran *et al.*, 2019), many security breaches are the consequence of human error. Human users are the weakest link in cybersecurity, according to research. According to Verizon's 2021 report, human error was a factor in 85% of breaches (Lee, 2021). Eight event classification patterns, including social engineering, mistakes, lost and stolen property and others, are listed in the report. Similarly, the Kaspersky study cites 'inappropriate IT resource use by employees' and 'inappropriate data sharing through mobile devices' as the main reasons why data breaches occur (Leahovcenco, 2021).

2.2. User Cybersecurity Practises

According to Mashiane and Kritzinger (2018), cybersecurity behaviours are the behaviours, responses, mannerisms and general online conduct of an individual. Establishing defensive routines, utilising security tools and the avoidance of destructive behaviour by end users are all examples of cybersecurity behaviour. It is frequently referred to as 'cyber hygiene'. Cyber hygiene aids in reducing the human factor-related cybersecurity risk (Neigel *et al.*, 2020). There is no agreement on what cyber hygiene entails, nevertheless (Vishwanath *et al.*, 2020). Depending on the situation, cyber hygiene has many meanings. To reduce the risk associated with enterprise IT systems, the National Institute of Standards and Technology (NIST) classified patching as a crucial aspect of cyber hygiene (Ferdausi, 2020). According to the European Commission (2018), the main goal of cyber hygiene for kids is to stop fake news from spreading and cyberbullying.

2.3. The Variances among National Cybersecurity Strategies and Their Distinctions

National cybersecurity strategies are comprehensive plans developed by governments to protect their countries' critical infrastructure, data and networks from cyber threats. While the specific details and approaches vary from country to country, these strategies generally share common goals of ensuring cybersecurity, promoting resilience and mitigating cyber risks (Sabillon *et al.*, 2016). Different countries may have varying scopes and objectives for their cybersecurity strategies based on their national priorities, threat landscape and technological capabilities. The emphasis of national cybersecurity strategies can vary depending on the country's priorities. For example, some strategies may prioritise international cooperation, information sharing and collaboration with other nations to combat cyber threats (Štitilis *et al.*, 2017). The organisational structure and governance mechanisms for implementing cybersecurity strategies can differ among countries. Some nations may establish dedicated national cybersecurity agencies or central coordinating bodies responsible for formulating and implementing policies, coordinating incident response and overseeing cybersecurity initiatives (Hills, 2019).

3. METHODOLOGY

The data from qualitative and quantitative research are combined or integrated in this study to offer a comprehensive picture of the research problem. The main tenet of this method is that integrating qualitative and quantitative methodologies produces a more thorough and accurate understanding of a research topic than each way by itself. Studies of this nature follow the pragmatic paradigm.

3.1. Sampling Method

The researcher utilised a two-step approach, starting with a qualitative method followed by a quantitative method. The purpose of this approach was to assess whether findings from a small sample of individuals (Study 1) could be generalised to a larger population sample (Study 2). To facilitate data collection through participants' smartphones whenever feasible, a practical sampling technique was employed. The number of participants was determined, taking into account the study's objectives and theoretical saturation, as outlined by Blandford (2018). Due to the aim of comprehending users' security practices, the researcher opted to limit the number of participants to ensure a thorough understanding of the interviews.

3.2. Mixed Method Approach

There are incredibly few research studies on Indian users of key assets' cybersecurity behaviour. Consequently, a mixed-methods approach was employed to fully investigate the cybersecurity behaviour of smart device users. This study approach combines the advantages of qualitative and quantitative methodologies. While very little qualitative empirical research is undertaken, quantitative empirical research is overwhelmingly used to analyse security awareness and behaviour (Omidosu and Ophoff, 2016).

4. RESULT AND DISCUSSION

The methods mentioned in the previous section are used to arrange the results of research studies. The study issue of cybersecurity behaviour and practices is addressed in the subsequent sections.

4.1. Protection via Settings and Add-On Utilities for Critical Assets

These included employing the authentication system, performing backups, updating software and so forth. These built-in features are easy to use and may be enabled or disabled by the user with just one action. Figure 1 displays the proportion of those polled who have the screen auto-lock feature enabled. As can be seen in the graph, a strong adoption of authentication systems is indicated by the fact that 89% of respondents have turned on-screen auto-lock mechanisms.

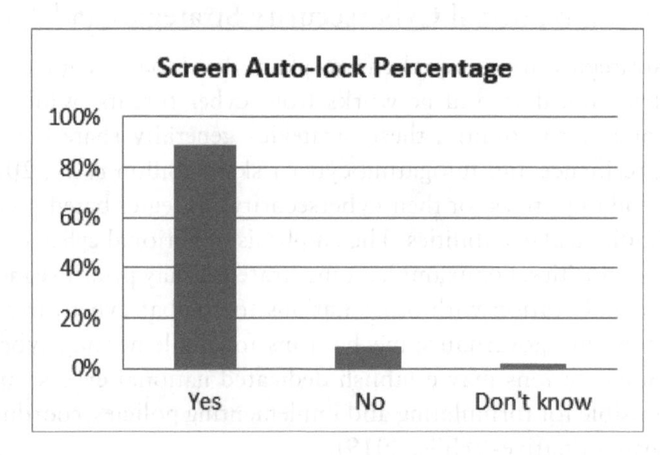

Figure 1. Screen auto-lock. Source: Author's compilation.

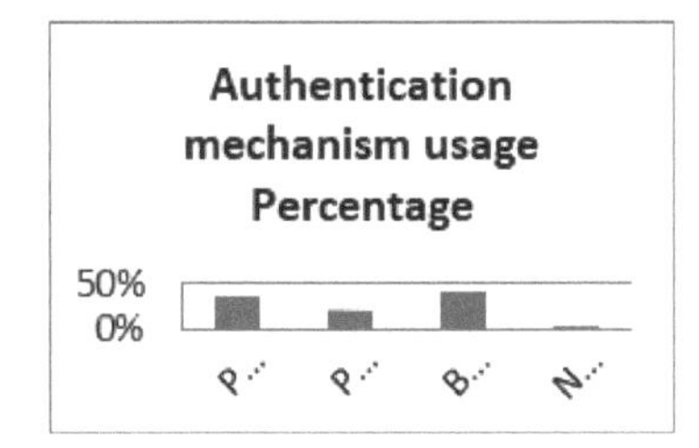

Figure 2. Authentication mechanism usage. Source: Author's compilation.

Figure 2 illustrates how the study respondents chose the biometric authentication technique first, followed by the PIN/password and pattern system. Developers of mobile operating systems and applications publish security updates to close any holes in their software.

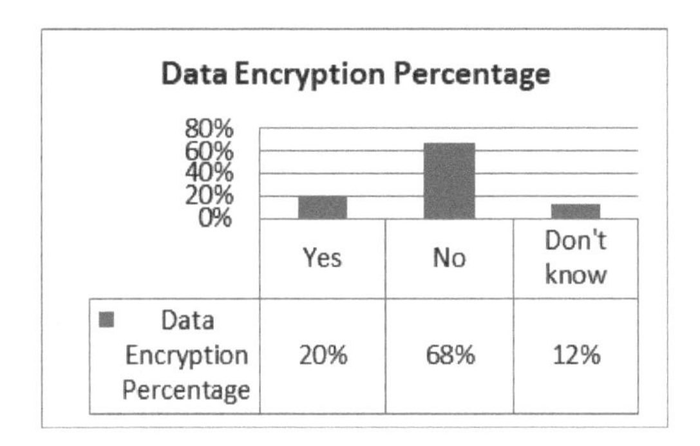

Figure 3. Data encryption percentage. Source: Author's compilation

By locking the device and encrypting the data, it may be possible to stop simple access to information about essential assets that have been lost or stolen. Figure 3 shows that approximately 12% of participants do not know about encryption and approximately 68% of participants do not have data secured on their critical assets.

4.2. Avoid Negative Behaviour

The survey asked about potentially hazardous habits that users of key assets should avoid to increase their security. Acceptance of these points signifies a riskier course of action. Figure 4 displays respondents' opinions regarding the authentication information sharing that online users should avoid.

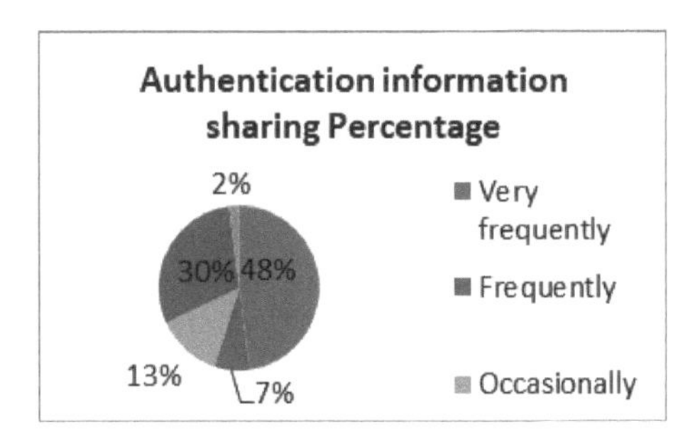

Figure 4. Authentication information sharing. Source: Author's compilation.

According to Figure 4, the majority of respondents (about 77%) do not divulge their PIN or password to anyone. Another 23% of respondents, however, divulge their PIN or password to others. Due to the exchange of authentication data, they become open to several cyber security risks, including unauthorised access to cell phones and the data they hold on them.

5. CONCLUSION

Organisations must become more aware of the changing cybersecurity landscape and act rapidly to address it as a result of the increasing cybersecurity dangers offered by adversaries and cybercriminals. The difficulty of estimating the costs and benefits of cybersecurity risk management, like in many other IT initiatives, is one of the obstacles to investment in this area. The organisation is in charge of determining the best technology to address the demand for cyber-acquisition and the requirement itself. Organisations may lessen the impact of cybercrime and unlock future economic potential by giving priority to technologies that enhance cybersecurity protection since higher levels of trust attract customers to do more business with them.

REFERENCES

Attaran, M., A. Gunasekaran, M. Attaran, and A. Gunasekaran (2019). "Blockchain and cybersecurity," *Applications of Blockchain Technology in Business: Challenges and Opportunities*, 67–69.

Corallo, A., M. Lazoi, and M. Lezzi (2020). "Cybersecurity in the context of Industry 4.0: A structured classification of critical assets and business impacts," *Computers in Industry*, 114, 103165.

Ferdousi, B. (2020). "Cyber security risks of bring your device (BYOD) practice in workplace and strategies to address the risks."

Greifeneder, E., S. Pontis, A. Blandford, H. Attalla, D. Neal, and K. Schlebbe, (2018). "Researchers' attitudes towards the use of social networking sites," *Journal of Documentation*, 74(1), 119–136.

Hills, M. (2019). "Hybrid threats a strategic communications perspective: 2007 cyber attacks on Estonia," *Hybrid Threats: A Strategic Communications Perspective* (pp. 52–53). NATO Strategic Communications Centre of Excellence.

Leahovcenco, A. (2021). "Cybersecurity is a fundamental element of the digital economy," *MEST Journal*, 9(1), 97–105.

Lee, I. (2021). "Cybersecurity: Risk management framework and investment cost analysis," *Business Horizons*, 64(5), 659–671.

Mashiane, T., and E. Kritzinger (2019). "Cybersecurity behaviour: a conceptual taxonomy," *Information Security Theory and Practice: 12th IFIP WG 11.2 International Conference, WISTP 2018, Brussels, Belgium, December 10–11, 2018, Revised Selected Papers 12* (pp. 147–156). Springer International Publishing.

Neigel, A. R., V. L. Claypoole, G. E. Waldfogle, S. Acharya, and G. M. Hancock (2020). "Holistic cyber hygiene education: accounting for the human factors," *Computers & Security*, 92, 101731.

Omidosu, J., and J. Ophoff (2016). "A theory-based review of information security behavior in the organization and home context," *2016 International Conference on Advances in Computing and Communication Engineering (ICACCE)* (pp. 225–231). IEEE.

Sabillon, R., V. Cavaller, and J. Cano (2016). "National cyber security strategies: global trends in cyberspace," *International Journal of Computer Science and Software Engineering*, 5(5), 67.

Štitilis, D., I. Rotomskis, M. Laurinaitis, S. Nadvynychnyy, and N. Khorunzhak (2020). "National cyber security strategies: management, unification, and assessment."

Štitilis, D., P. Pakutinskas, M. Laurinaitis, and I. Malinauskaitė-van de Castel (2017). "A model for the national cyber security strategy. The Lithuanian case," *Journal of Security and Sustainability Issues*, 6, 357–372.

Treaty, N. A. (2020). *Human Systems Integration Approach to Cyber Security.*

Vishwanath, A., L. S. Neo, P. Goh, S. Lee, M. Khader, G. Ong, and J. Chin (2020). "Cyber hygiene: the concept, its measure, and its initial tests," *Decision Support Systems*, 128, 113160.

35. Use of Blockchain Technology in Banking Industry and Evolution of Digital Currency

Imran Qureshi, Mohammed Abdul Habeeb, Burhanuddin Mohammad and
Syed Ghouse Mohiuddin Shadab

Lecturer, University of Technology and Applied Science, IT Department, Al Musanna, Sultanate of Oman, Pin: 314, Muscat, South Al Batinah, Almuladdah Sultanate of Oman

ABSTRACT: This article explores the rising development of blockchain, which upholds Bitcoin and other digital currencies look at what the innovation is and its capability to disturb and change the financial administration industry. The article likewise features the work that the business needs to do to make blockchain applications a standard piece of the financial scene. It focuses on the fact that this is not an innovation that a solitary association can expect to be wonderful to acquire a benefit over rivals.

KEYWORDS: Blockchain, Digital Currency, Finance, Digital Technology

1. INTRODUCTION

Blockchain, by and large, known as the spine improvement behind Bitcoin, is one of the arising movements as of now in the market drawing in a great deal of thought from attempts, new associations and media. As everyone has started to understand the risky potential of this advancement, several use cases are also being examined across projects. Despite the fact that this progress is still in its early stages, financial players are the primary movers to profit from it. According to a prediction of the World Monetary Organisation, banks and regulators worldwide will be ready to explore several paths in relation to different blockchain models in 2017 (Pal *et al.*, 2021). With 2500+ licences recorded over the past few years, 90+ public banks participating in blockchain discussions worldwide, and 80% of banks planning to launch blockchain and distributed ledger technology (DLT) projects by 2017, blockchain development is on track to become the new norm in the world of financial organisations (Garg *et al.*, 2021).

1.1. Objectives and Hypothesis

- To investigate how 'blockchain technology' is used in the financial sector and how it affects different banking procedures.
- To examine the development of digital currency and its effects on the banking industry.
- To evaluate the advantages, difficulties and prospects related to the implementation of blockchain technology in the banking sector.
- To explore possible uses of blockchain technology outside of the realm of cryptocurrencies, such as improving the security, openness and effectiveness of banking transactions.

Hypothesis 1: *The security, transparency and effectiveness of banking transactions are all improved by the implementation of blockchain technology.*

Hypothesis 2: *With unprecedented options for financial inclusion, international trade and decentralised financial institutions, the development of digital currency has the potential to transform the traditional banking industry.*

DOI: 10.1201/9781003532026-35

Hypothesis 3: *The establishment of enabling infrastructure and regulatory frameworks, coupled with successful cooperation between banks, regulators and technology suppliers, will determine the success of digital currency in the banking sector.*

2. LITERATURE REVIEW

Blockchain innovation and computerised currencies have been hotly debated issues in the financial business throughout the course of recent years. The development of advanced currencies has changed conventional instalment and banking frameworks. Blockchain innovation is essential in making secure and straightforward computerised instalment frameworks (Xu, 2022).

Blockchain is at this point a thought that has gotten basic thought in monetary advancement (FinTech). It joins a couple of PC developments, including passed-on data limit, feature point transmission, understanding frameworks and encryption estimations. It has moreover been recognised as an irksome improvement of the Internet time frame (Zachariadis *et al.*, 2019). Regardless, as blockchain is a critical jump forward in data limit and information transmission, it could basically change the ongoing working models of money and economy, which could provoke one more round of mechanical turns of events and present-day change inside the FinTech business. Blockchain innovation can possibly change the banking business by diminishing the requirement for mediators, further developing straightforwardness, and improving security (Wu and Duan, 2019).

As per Deloitte, blockchain innovation can assist banks with diminishing the gamble of misrepresentation, upgrading security and working on administrative consistence (Polyviou *et al.*, 2019). The utilisation of blockchain innovation can likewise work on the effectiveness of the KYC (Know Your Client) process (Zhao and Meng, 2019).

Computerised cash, otherwise called cryptocurrency, is a computerised resource that utilises cryptography to get and check exchanges. The rise of advanced money has changed conventional instalment and banking frameworks. Bitcoin, the primary decentralised advanced money, was made in 2009 (Zhao, and Meng, 2019). Computerised currencies can likewise be utilised for cross-line instalments, and the expense of exchanges can be essentially decreased. Computerised currencies are likewise decentralised and that implies they are not constrained by any focal power. This makes computerised currencies safer and more straightforward (Khalil *et al.*, 2021).

The future of blockchain innovation and advanced money is dubious. Notwithstanding, a few specialists accept that blockchain innovation and computerised cash will change customary banking and instalment frameworks. As per McKinsey, blockchain innovation can make another financial framework that is safer, more proficient and more straightforward. Blockchain innovation can likewise make new plans of action and income streams for banks (Muminova *et al.*, 2020). As indicated by a report by Exploration and Markets, the worldwide computerised cash and the market should create at a CAGR of 32.31% someplace in the reach somewhere in the range of 2021 and 2026. A peaking reception of advanced currencies by organisations and people is supposed to drive the development of the computerised money market (Chen and Wang, 2020).

Blockchain innovation and computerised cash can possibly change customary banking and instalment frameworks. Blockchain innovation can make secure and productive instalment frameworks that can cycle exchanges progressively (Trivedi *et al.*, 2021). The utilisation of blockchain innovation can likewise decrease the expense of exchanges and work on the general productivity of the instalment framework (Zachariadis *et al.*, 2019). On October 18, 2016, the Assistance of Industry and Data Advancement scattered the 'Chinese Blockchain Headway and Application Improvement White Paper' which explores the consistent status of blockchain advancement and proposes suggestions for the future turn of events (Muminova *et al.*, 2020).

The banking business Universally is as of now confronting various tensions, remembering a decay for benefits and an expansion in risk, and has entered another condition of progress and improvement. The unexpected Web finance blast has likewise prompted various difficulties in the customary banking business (Kim *et al.*, 2020).

3. RESEARCH METHODOLOGY

The methodology that will be utilised to gather and analyse data should be specified in the study design. A combination of qualitative and quantitative research techniques, including as interviews, surveys, case studies and data analysis of blockchain transactions, may be used in the study design. The people or instances that will be included as part of the research should be chosen using the sampling approach. The sample might be chosen based on particular standards such bank size, location and kind. To choose people with knowledge or experience in blockchain technology for use in the financial sector, a purposive sampling approach might be utilised. The study design and sample plan should be taken into consideration while choosing the data gathering techniques. Interviews with banking managers and technology specialists, surveys of banking clients and data evaluations of blockchain transactions might all be used as data collecting techniques. The total sample size of this research study was 25. The information gathered should be pertinent to the study issue and include details on the advantages and difficulties of using blockchain technology to the financial services industry.

4. ANALYSIS AND DISCUSSION

A blockchain is an electronic, extremely durable, dispersed record that consecutively keeps trades in close to constant. The essential for each ensuing exchange to be added to the record is the particular agreement of the organisation members (called hubs), subsequently making a persistent instrument of control in regard to control, mistakes and information quality (Zhao and Meng, 2019).

$$\frac{\sum_{i=0}^{32} 210000 * \left[\frac{50*10^8}{2^i}\right]}{10^8} \approx 21000000$$

Blockchain unlike traditional structures is adequately unique to transform into a harbinger of execution in unpredictable market circumstances. In a blockchain, the superior advantage it ensures is that each party has a record which is kept up in a record open to everybody (Karim *et al.*, 2022).

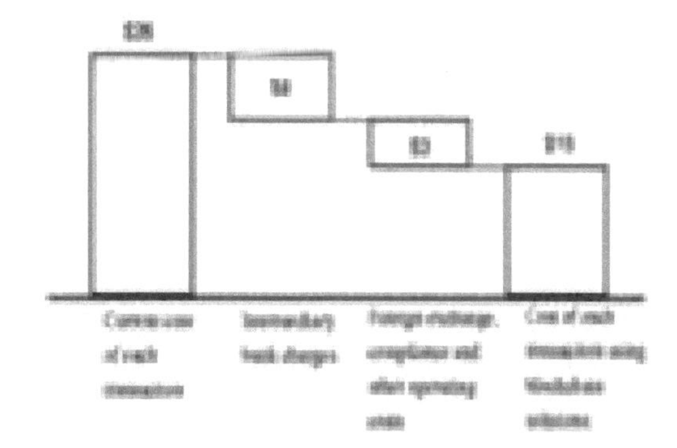

$$E(\sigma_t) = f\ \left(w\left(\sum_{i=1}^{n}\frac{\left(r_i - \Sigma\frac{r_i}{n}\right)^2}{n-1}\right) + (1-w)(\sigma'_{t-1})\right) | \sigma'_{t-1} = \sqrt{\sum_{i=1}^{n}\frac{(r_i - \Sigma\frac{r_i}{n})^2}{n-1})}$$

where 'σ_t is the standard deviation in time t, σ'_{t-1} is the standard deviation in a previous period with low volume, r_i is the price returns, n is the total number of price-time integer observations and w is the percentage factor weighting of σ.'

As portrayed here the blockchain is only a circulated shared decentralised public record which is available to all. It is an information structure which is seen to be powerful and permanent (Han *et al.*, 2019). In the Bitcoin organisation, most of the members are supplanted by significant individuals assigned as the validators in the organisations (Xu, 2022).

$$\frac{\sum_{i=0}^{32} 210000 * \left[\frac{50*10^8}{2^i} \right]}{10^8}$$

The overall financial industry today is going up against issues, for instance, expanding costs of exercises, extending feebleness to underhanded attacks on bound-together servers and troubles in ensuring straightforwardness (Kuzior and Sira, 2022). Banks are constantly researching better ways to deal with performing trades quicker for overhauled client support while ensuring cost efficiency in their exercises and ensuring straightforwardness to clients and regulators (Pashkevych *et al.*, 2020).

Considering the above discussion of what are the continuous pain points of the Financial Business and the benefits of blockchain, a Blockchain Assessment Framework is made to evaluate whether a particular cycle or use case is an ideal decision for a blockchain-based plan (Gao and Li, 2022).

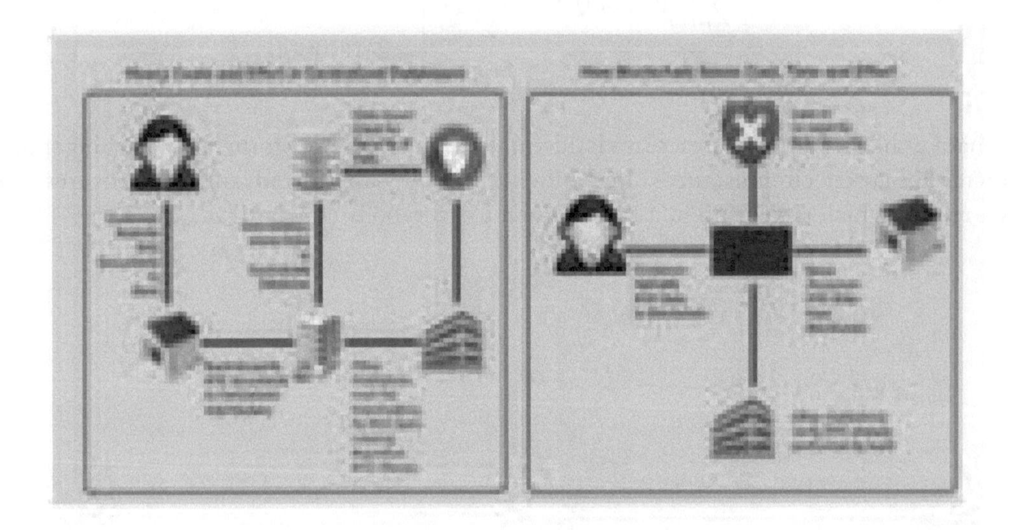

KYC - using Blockchain Technology

Partnerships attempt different huge activities, for example, the improvement of roads, train structures, air terminals, creation lines, new business territories, etc., which require a huge extent of help. The trailblazer bank or people by upset could achieve made by assessment and affirmation of assurance. Different managerial cures concerning the lead of consortium/different banking/accomplice approaches were taken out by the Hold Bank of India in October 1996 with the ultimate objective of introducing versatility in the credit transport system and working with the smooth progression of credit (Wang *et al.*, 2020).

The financial region has been growing actually, progressing and endeavouring to embrace and execute electronic portions to further develop the financial system (Arjun and Suprabha, 2020).

Feature point portion can similarly be executed using blockchain advancement, as such taking out the go-between association of untouchable monetary associations, which will remarkably additionally foster organisation efficiency and diminish the trade costs of banks.

Each individual produces enormous measures of information on the Web, which is incredibly important as evidence of their credit circumstance. By the by, this information is as of now being hoarded by enormous Web organisations. Subsequently, people cannot lay out their own or use this information (Harris and Wonglimpiyarat, 2019). This can additionally ensure that the data are veritable and solid, while likewise lessening the expenses of information assortment by credit organisations.

Banks may utilise DLT to build a shared ledger that is transparent, secure and impenetrable. Ethereum, a decentralised platform that enables the construction of smart contract frameworks and the building of decentralised apps is one of the most well-known blockchain platforms for DLT. In order to build their own private blockchain networks and use smart contracts that automate numerous financial procedures, banks can utilise Ethereum. Self-executing software, or 'smart contracts', automates the conditions of a contract between two parties (Wang *et al.*, 2020).

5. CONCLUSION

But the ability of blockchain is extensively pronounced to be at standard with early business Web, banking firms need to understand the imperative components of the development and how it can address the continuous business issues on one hand, web-engaged exchange of data while on the other, the blockchain can incorporate the exchanging of critical worth. This will allow them to get a handle on the development, assess the bet and enable tailor-made deals with serious consequences regarding their specific hindrances. This research article provides an in-depth analysis, weighs advantages and disadvantages and offers practical insights, going beyond the cursory review of blockchain and digital currency in banking. The study enhances knowledge and understanding in the field by making these original contributions, directing the financial sector and decision-makers toward the efficient application of blockchain technology and the advancement of digital currency.

Blockchains have the potential to disrupt the fundamental evolution of banks' credit and portion clearance systems, modernising and developing them. Blockchain applications also progress the 'multi-concentrate, sadly intermediated' scenario design, which will improve financial business capabilities.

REFERENCES

Arjun, R., and K. R. Suprabha (2020). "Innovation and challenges of blockchain in banking: a scientometric view."

Chen, T., and D. Wang (2020). "Combined application of blockchain technology in fractional calculus model of supply chain financial system," *Chaos, Solitons & Fractals*, 131, 109461.

Gao, H., and G. Li (2022). "Fintech, digital currency, and the restructuring of the global financial system," *China Economic Transition= DangdaiZhongguoJingjiZhuanxingYanjiu*, 5(2), 262–276.

Garg, P., B. Gupta, A. K. Chauhan, U. Sivarajah, S. Gupta, and S. Modgil (2021). "Measuring the perceived benefits of implementing blockchain technology in the banking sector," *Technological Forecasting and Social Change*, 163, 120407.

Han, X., Y. Yuan, and F. Y. Wang (2019). "A blockchain-based framework for central bank digital currency," *2019 IEEE International conference on service operations and logistics, and informatics (SOLI)* (pp. 263–268). IEEE.

Harris, W. L., and J. Wonglimpiyarat (2019). "Blockchain platform and future bank competition," *Foresight*.

Karim, S., M. R. Rabbani, and H. Bawazir (2022). "Applications of blockchain technology in the finance and banking industry beyond digital currencies," *Blockchain Technology and Computational Excellence for Society 5.0* (pp. 216–238). IGI Global.

Khalil, M., K. F. Khawaja, and M. Sarfraz (2021). "The adoption of blockchain technology in the financial sector during the era of fourth industrial revolution: a moderated mediated model," *Quality & Quantity*, 1–18.

Kim, K., G. Lee, and S. Kim (2020). "A study on the application of blockchain technology in the construction industry," *KSCE Journal of Civil Engineering*, 24(9), 2561–2571.

Kuzior, A., and M. Sira (2022). "A bibliometric analysis of blockchain technology research using VOS viewer," *Sustainability*, 14(13), 8206.

Muminova, E., G. Honkeldiyeva, K. Kurpayanidi, S. Akhunova, and S. Hamdamova (2020). "Features of introducing blockchain technology in digital economy developing conditions in Uzbekistan," *E3S Web of Conferences* (Vol. 159, p. 04023). EDP Sciences.

Pal, A., C. K. Tiwari, and A. Behl (2021). "Blockchain technology in financial services: a comprehensive review of the literature," *Journal of Global Operations and Strategic Sourcing*.

Pashkevych, M., L. Bondarenko, A. Makurin, I. Saukh, and O. Toporkova, (2020). "Blockchain technology as an organization of accounting and management in a modern enterprise," *The International Journal of Management*, 11.

Polyviou, A., P. Velanas, and J. Soldatos (2019). "Blockchain technology: financial sector applications beyond cryptocurrencies," *Multidisciplinary Digital Publishing Institute Proceedings*, 28(1), 7.

Trivedi, S., K. Mehta, and R. Sharma (2021). "Systematic literature review on application of blockchain technology in E-finance and financial services," *Journal of Technology Management & Innovation*, 16(3), 89–102.

Wang, Y., D. K. Kim, and D. Jeong (2020). "A survey of the application of blockchain in multiple fields of financial services," *Journal of Information Processing Systems*, 16(4), 935–958.

Wu, B., and T. Duan (2019). "The application of blockchain technology in financial markets," *Journal of Physics: Conference Series* (Vol. 1176, No. 4, p. 042094). IOP Publishing.

Xu, J. (2022). "Developments and implications of central bank digital currency: the case of China e-CNY," *Asian Economic Policy Review*, 17(2), 235–250.

Zachariadis, M., G. Hileman, and S. V. Scott (2019). "Governance and control in distributed ledgers: understanding the challenges facing blockchain technology in financial services," *Information and Organization*, 29(2), 105–117.

Zhao, C., and X. Meng (2019). "Research on innovation and development of blockchain technology in financial field," *2019 International Conference on Pedagogy, Communication and Sociology (ICPCS 2019)* (pp. 421–424). Atlantis Press.

36. A Detailed Discussion about Managing Security Risks Associated with Cloud Computing

Pradeep Kumar Bharadwaj[1] and G. Vinodini Devi[2]

[1]Research Scholar, Department of Law, Koneru Lakshmaiah Education Foundation, Green Fields, Vaddeswaram, Guntur, Andhra Pradesh, 522302

[2]Assistant Professor, Department of Law, Koneru Lakshmaiah Education Foundation, Green Fields, Vaddeswaram, Guntur, Andhra Pradesh

ABSTRACT: In recent years, cloud computing has become ubiquitous, and the safety issues related to the cloud model are also increasing proportionally. The industry's concept of infrastructure, service delivery and development methods has been drastically altered by cloud computing. From straightforward data storage to online banking to data analytics and/or predictive models, commercial uses of cloud-based services are growing. Concerns about important issues like data confidentiality, access control, network security, malicious attacks, denial-of-service attacks and challenges like encryption time and computational burden associated with effective implementation of security models while handling the data in cloud architecture have been raised by this shift to cloud architecture. Therefore, the primary goal of the research project is to evaluate the current security mechanisms across the cloud ecosystem, from the fundamental cloud storage architecture to more advanced services like web application services, in order to give insight into the security system that accomplishes the desired benefits.

KEYWORDS: Cloud computing, security threats and attacks, encryption, network security

1. INTRODUCTION

The most effective innovations that has captured the attention of the technologists all across the world is cloud computing. While there are numerous advantages to cloud computing, including scalability, rapid elasticity, quantifiable offerings and – most importantly – the ability to reduce costs for organisations, there are also security risks that the companies cannot ignore. Despite cloud computing being an effective setting overall, enterprises are hesitant to utilise it because there are no reliable security requirements. The numerous weaknesses present in every type of cloud computing infrastructure are the source of the security issues. For all cloud-based business operations, it is now essential to identify the best approach guidelines to boost safety and confidentiality in the cloud environment. The research study's topic, 'Security Threats and Attacks on Cloud Computing System: An Empirical Study', is not only highly pertinent and current but also presents an intriguing challenge for boosting organisations' confidence and assurance levels by successfully lowering security risks and dangers in this newly developed area of cloud computing. In this article, we investigate and examine well-known data and network security attacks on cloud system. DoS (HDoS, XDoS and DDoS) assaults and man-in-the-middle attacks are more common attacks on cloud networks, according to research already conducted. Additionally, the most frequent and well-known data security threats on the web of cloud networks fall under the categories of SQL injection and Cross Site Scripting (XSS) assaults, which are both malware injection attacks (Sun, 2020).

1.1. Research Gap and Problem Statement

The cloud determines the specific security requirements. To improve cloud environment security and privacy, it is also necessary to identify applicable solution instructions, embrace cloud computing and identify vulnerabilities and threats.

DOI: 10.1201/9781003532026-36

An organisation must conduct thorough planning and identify emerging risks, hazards, weaknesses and potential countermeasures in order to successfully implement cloud computing. A key component of secure cloud computing would continue to be creating trusted applications from untrusted components (Butt *et al.*, 2020).

1.2. Research Objectives

- Objective 1: study cloud assaults and provides methods of mitigation or prevention.
- Objective 2: examine data security attacks on cloud networks and provide methods for prevention and mitigation.

2. LITERATURE REVIEW

2.1. Cloud Deployment

Administrations of cloud computing might be open, closed or hybrid. The corporate server provides private cloud management to the client. The deployment approaches maintain the structure, protection and regular impact on local server farms while offering flexibility and cloud services. Cloud services are delivered online by third-party cloud providers. Public cloud providers are available on demand, typically every hour or continuously, however some administrations offer long-haul obligations. The client uses the CPU utilisation and data transfer capability by paying for the services in accordance with their requirements. The public cloud includes the Google Cloud Platform, Microsoft Azure, IBM and Amazon Web Services (Tabrizchi and Kuchaki Rafsanjani, 2020).

2.2. Cloud Computing Security

Security continues to be the primary worry for organisations since cloud computing, particularly open cloud adoption. As the open cloud environment has multiple tenants, open cloud authority divides their veiled gear establishment among various consumers. This condition's requests for solitude from cogent process resources are overflowing. In the interim, to prevent unauthorised attempts to open shared databases and digital assets, account login information are used (Jannati *et al.*, 2021). The risks to cloud-based data resources, or cloud security threats, can change depending on the cloud delivery patterns used by cloud client associations. Distributed computing is defenceless against some security risks (Kumar *et al.*, 2018). Review of the hazards for cloud users as categorised by the Confidentiality, Integrity, and Availability/Accessibility (CIA) security display, as well as their applicability to all cloud benefit delivery scenarios, are provided in Table 1 (Alouffi *et al.*, 2021).

Table 1

Thread	Confidentiality	Integrity	Availability
Description	• Malicious cloud – Supplier client. • Remote programming assault of cloud applications. • Among many fields, there is a lack of security access rights.	• Security attributes are wrongly defined. • Infrastructure segments or faulty applications are introduced.	• Many cloud users are suffering from other customer saturation testing. • Network DNS DOS. • Disturbance of outsider WAN suppliers' administrations.

2.3. Compared Studies from Related Fields

Previous studies in the literature have predominantly employed one or two machine learning (ML) techniques to address cloud security concerns and threats. In contrast, this article explores the utilisation of multiple ML algorithms to tackle various cloud security issues. The comparison between our review paper and related works is presented in Table 2, where we assess different algorithms to determine the most effective approaches for resolving these issues. Moreover, we employ distinct supervised and unsupervised algorithms for conducting comprehensive problem analysis and address the legal aspects, making our study distinct from previous surveys and articles.

Table 2

Ref.	Area of interest	Techniques	Issues related to security	Cloud computing
Liu et al., (2019)	Security maintained encrypted data	Supervised and unsupervised learning	Limited	Small or Intermediate Problems
Selamat and Ali, (2019)	Debate over trust-based access	Unsupervised learning	No	Several options are available
Shamshirband et al., (2020)	Security issues	Supervised and unsupervised learning	Limited	Small issues
Shukla andMaheshwari, (2019)	Malware security risks and defence	Supervised learning	Yes	Long-term problems
Alsolami (2018)	Threats and security concerns	Supervised learning	Limited	Small or Intermediate Problems

3. RESEARCH METHODOLOGY

We developed the data collecting method using the literature review as a foundation. For this investigation, both primary and secondary data were used.

3.1. Research Design

Our questionnaire is set up with a total of seven items. The questions are intended to assess the security risks associated with the hesitation to use cloud computing. In addition, the examination of replies to two of the seven questions' elements will assess how well-informed different industries are about cloud computing. The following paragraphs provide a detailed study of the statistical analysis of the responses to the seven items with the aid of graphs and SPSS software.

3.2. Primary Data and Secondary Data

To examine the causes of the restricted adoption of cloud computing, we have developed a questionnaire for primary data collecting that is based on practical sampling into three groups, including the IT industry, the educational sector and non-IT individual customers from diverse companies in terms of security threats. Our questionnaire is set up with a total of seven items. The questions are intended to assess the security risks linked with the hesitation to use cloud computing. Additionally, the examination of responses to two questions will assess how well-informed different industries are about cloud computing. The following results are a statistical interpretation of the responses and their analysis using graphs and SPSS software.

Secondary data are gathered from printed or published sources like books, journals, research papers, articles and newspapers.

4. RESULT AND DISCUSSION

4.1. Survey Analysis

Since the primary focus of this survey is to evaluate the variables influencing cloud computing adoption in enterprises. We created a survey questionnaire to find out how different security measures affect cloud computing adoption. Unexpectedly, we discovered that 58% of industrialists select 'cloud computing is a type of outsourcing of IT, 27% – cloud computing is an interesting technology', followed by 'cloud computing is an unknown or unclear subject', 9% and 'cloud computing is something else for me', according to 2% of respondents. Figure 1 illustrates these results.

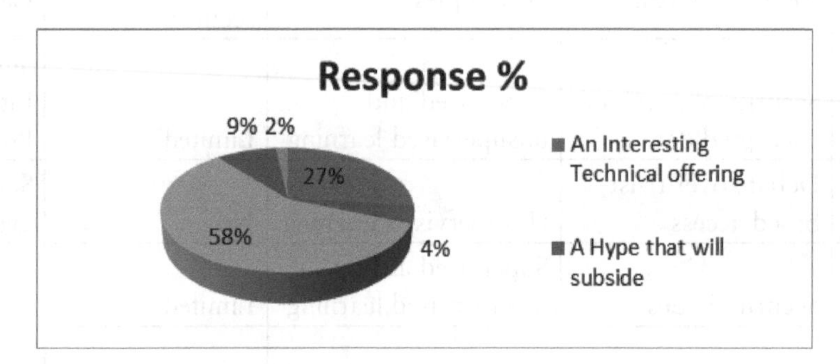

Figure 1. Cloud computing understanding for IT sector. Source: Author's compilation.

When we asked questions to high- and medium-level employees in the education sector, including teachers, non-teaching (directors) and the IT department then we discovered that, as shown in Figure 2, 41% of educators say that 'cloud computing is an unknown or unclear subject', 30% say that 'cloud computing is an interesting technical offering', 29% say that 'cloud computing is a hype that will subside', 13% say that 'cloud computing is a type of outsourcing of IT' and 4% say that 'cloud computing is something else'.

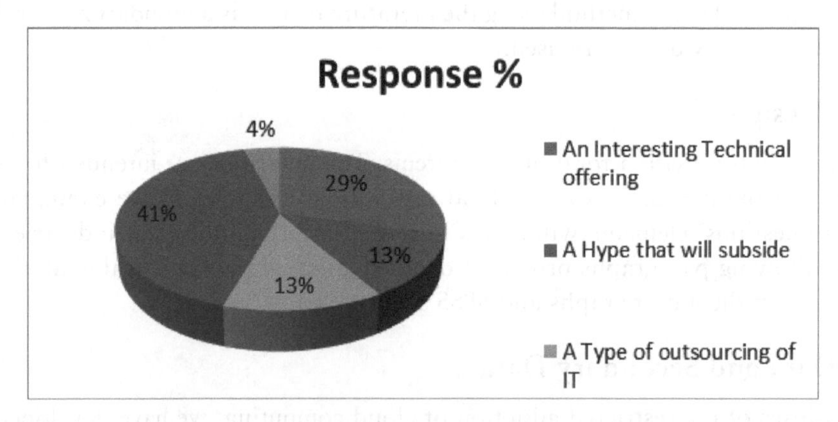

Figure 2. Cloud computing understanding for education sector. Source: Author's compilation.

From the IT department and non-IT department, we posed the query to non-IT industries. As shown in Figure 3, 34% of respondents said that 'cloud computing is an uncharted or unclear subject', 26% said that 'cloud computing is a type of outsourcing of IT', 19% said that 'cloud computing is an interesting technical offering', 13% said that 'cloud computing is a hype that will subside' and 8% said that 'cloud computing is something else'.

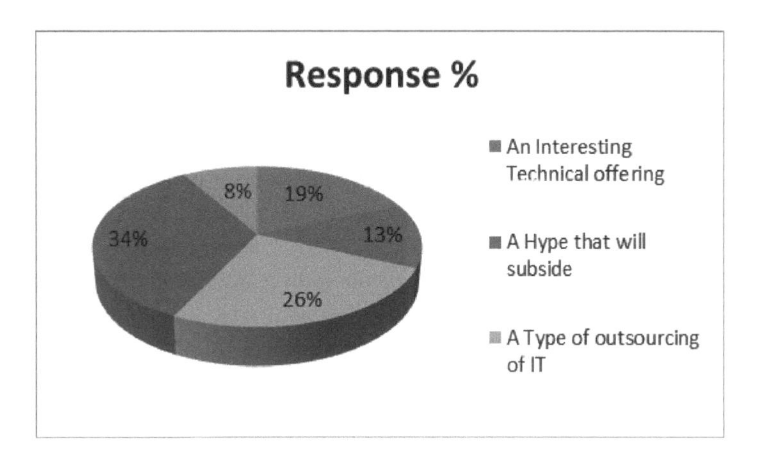

Figure 3. Cloud computing understanding for non-IT sector. Source: Author's compilation.

According to the survey's interpretation, while most IT industry customers have a sufficient degree of cloud computing knowledge, personnel in the education sector and other non-IT businesses are less knowledgeable about the technology. Employees in the education sector and other non-IT industries do not clearly understand the capabilities and principles of cloud computing, which is bad news for both the commercial plans for providers of cloud services in general and the future growth of this powerful technology.

4.2. Correlation Analysis

The symbol for correlation, r, denotes the direction of covariance. The range of r's value is between 1. Positive values of r denote positive correlation between two variables that means there is a change in both the variables which is followed by the statement direction, whereas negative values r denotes the negative correlation, meaning that variations in the 2 variables follow the opposite directions. If the value of r is zero, there is no correlation between the two variables. r = (+) 1 shows the positive correlation and r = (–) 1 indicates the negative correlation. This means that changes in the independent variable (X) account for all changes in the dependent variable (Y) completely. We may also state that a correlation will be referred to as perfect positive if there is an ongoing shift in the dependent variable of same direction for a unit change in the independent variable. However, if the shift is the other way around, the correlation will be referred to be perfect negative. A higher degree of correlation b/w the two variables is indicated by a *r* value that is closer to +1 or –1.

5. CONCLUSION

For all organisations, cloud computing is a game-changing technology. This study focuses on a particular knowledge gap about security threats and assaults on cloud computing networks. After a thorough assessment of the literature, we decided to conduct a survey using a questionnaire in order to meet our research goals. There is a noticeable hesitation on the part of organisations to use cloud computing, apparently due to the lack of trustworthy security guidelines and other considerations. It has been discovered that there is a significant knowledge gap among consumers about cloud computing. Particularly, most sectors in the education sector and non-IT sectors lack knowledge and understanding of cloud computing.

REFERENCES

Alouffi, B., M. Hasnain, A. Alharbi, W. Alosaimi, H. Alyami, and M. Ayaz (2021). "A systematic literature review on cloud computing security: threats and mitigation strategies," *IEEE Access*, 9, 57792–57807.

Alsolami, E. (2018). "Security threats and legal issues related to cloud based solutions," *IJCSNS International Journal of Computer Science and Network Security*, 18(5).

Butt, U. A., M. Mehmood, S. B. H. Shah, R. Amin, M. W. Shaukat, S. M. Raza, D. Y. Suh, and M. J. Piran (2020). "A review of machine learning algorithms for cloud computing security," *Electronics*, 9(9), 1379.

Jannati, H., E. Ardeshir-Larijani, and B. Bahrak (2021). "Privacy in cross-user data deduplication," *Mobile Networks and Applications*, 1–13.

Kumar, P. R., P. H. Raj, and P. Jelciana (2018). "Exploring data security issues and solutions in cloud computing," *Procedia Computer Science*, 125, 691–697.

Liu, Y., Y. L. Sun, J. Ryoo, S. Rizvi, and A. V. Vasilakos (2019). "A survey of security and privacy challenges in cloud computing: solutions and future directions," *Journal of Computing Science and Engineering*, 9(3), 119–133.

Selamat, N., and F. Ali (2019). "Comparison of malware detection techniques using machine learning algorithm," *Indonesian Journal of Electrical Engineering and Computer Science*, 16, 435.

Shamshirband, S., M. Fathi, A. T. Chronopoulos, A. Montieri, F. Palumbo, and A. Pescapè (2020). "Computational intelligence intrusion detection techniques in mobile cloud computing environments: review, taxonomy, and open research issues," *Journal of Information Security and Applications*, 55, 102582.

Shukla, S., and H. Maheshwari (2019). "Discerning the threats in cloud computing security," *Journal of Computational and Theoretical Nanoscience*, 16(10), 4255–4261.

Sun, P. (2020). "Security and privacy protection in cloud computing: discussions and challenges," *Journal of Network and Computer Applications*, 160, 102642.

Tabrizchi, H., and M. Kuchaki Rafsanjani (2020). "A survey on security challenges in cloud computing: issues, threats, and solutions," *The Journal of Supercomputing*, 76(12), 9493–9532.

37. Applications of the Internet of Things for Cyber Security and Data Privacy and Managing Future Trends

K. Seshadri Ramana[1], R. Deepthi Crestose Rebekah[2], Kommireddy Srividya[3], Kapu Hemalatha[3], K. Divya Nethri[3], and B. V. Baby Sweshika Reddy[3]

[1]Professor & HoD, Department Computer Science and Engineering, Ravindra College of Engineering for Women, Kurnool

[2]Assistant Professor, Department of CSE, Ravindra College of Engineering for Women, Kurnool

[3]B.Tech IV Year Student, Department of CSE, Ravindra College of Engineering for Women, Kurnool

ABSTRACT: This study describes the current state of the Internet of things (IoT) and how it is being used to address cyber security and data privacy issues, as well as to manage future trends. It highlights the need for a comprehensive approach to security and privacy to ensure that the technology is used responsibly and securely. It also looks at the potential benefits that the IoT can bring, from improved efficiency and cost savings to enhanced customer experiences and improved public safety. Finally, it examines the challenges that need to be addressed for the secure and privacy-preserving use of IoT technology, including the need for effective governance, the need for secure authentication and authorisation protocols and the need for new technological developments. This abstract provides a snapshot of the current state of the IoT and how it can be used to improve cyber security, data privacy and the management of future trends.

KEYWORDS: Internet of things, cyber security, data privacy

1. INTRODUCTION

Due to the cutting-edge innovation of the Internet of things, cities have grown increasingly involved in enhancing the efficiency of emergency services' readiness, improve accuracy, reduce costs, developing infrastructure and much more (IoT). The usage of IoT systems will lead to the emergence of more smart cities in the future years (Arasteh *et al.*, 2016). Cities are moving towards IoT technology primarily because it provides wireless connectivity. Expense is the primary factor in the choice to switch from wired to wireless systems (Qian *et al.*, 2019). Landlines are costly to establish and maintain, and cellular data subscriptions are becoming more affordable (Al-Turjman, 2018). As a result, new scenarios are now possible that were previously cost-prohibitive owing to wireless technology. Reliability is a crucial additional factor (Cvar *et al.*, 2020). To maintain the communications network, service professionals must manually travel to the installation site, which is time-consuming and expensive.

IoT implementations can be remotely managed and watched over thanks to wireless connections. Additionally, it enables administrators to apply firmware and security patch upgrades during the installation. It can send out automated alerts for all the problems and consume less energy overall (Muhammed *et al.*, 2019). Additionally, IoT systems ensure to use of sensors to collect data and wireless devices to manage how a service is used (Ahmed and Rani, 2018). IoT application development is a rapidly expanding business. A group of industries, such as emergency services, sewage treatment, city lighting and transit and others, can be grouped together as smart cities. Municipalities and smart cities that use wireless technology for lighting in order to conserve resources and money are common examples of IoT applications.

1.1. Objectives

DOI: 10.1201/9781003532026-37

The goal of this investigation is to perform a comprehensive examination of IoT utilisation in smart cities. The goal of this research is to explore the applications of the IoT in the context of cyber security and data privacy and investigate its potential for managing future trends.

1.2. Hypothesis

- Implementing IoT technologies enhances cyber security and data privacy measures.
- Utilising IoT for managing future trends in cyber security and data confidentiality leads to more efficient and effective protection strategies.

2. LITERATURE REVIEW

Talari *et al.* (2017) Owing to the advancement of smart metres, the Internet of things, every smart city today contains various electronic devices. As a result, technologies and techniques make us smarter and hasten the availability of various smart city characteristics. The goal of the most current study is to offer a description of the smart city idea, along with its various uses and benefits. The bulk of the prospective IoT techniques are also discussed, along with how they could be used for various smart city characteristics. How upcoming smart cities could be used to improve technologies is another brilliant idea covered in this essay. Various real-life instances from throughout the world are provided while this is going on, along with the primary barriers to its acceptance.

Alsamhi *et al.* (2019) Intelligent objects can routinely and cooperatively improve the standard of existence, preserve human life and serve as a responsible resource environment in smart cities. AI, IoT, drones and robotics are needed to implement these cutting-edge collaborative innovations in order to boost the intellect of smart cities by enhancing connection, power efficiency and service quality (QoS).

Tragos *et al.* (2014) In recent years, smart cities have been viewed as a viable way to leverage information and communication technologies to deliver effective services to inhabitants. With the most recent developments in the IoT, a novel period in the field of smart cities has been created, creating novel ideas for the development of effective and affordable uses intended to enhance urban quality of life. Even though a lot of studies in these fields and numerous commercial brands have been developed as a result, critical factors like accuracy, security and privacy have not previously been seen to be highly relevant.

Kshetri (2017) This article explores the potential of blockchain technology in enhancing IoT security and privacy. It discusses how blockchain can ensure secure data exchange, decentralised identity management and transparent auditing in IoT systems.

Roman *et al.* (2013) This research article focuses on security and privacy challenges in distributed IoT systems. It examines authentication, access control, and data privacy mechanisms and proposes a secure framework for IoT devices based on these considerations.

3. RESEARCH METHODOLOGY

The term 'quantitative approach' refers to the gathering of numbers from diverse demographic categories within the study goal. In contrast to qualitative data, it is therefore more impartial and rational. Typically, structured questionnaires are used in the quantitative approach.

The sample size of 500 is a typical choice for surveys and studies involving professionals in a specific field. The sampling technique involved in this method is stratified random sampling. The target population for this study are professionals working in the field of cybersecurity and data privacy, specifically those involved in managing Internet of things (IoT) devices and systems, and individuals responsible for identifying and addressing future trends in this domain.

Using a quantitative data collection method using a questionnaire and a quantitative analysis approach using the SPSS software package, based on the study purpose, and the characteristic of the descriptive study.

A structured questionnaire is used to collect data, enabling researchers to reach a broad population fast and effectively using the Bristol online survey tool. The questionnaire can also be simply coded and exported to SPSS for statistical analysis.

This research utilises a statistical sampling method while considering the feasibility, research goal and issue, and the population characteristics that are needed to evaluate statistically. The study question's nature essentially consists of opinion variables, which track how participants to surveys feel about certain topics or what they consider to be true or false. The four algorithms SAAPAES, DES, BLOWFISH and AES were contrasted for finding the best one.

4. RESULT AND DISCUSSION

Figure 1 compares the suggested algorithm's security level, in which BLOWFISH is approximately 89%, to that of the other techniques, which have security levels of 81%, 74% and 67%.

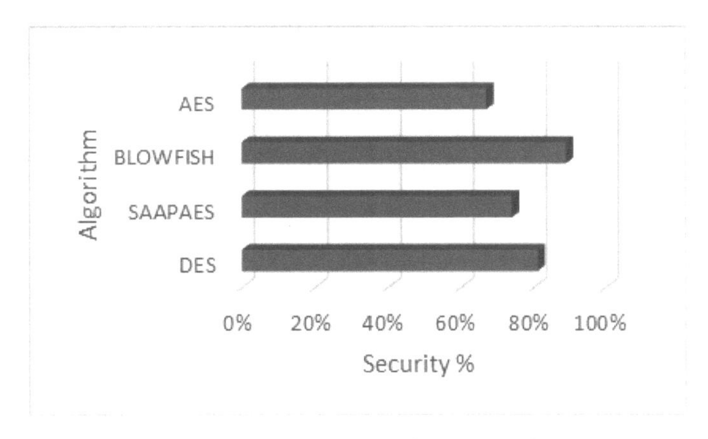

Figure 1. *Security. Source: Author's compilation.*

4.1. Cronbach's Alpha Test of Privacy and Security

To assess the privacy choices of people, this research use Cronbach's Alpha assessment to evaluate the consistency instead of testing reliability (see Tables 1 and 2)

Table 1: Cronbach's Alpha test.

	N	%
Valid	76	71.02
Excluded[a]	31	28.97
Total	107	99.99

Source: Author's compilation.

4.2. Consistency of Statistics

Table 2: Consistency of statistics.

Cronbach's alpha	Cronbach's alpha relies on standardised items	N of items
.518	.518	4

Source: Author's compilation.

5. CONCLUSION

The Internet of things is a powerful tool for managing cyber security, data privacy, and future trends. By leveraging the power of the IoT, companies can gain visibility into their network traffic, monitor user activity, detect threats and protect their data. Additionally, using predictive analytics, companies can use the data collected by the IoT to make informed decisions about their future. With the right strategy and implementation, businesses can use the Internet of Things to effectively secure their data and protect their privacy. One major contribution of this paper is the identification of key IoT applications for cyber security. These include real-time threat detection and response, secure authentication and access control and data encryption and integrity. By leveraging IoT technologies, organisations can better protect their systems from cyber-attacks and mitigate potential risks.

6. CHALLENGES

The main challenge of IoT for cyber security and data privacy is to ensure that all devices connected to the network are secure and not vulnerable to cyber-attacks. This requires the implementation of industry-standard security protocols and encryption techniques that can protect user data from unauthorised access. Additionally, the challenge of managing future trends in the IoT field requires a comprehensive understanding of the technology and its implications for cyber security and data privacy.

REFERENCES

Ahmed, S. H., and S. Rani (2018). "A hybrid approach, smart street use case and future aspects for Internet of Things in smart cities," *Future Generation Computer Systems*, 79, 941–951.

Alsamhi, S. H., O. Ma, M. S. Ansari, and F. A. Almalki (2019). "Survey on collaborative smart drones and internet of things for improving smartness of smart cities," *IEEE Access*, 7, 128125–128152.

Al-Turjman, F. (2018). *Intelligence in IoT-enabled Smart Cities*. CRC Press.

Arasteh, H., V. Hosseinnezhad, V. Loia, A. Tommasetti, O. Troisi, M. Shafie-Khah, and P. Siano (2016). "IoT-based smart cities: a survey," *2016 IEEE 16th International Conference on Environment and Electrical Engineering (EEEIC)* (pp. 1–6). IEEE.

Cvar, N., J. Trilar, A. Kos, M. Volk, and E. Stojmenova Duh (2020). "The use of IoT technology in smart cities and smart villages: similarities, differences, and future prospects," *Sensors*, 20(14), 3897.

Kshetri, N. (2017). "Can blockchain strengthen the internet of things?" *IT Professional*, 19(4), 68–72.

Muhammed, T., R. Mehmood, and A. Albeshri (2017). "Enabling reliable and resilient IoT based smart city applications," *International conference on smart cities, infrastructure, technologies and applications* (pp. 169–184). Springer, Cham.

Qian, L. P., Y. Wu, B. Ji, L. Huang, and D. H. Tsang (2019). "Hybrid IoT: integration of hierarchical multiple access and computation offloading for IoT-based smart cities," *IEEE Network*, 33(2), 6–13.

Roman, R., J. Zhou, and J. Lopez (2013). "On the features and challenges of security and privacy in distributed internet of things," *Computer Networks*, 57(10), 2266–2279.

Talari, S., M. Shafie-Khah, P. Siano, V. Loia, A. Tommasetti, and J. P. Catalão (2017). "A review of smart cities based on the internet of things concept," *Energies*, 10(4), 421.

Tragos, E. Z., V. Angelakis, A. Fragkiadakis, D. Gundlegard, C. S. Nechifor, G. Oikonomou, H. C. Pöhls, and A. Gavras (2014). "Enabling reliable and secure IoT-based smart city applications," *2014 IEEE International Conference on Pervasive Computing and Communication Workshops (PERCOM WORKSHOPS)* (pp. 111–116). IEEE.

38. An Analysis of Risks Associated with Information Security Management in Computer Networks

Pradeep Kumar Bharadwaj[1] and G. Vinodini Devi[2]

[1]Research Scholar, Department of Law, Koneru Lakshmaiah Education Foundation, Green Fields, Vaddeswaram, Guntur, Andhra Pradesh, 522302

[2]Assistant Professor, Department of Law, Koneru Lakshmaiah Education Foundation, Green Fields, Vaddeswaram, Guntur, Andhra Pradesh

ABSTRACT: Organisations have traditionally been extremely concerned with managing information security (IS) in computer networks. The risk involved with IS management has increased dramatically with the intricacy of technologies. This essay tries to evaluate the dangers connected to computer network IS management. It gives a general outline of the numerous dangers and weaknesses that organisations may run into while addressing IS. The interviewing methods used in this research are both quantitative and qualitative. A request for participation in the research was addressed through email to 200 CISSPs to start the data collection process. The article offers a thorough analysis of both the hazards connected to IS management in networks of computers and the countermeasures. To properly safeguard their data and systems, organisations must have a thorough awareness of the dangers involved with IS management.

KEYWORDS: Information security management, risk factors

1. INTRODUCTION

The virus issue has long baffled the industrial control system. It is a difficult challenge. According to Akbanov *et al.* (2019), the frequency of industrial control security instances has increased steadily over the past three years, from 290 in 2016 to 305 in 2017 to 320 in 2018, which is more than eight times the frequency of the Iran earthquake network incident in 2010. The most significant security events between 2008 and 2017 include apt attacks, worms, Trojan horses, malware, blackmail viruses and others. The industrial control network will almost certainly become the next attack target if no action is taken against it. Unlike conventional IT, it will harm people and bring down systems (Farn *et al.*, 2004).

Although the term 'network security' most often relates to computer network security, it can also refer to computer communication network security. Computer communication networking is a collection of interlinked systems that use communication instruments and transfer media to support data exchange and transfer between systems (Xiong *et al.*, 2019).

The administration of IS is a crucial part of any computer network system. Since the beginning of computing, organisations and IT professionals have given careful thought to the possibility of unauthorised access to critical data. The potential dangers related to IS management have expanded along with the complexity of computer networks and the number of data stored on them (Singh *et al.*, 2013).

In addition, malicious actors can use a network to launch attacks on other systems, either within an organisation or outside of it. Furthermore, the proliferation of Internet-connected devices, such as smartphones and tablets, has increased the surface area for potential attacks. The risks associated with information security management can be split into 2 broad classes: exterior threats and interior threats (Rutledge and Hoffman, 1986). Exterior threats include attacks from malicious actors, who may be motivated by financial gain, espionage or simply amusement.

DOI: 10.1201/9781003532026-38

Passive threats involve attempts to gain unauthorised access to a system, such as through the exploitation of a vulnerability or the use of social engineering techniques, while active threats comprise the application of malicious code, like Trojan horses, viruses, and worms, to disrupt or control a system's operations (Zaini *et al.*, 2020). Organisations must take appropriate steps to guarantee the safety of their networks and data.

In conclusion, information security administration is an important part of any computer network infrastructure. This article has analysed the risks associated with information security administration in computer systems, exploring the potential threats posed by malicious actors, as well as the steps that organisations can take to protect their networks and data (Nyanchama, 2005).

1.1. Objective

This analysis aims to determine and evaluate the risks linked with information security administration in computer networks and to provide a comprehensive understanding of the potential threats and vulnerabilities that organisations face in maintaining the confidentiality, integrity and availability of their data and systems. Through this analysis, we aim to develop insights into the key risk areas, assess their potential impact and propose strategies and measures to mitigate and manage these risks effectively.

1.2. Hypothesis

- The increasing complexity and interconnectedness of computer networks pose a higher risk of information security breaches.
- Inadequate implementation and adherence to information security management practices increase the vulnerability of computer networks to various security risks.

2. LITERATURE REVIEW

To guarantee the dependable function of information systems and the amenities they offer, this article examines the issues related to IS in corporate networks (Lavrov *et al.*, 2021). IS is described as the security of data resources, information systems and accompanying architecture from unintentional or intentional external factors of a natural or artificial type that could pose a risk to the information resources' owners or users. IS concerns are presently undergoing significant developments.

IS is now a global concern (Von Solms, 1999). The IS of one organisation undoubtedly influences its business partners in this era of Internet commerce. Business partners must now demand an acceptable standard of IS from one another as a result of this. Standards for IS management should unquestionably be crucial in this regard.

In today's industry, IS is a crucial concern (Dey, 2007). Only a set of hardware and software can no longer handle IS. Instead, an entire end-to-end system is needed. Information Security Management System (ISMS) is the name of such a system. Developing and executing such a system within the organisation calls for special attention, engagement and full commitment from all levels of personnel. Government compliance laws and ISO security standards direct and compel organisations to adhere to norms and demands. By assembling all the components according to their business requirements, organisations must create an ISMS.

Technology has ingrained itself into every aspect of our life in the modern day (Damodharan and Kurmi, 2023). To interact, purchase and obtain information, we utilise the Internet. Technology usage has increased, and security precautions are now of utmost importance. Network security management is the procedure of preventing unauthorised access, misuse and destruction of systems and the statistics they comprise. This research proposes a network security administration paradigm for high-speed computer networks and analyses its effectiveness.

This study identified various network security risks, including unauthorised access, malware attacks and data breaches (Smith, 2018). It highlighted the importance of implementing effective security measures to mitigate these risks.

This research examined the risks posed by insider threats to information security (Johnson, 2019). It discussed the various kinds of internal threats and their potential impact on organisational networks. Strategies for detecting and preventing insider attacks were also discussed.

3. RESEARCH METHODOLOGY

The interviewing methods used in this research study are both quantitative and qualitative. A request for participation in the study was addressed through email to 200 CISSPs to start the data collection process. Of the 200 people, 150 who were contacted responded to the semi-structured poll, yielding a 12.58% response percentage. These researches, distinguished by Likert-scale queries, gather data in the form of numbers, but this does not make them quantitative.

The development of a survey instrument to examine the convoluted ISRA procedure marked the start of the first phase of this methodology. The lead researcher developed an apparatus to do this. Four Certified Information Systems Security Professionals (CISSPs) and two prominent university researchers made up the next expert group that was contacted. Over two months, recommendations were made, adjustments were made and feedback was provided in phases. The instrument was declared ready for data gathering after this repeated improvement process.The sample was noteworthy for several factors.

4. RESULT AND DISCUSSION

Table 1 demonstrates how many organisations incorporate some or all the risk categories when formulating their individual IS plans. These responses were more industry-specific when asked about the other category. Respondents expressed worry about downstream liability in numerous businesses as well as breaches of patient anonymity in health care and regulatory standards for the financial services sector.

Table 1: Risk factors by percentages.

Which aspects does your organisation prioritise the most when creating risk factors for its risk analysis?	YES	NO
Legal, regulatory, or statutory standards	45.21%	28.03%
Loss of customer confidence	61.37%	12.76%
Image destruction to the organisation	81.09%	6.78%
Dangers to infrastructure	31.56%	2.32%
Financial losses	27.12%	17.3%

Source: Author's compilation

It has historically been challenging to estimate the financial return on investment for IS remedies. FUD, or dread, uncertainty, and doubt, is a frequent tactic used to sell investments utilising anecdotes from actual worst-case situations. Another way is to calculate the IS return on investment (ROI) relying on the price of countermeasures. Utilising indirect calculations of the potential costs related to security breaches is another technique. Using a conventional risk architecture is a more conventional strategy. In this study, participants were merely asked if their company used any approach to determine the return on investment for IS investments (Table 2).

Table 2: Insurance and ROI for IS.

Does your business track the ROI of its IS expenditures and investments?	**Response Percent**
Yes	27.89%
No	44.21%
Does your company have insurance to protect its data assets?	**Response Percentage**
Yes	67.34%
No	34.12%

Source: Author's compilation.

In all, 25% of respondents who were asked how often their departments and organisations used the ISRA process responded never or rarely. It is alarming that these many organisations are carrying out their ISRA process so haphazardly and seldom. For their department and organisation, nearly half (40.6%) opted for quarterly or annually. The remaining individuals (34.4%) selected continuously or weekly/monthly for the rate with which their ISRA procedures. Individuals from this category responded when queried regarding the regularity of the procedure at their firms, noting that it is a continuous procedure with committees that satisfy frequently all year round (Table 3).

Table 3: ISRA Frequency.

Frequency	Percentage
Never or rarely for their organisation	13%
Annually or quarterly for their department or organisation	57%
Weekly/monthly or continuously for the organisation	24.51%

Source: Author's compilation.

5. CONCLUSION

Information security administration in computer systems is an important process for preventing unauthorised access to confidential data and protecting the integrity of a network system. Risks associated with information security management in computer networks can be split into 2 classless: exterior and interior threats. Exterior threats include malicious attacks from hackers, viruses and malware. Internal threats include negligence of users, weak passwords and system vulnerabilities. Businesses ought to inspect their networking on frequently for any unusual activities and take appropriate measures to mitigate any potential risk. By following these excellent practices, firms can protect their networking from external and internal threats and ensure that their data is secure.

REFERENCES

Akbanov, M., V. G. Vassilakis, and M. D. Logothetis (2019). "Ransomware detection and mitigation using software-defined networking: the case of WannaCry," *Computers & Electrical Engineering*, 76, 111–121.

Damodharan, D., and R. S. Kurmi (2023). "The performance analysis of network security management model in high-speed computer networks," *2023 2nd International Conference for Innovation in Technology (INOCON)* (pp. 1–6). IEEE.

Dey, M. (2007). "Information security management practical approach," *AFRICON 2007* (pp. 1–6). IEEE.

Farn, K. J., S. K. Lin, and A. R. W. Fung (2004). "A study on information security management system evaluation—assets, threat, and vulnerability," *Computer Standards & Interfaces*, 26(6), 501–513.

Johnson, A. (2019). "Assessing the impact of insider threats on information security," *International Journal of Cybersecurity Research*, 15(4), 123–140.

Lavrov, E. A., A. L. Zolkin, T. G. Aygumov, M. S. Chistyakov, and I. V. Akhmetov (2021). "Analysis of information security issues in corporate computer networks," *IOP Conference Series: Materials Science and Engineering* (Vol. 1047, No. 1, p. 012117). IOP Publishing.

Nyanchama, M. (2005). "Enterprise vulnerability management and its role in information security management," *Information Security Journal: A Global Perspective*, 14(3), 29–56.

Rutledge, L. S., and L. J. Hoffman (1986). "A survey of issues in computer network security," *Computers & Security*, 5(4), 296–308.

Singh, A. N., A. Picot, J. Kranz, M. P. Gupta, and A. Ojha (2013). "Information security management (ISM) practices: lessons from select cases from India and Germany," *Global Journal of Flexible Systems Management*, 14, 225–239.

Smith, J. (2018). "A comprehensive study of network security risks," *Journal of Information Security*, 20(2), 45–62.

Von Solms, R. (1999). "Information security management: why standards are important," *Information Management & Computer Security*, 7(1), 50–58.

Xiong, Z., G. Hao, H. Xiaoyun, L. Zhou-Bin, S. Xue-Jie, and C. Hong-Song (2019). "Research on security risk assessment methods for state grid edge computing information systems," *Computer Science*, 46, 428–432.

Zaini, M. K., M. N. Masrek, and M. K. J. Abdullah Sani (2020). "The impact of information security management practices on organizational agility," *Information & Computer Security*, 28(5), 681–700.

39. The Critical Impacts of Technology on Various Organisational Structures: A Study Focusing on Current Digital Practices in This Digital Age

Nukalapati Naresh Kumar[1] and Sailaja Petikam[2]

[1]Research Scholar, Department of Law, Koneru Lakshmaiah Education Foundation, Green Fields, Vaddeswaram, Guntur, Andhra Pradesh, 522302

[2]Associate Professor, Department of Law, Koneru Lakshmaiah Education Foundation, Green Fields, Vaddeswaram, Guntur, Andhra Pradesh, 522302

ABSTRACT: The development of subsequent business model and digital transformation have completely upset various business markets and significantly modified customer expectations and behaviour. Utilising a corpus of late literature, we distinguish three phases of the digital transformation: digital transformation, digitalisation and digitisation. We explore the tools and knowledge required to successfully manage the digital transformation as well as build and describe growth strategies for digital enterprises. We contend that the calibrated performance measurements are influenced by the organisational frameworks required for digital transformation. In our conclusion, we offer a research agenda to support and guide additional investigation into the digital transformation.

KEYWORDS: Digital ages, digital transformation, organisational structure

1. INTRODUCTION

New digital technologies, for example, long range interpersonal communication, mobile data analytics and cloud computing, represent a danger to lay out organisation structures. The ability of an organisation to strategise will determine whether it succeeds or fails in this new, intensely competitive climate. What's going on has been marked as the fourth industrial revolution, which is presently moving and is based on combination of technologies fusing biological, physical and digital environment. Robots, nanotechnology, 3D printing, cloud computing, artificial intelligence and machine learning are examples of separate technologies that have come together and built upon one another. These changes lead to the development of digital ecosystems where traditional businesses and those born in the digital age cooperate and compete to produce and capture value. This presents both academics and practitioners with brand-new strategic management issues. Additionally, digital technologies, in accordance with Rossi *et al.*, (2020), enable businesses to modify how they internally organise and structure their activities in order to attain the fluidity and regularity required in digital settings. All participants are impacted by these new emergent standards, including huge 'born digitals' like Google and Amazon as well as smaller start-ups. Due to the vastly different set of structures and practises that they acquired from the early twentieth-century industrial age; it can be extremely difficult for current enterprises to embrace these new working practises. As per various reviews, these organisations' digital transformation, or a shift from a bunch of standards and norms acquired from that previous time to ones that are appropriate for the contemporary digital environment is the main thing on their essential plan (Autio *et al.*, 2021).

To more readily grasp the peculiarity of digital transformation and to invigorate future examination by laying out essential objectives and an exploration plan, we look to consider the peculiarity and the writing from various disciplines in this article. The accompanying three objectives are our own: first, to decide the external reasons that have made the requirement for digital transformation more earnest. As

DOI: 10.1201/9781003532026-39

far as (1) the digital resources that should be accessible, (2) the organisational structure that should be set up, (3) the development targets and (4) the metrics that should be utilised, the essential goals that come from the digital transformation will next be inspected. An examination plan for future (inter)disciplinary concentrates on digital transformation is the third objective.

In our discussion, we identify the motivators, stages and requirements of digital transformation using a standard flow model (see Figure 1).

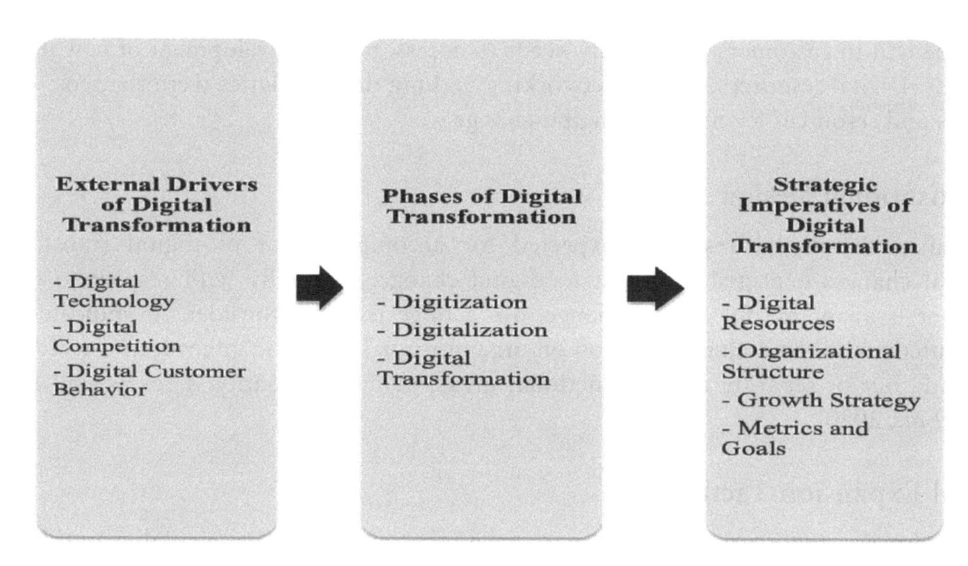

Figure 1. *Digital transformation. Source: Author's compilation.*

1.1. Objective of Study

RO 1: To evaluate the factors that has increased the urgency of the need for digital transformation.

RO 2: The goal is to develop a plan for future (inter) disciplinary studies that focus on digital transformation.

2. LITERATURE REVIEW

Digital technology advancements are opening up new opportunities and challenges for strategy in a variety of ways (Hanelt *et al.*, 2020). They frequently have an impact on every aspect of the organisation and even extend outside of its walls, affecting ecosystems, supply chains, sales channels and business processes. First and foremost, digital technologies alter the content of corporate strategy, presenting new challenges and options to strategists. By empowering opponents to take part in new sorts of competition, for example, stage and environment-based plans of action, these advances have changed the collaboration elements and competition inside and between businesses (Kretschmer *et al.*, 2020). Rising industry convergence is redefining how firms collaborate and compete with one another as they participate in atypical organisations from previously unrelated industries in digital ecosystems.

Second, numerous established companies are being compelled to shift as a result of these new competitive dynamics and the issues they pose (Sony and Naik, 2020). For academics, this presents an environment that is ripe for the study of strategic change. Due to the pace, size and scope of these most recent changes as well as the high amount of uncertainty in digital ecosystems, it is impossible for us to solely rely on prior research (Khanagha *et al.*, 2018).

Third, as a result of the influence of new technology, organisational change and the process of developing strategies have changed. New technologies, such big data, offer fresh approaches to managing issues (George *et al.*, 2016). Open and comprehensive ways to deal with system are turning out to be more well-known while managing the new social media technologies and interconnected environments. The traditional job of a strategist is put to the test by digital technology, which transforms them from strategists into process coordinators.

3. RESEARCH METHODOLOGY

3.1. Digital Resources

According to Barney, resources were the assets and capabilities that a company owns and controls. Resources are the asset blessings in intellectual and physical assets, while abilities are habitually seen as in the company's human, educational or organisational capital and connection resources to support the successful deployment. At the point when an organisation needs to re-examine how it creates and offers some benefit to customers, it normally needs to access, buy or development of new digital assets as well as talents. Digital resources, agility, networking and big data analytics were the most crucial digital competencies and resources for adopting digital change.

3.2. Organisational Design

Notwithstanding the digital resources expected for accomplishment of digital transformation, the organisational changes expected to adjust to digital change, especially with respect to organisational structures that is adaptable for digital change, are a huge issue to consider. As indicated by a recent report, the outcomes of the digital revolution on organisational structure might lean toward an adaptable structure made out of separate digital functional areas, agile organisational forms and business units (Bonanomi *et al.*, 2020).

3.3. Digital Expansion Tactics

For digital enterprises, there were many alternative digital growth strategies available, but the most widely employed one made use of digital platforms. Table 1 highlights the variety of growth strategies utilised at various stages of the digital transformation and how platform strategies become more and more common in the latter, broader stages. Their remarkable growth rates were an almost universal trait of digital businesses, and particularly digital platforms.

Their remarkable growth rates are an almost universal trait of digital businesses in general and digital platforms in particular. As Google, which saw a surge in look from 1 billion every year in 1999 to 2 trillion every year in 2016 (Digital, 2017). From 2.7 million rides in 2013 to 162.6 million outings in 2016, Lyft experienced yearly development of practically 300%. Similar to this, between 2009 and 2017 there were around 25% more active Facebook users (Statista, 2018).

Table 1: Strategic objectives in relation to digital transformation phases.

Type	Example	Digital Resources	Organisational Structure	Digital Growth
Digitisation	Automated processes and duties; transformation of analogue information into digital	Digital assets	Standard top-down hierarchy	Market penetration, market expansion and product creation
Digitalisation	The use of digital components in goods or services; introduction of digital communication and distribution methods: the employment of robots in production.	Networking ability and Digital agility + Above	Separate agile units	Market penetration via platforms, platforms for co-creation + above

Type	Example	Digital Resources	Organisational Structure	Digital Growth
Digital transformation	New business models are being introduced, such as 'product-as-a-service', 'digital platforms' and 'pure data driven business models'	Big data analytics + Above	Separate units with adaptable organisational structures, internalisation of the functional domains of IT and analysis	Platform diversification + Above

Source: Author's compilation.

3.4. Goals and Metrics

To completely understand the capability of the digital transformation, digital firms should screen improvements in performance on Key Performance Indicators (KPIs) to drive learning and adjust the business model. At certain stages of the digital transformation process, KPIs may be more or less significant or useful.

4. RESULT AND DISCUSSION

To take advantage of digital opportunities, new digital routines can be created, and new organisational structures can be implemented to set up and integrate digital processes. These adjustments might be accomplished by reevaluating cognitive management methods and developing fresh digital business models. We are largely uninformed of how digital changes truly occur, despite the importance of ensuring that these three key elements are tightly matched. Varied industry sectors have experienced the digital transformation process at varying rates and to varied degrees. Academic writing frequently focuses on industries where new competitors have experienced the greatest success and where long-standing, firmly established businesses have experienced the greatest difficulty, such as technology, media, telecom, or retail (Hastig and Sodhi, 2020). Digital technology' effects on company strategies have been less disruptive in many other sectors, including consumer goods, engineering and energy. For our experiment, we conduct a survey of 450 consumers and get their responses through questionnaires. How digital transformation is driving the consumer experience is given in Table 2 and Figure 2.

Table 2: Outcome in adoption of digital model.

Outcomes	Responses (%)
Improves operational efficiency	40
Meet the changing customer expectation	35
Improves new product quality	26
Increases design reuse	25
Reduction of cost in product development	24
Introduce new revenue streams	21
Reduction of cost of poor quality	14
Increase first pass yield	5

Source: Author's compilation.

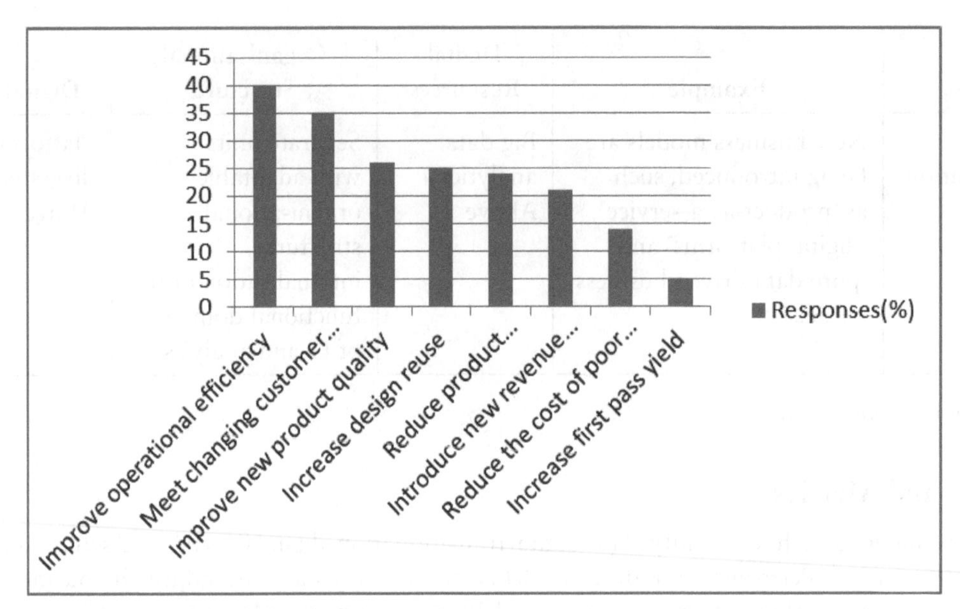

Figure 2. *Graphical representation of digital model. Source: Author's compilation.*

However, reforms at this time seemed unimportant to outsiders due to the steadiness of its economic structure. For better understanding of various digital transformation routes that a company may take, based on its specific situation and the decisions made by its senior executives, conceptualise them. The sort of change, as indicated by its speed and extent, is one crucial factor. This may have revolutionary effects on some businesses because it changes their routines, structures and frames of thought in fundamental ways. Others may see subtle changes connected to evolution in each of these areas. The business's strategic orientation towards the ecosystem, or how it sees its relationships with other ecosystem actors, is the second component. In certain digital journeys, the company actively shapes a new ecosystem, while in others, it just adapts to existing ecosystems.

5. CONCLUSION

The major goal of this essay is to give many disciplinary perspectives on digital revolution. We began by examining the reason why organisations need to go through a digital transformation, and we reached the resolution that it occurs because of advances in digital technology, raising digital competition and coming about digital customer behaviour. The three stages of digital transformation digitalisation, transformation and digitisation are then separated using literature analysis. A company's digital assets, organisational structure, growth strategies and measures must meet specific requirements at each level. In addition to having digital assets, businesses intending to undergo a digital transformation also need to establish or purchase capabilities for digital networking, big data analytics and agility. IT and analytical functional responsibilities must be internalised by organisations, which must also develop agile frameworks with a flat hierarchy. We identify particular growth strategies for businesses using platform-based techniques, which are inspired by the Ansoff matrix notably, platform diversity, customer co-creation and market development based on platforms. We conclude by talking about how crucial it is for digital organisations to create new (intermediate) digital measures and goals.

REFERENCES

Autio, E., R. Mudambi, and Y. Yoo (2021). "Digitalization and globalization in a turbulent world: centrifugal and centripetal forces," *Global Strategy Journal*, in press.

Bonanomi, M. M., D. M. Hall, S. Staub-French, A. Tucker, and C. M. L. Talamo (2020). "The impact of digital transformation on formal and informal organizational structures of large architecture and engineering firms," *Engineering, Construction and Architectural Management*, 27(4), pp.872–892.

George, G., E. C. Osinga, D. Lavie, and B. A. Scott (2016). "Big data and data science methods for management research," *Academy of Management Journal*, 59(5), 1439–1507.

Hanelt, A., R. Bohnsack, D. Marz, and C. Antunes Marante (2020). "A systematic review of the literature on digital transformation: insights and implications for strategy and organizational change," *Journal of Management Studies*. in press.

Hastig, G. M., and M. S. Sodhi (2020). "Blockchain for supply chain traceability: Business requirements and critical success factors," *Production and Operations Management*, 29(4), 935–954.

Khanagha, S., M. T. Ramezan Zadeh, O. R. Mihalache, and H. W. Volberda (2018). "Embracing bewilderment: responding to technological disruption in heterogeneous market environments," *Journal of Management Studies*, 55(7), 1079–1121.

Kretschmer, T., A. Leiponen, M. Schilling, and G. Vasudeva (2020). "Platform ecosystems as metaorganizations: implications for platform strategies," *Strategic Management Journal*, in press.

Rossi, M., J. Nandhakumar, and M. Mattila (2020). "Balancing fluid and cemented routines in a digital workplace," *The Journal of Strategic Information Systems*, 29(2), 101616.

Sony, M., and S. Naik (2020). "Critical factors for the successful implementation of Industry 4.0: a review and future research direction," *Production Planning & Control*, 31(10), 799–815.

40. Interaction of Human Workers and AI with Strategies of HRM: Assessment of Potentialities and Challenges in HRM

Neha Tomer[1] and Pallavi Tandon[2]

[1]Assistant Professor, ICFAI Business School, ICFAI University, Dehradun

[2]Assistant Professor, Department of Management, Noida International University

ABSTRACT: Automated processes, machine learning and artificial intelligence have been quickly advancing, considerably boosting the significance of information technology (IT) within corporate procedures. Rising AI-based responses in human resource management (HRM) have been rapidly being used to handle time-consuming and difficult activities within HRM capabilities. This study attempts to concentrate in particular on the challenges that HRM employees and divisions in modern firms confront as an outcome of close relationships between AI and human employees, specifically at the group level. Depending on an analysis of existing research, then highlights major potential solutions that might be used to solve these challenges. The secondary source of the HRM Association collected the study sample of 186 top HR workers. The findings showed that top management support and performance expectations are important determinants of the intent to embrace AI–human interaction, although competitive pressure did not appear to have any significant connection with such an intention.

KEYWORDS: Human resource, artificial intelligence, potential, and challenges, human resource management

1. INTRODUCTION

Artificial intelligence (AI) has been introduced by the so-called 'Industry 4.0' (Kong *et al.*, 2021). As information and communication technologies (ICT) continue to advance, events like AI can have a significant impact on many facets of modern life, making them among the most important factors in all potential changes. There are examples discussed in recent studies like employees and managers who behave in complicated environment for technical intensity, quick cycle periods and quick digital contacts as well as gig economy (Glikson and Woolley, 2020). The environment in which functions of managers and employees change technological progress as an ideal storm. The advantages and drawbacks for using the contemporary technologies at work are pervasive. The next phase of HR is both AI and human, according to a recent Forbes article, indicating the critical role that contemporary technology play in HRM (Meister, 2019).

Although the technological development of human resource management begins with industrial revolution with few alterations either physical or mental services (Luo *et al.*, 2019). Nevertheless, modern advancements are ever more offers substitutes for human resources (HR) in tasks that historically required human communication and connection, altering organisational frameworks as well as the type of employment. AI and human assistance machines, for example, are gaining additional commercial emphasis (Larivière *et al.*, 2017). Conventional HR practices have been transformed through these smart 'beings', which not just present major challenges to HRM, such as career-specific breakdown, but also generate capabilities and potential. Simultaneously, ML techniques, Internet of things (IoT) and smart devices

DOI: 10.1201/9781003532026-40

can considerably assist cross-border organisations by promoting considerably better engagement and collaboration (Cooke *et al.*, 2019). Additionally, the advancement of digital human resource information systems (HRIS) and other technological advances offers various opportunities for reduce and improve the expenses of HRM processes, like the evaluation of employment applications and evaluations of worker performance (Abraham *et al.*, 2019). Thus, this research attempts to concentrate in particular on the challenges that HRM employees and divisions in modern firms confront as an outcome of close relationships between AI and human employees, specifically at the group level. Depending on an analysis of existing research, then highlights major potential solutions that might be used to solve these challenges.

2. LITERATURE REVIEW

Prior research (Coupe, 2019) has largely focused on the issues that AI poses for HRM in terms of jobs lost, transited professional requirements, requirements for novel abilities growth and talent administration dynamics. A significant number of CEOs, in contrast to the bulk of HR managers, feel that AI will generate more jobs than it will destroy. This was noted by KPMG (2019), in a recent survey. The HR workers within the organisation have historically viewed innovations, such as AI, from an operational viewpoint, and its primary emphasis has been on relearning and skills growth for the workers whose positions may be substituted by AI. This has played a significant role in the disparity in views regarding the role of AI (Libert *et al.*, 2020).

As per Driskell *et al.* (2018), improved interaction and cooperation between human workers (typically in addition to being a component of a group) has been a heavily studied subject in HRM investigations, given that cognitive prejudices, personality traits and senses perform an important function in the achievement or failure of these kinds of cooperation. These elements become increasingly complicated in situations and circumstances where AI-powered machines and humans collaborate as colleagues, since certain human employees may be resistant owing to the anxiety about losing their positions to technological advances, in addition to psychological issues with the adoption of innovations (Jurczyk-Bunkowska and Pawełoszek, 2015).

The integration of AI and human interaction presents distinct challenges for the HRM function in modern organisations. HRM needs to address employees' apprehension concerning AI's potential impact on job security and navigate the complexities of fostering trust between AI-enabled robots and human workers functioning as team members (Arslan *et al.*, 2020).

Artificial intelligence (AI) is reshaping human resource management (HRM) and, consequently, revolutionising the essence of work, the workforce and workplaces. The adoption of AI-assisted HRM is seen as a means to enhance organisational productivity (Malik *et al.*, 2022).

3. MATERIALS AND METHOD

The hypothesised correlations were empirically investigated using the survey technique. The HRM Association's secondary sources were used to construct the sampling structure, which included senior HR experts. Since this investigation is looking into the interaction of AI and humans in HRM practices, the respondents are HR workers with decision-making power who affect the company's adoption choices. As a result, senior HR workers in senior roles (for instance, HR head, Manager, CHRO and so on) were examined. A suitable number of 192 people were chosen to participate in this investigation. Following validation of response completion and lacking data, 186 replies have been retained for the ultimate evaluation, with six replies excluded due to mistakes or missing data. The poll was administered in English, replies were unidentified and privacy has been maintained.

The instrument measurements depend on past IT breakthrough interactions of human–AI investigations that have repeatedly demonstrated their reliability and validity. As a result, the items have been updated and reworded to meet the research setting. Behavioural intention (BI), top management support (TMS), performance expectancy (PE) and competitive pressure (CP) are adoption factors. Ulrich (1998) provided measurement items for the HR roles of administrative expert (AE), employee champion (EC), strategic partner (SP) and change agent (CA); however, the measurements have been simplified and just twenty criteria have been adopted.

H1: CP is positively connected to AI–human interaction in HRM.

H2: TMS is connected with a positive conduct intention to AI–human interaction in HRM.

H3: The behavioural intent of AI–human interaction in HRM is positively related to PE.

H4: SP HRM employment is positively connected to the AI–human interaction in HRM.

H5: The CA HRM function is positively connected to the AI–human interaction in HRM.

H6: The AE HRM function is positively connected to the AI–human interaction in HRM.

H7: The EC HRM function is negatively related to the behavioural intent of AI–human interaction in HRM.

4. RESULT AND DISCUSSION

By using the covariance-based structural equation modelling (CB-SEM) approach, the suggested model had been experimentally evaluated. A confirmatory factor assessment (CFA) has been used to assess the validity and reliability of the measurement scale (Table 1 and Figure 1).

Table 1: Measures of validity and reliability.

Constructs	Cronbach alpha	Average Variance Extracted	Composite Reliability
BI	0.926	0.827	0.935
TMS	0.898	0.749	0.899
PE	0.890	0.620	0.891
CP	0.716	0.635	0.914
CA	0.911	0.692	0.918
SP	0.927	0.730	0.931
EC	0.915	0.684	0.914
AE	0.934	0.745	0.936

Source: Author's compilation.

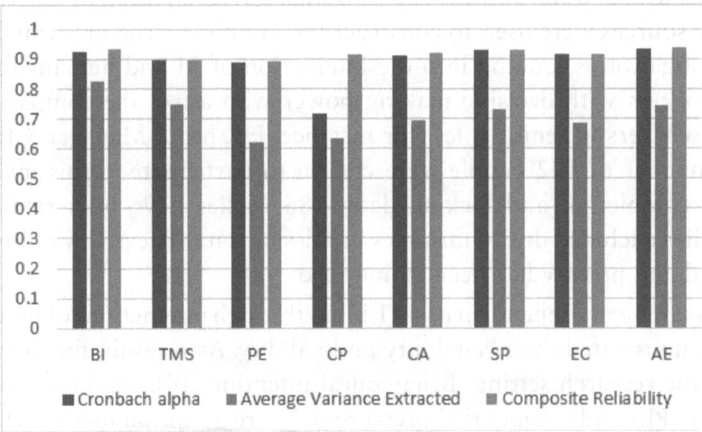

Figure 1. Validity and reliability measures. Source: Author's compilation.

The findings of the hypothesis testing are shown in Table 2.

Table 2: Results of hypothesis testing.

Hypothesis	Constructs	Scores	Critical Ratio	Standard Error	Significance Value	Outcome
H1	CP	0.09	2.02	0.04	0.043	Accepted
H2	TMS	0.1	1.68	0.06	0.092	Rejected
H3	PE	0.75	8.80	0.08	<0.001	Accepted
H4	EC	–0.21	–4.77	0.04	<0.001	Accepted
H5	CA	0.13	3.31	0.04	<0.001	Accepted
H6	AE	0.003	0.09	0.03	0.92	Rejected
H7	SP	0.009	0.26	0.03	0.78	Rejected

Source: Author's compilation.

The examination of statistical data provides scientific evidence and significant knowledge of the factors that influence of AI–human interaction in HRM, particularly CP, TMS and HR functions within the company, as indicators of HR employee's intention to implement AI. While a significant number of HRIS investigation has shown that HR workers are confident in the potential impact of technology, viewed it as strategically helpful and weighed its implementation as part of the procedure for strategic planning, this investigation indicates that broadening and illustrating this basis on AI–human interaction adoption is uncertain. Another possible argument is that AI-oriented HR services are nevertheless in the infancy stages of adoption, and CP continues to be young, which may reduce the need to strategically explore its implementation. This additionally seems in line with the lack of a link between CP and participants' adoption intentions of AI–human interaction. Furthermore, the considerable negative relationship between the EC position and the intention to embrace AI in HRM supports the hypothesis that HR workers who prioritise people above functions see HRIS generally as an obstacle to their function. Although information technology has increased corporate communication, it poses a danger to EC who prefer direct contact and employee sponsorship. AI has the potential to significantly increase the disparity in personal engagement by taking over a large portion of these direct interaction responsibilities. As a result, it raises concerns about the prospects of EC and additional HR employment. HR leaders who emphasised their function as agents of change through increased engagement in corporate transition and shift, on the other hand, saw AI in HRM as a potential and have a positive desire to utilise it. These technologies have been recognised for their influence on HRM strategies, including learning opportunities, decision making, collaboration between AI robots and humans and job replacement. Similarly, it affects the HRM practises like job performance, training and recruiting (Vrontis *et al.*, 2022).

5. CONCLUSION

The conclusions of this investigation assist in the spread of IT in HRM theory. The result was reached that HR leaders had a positive attitude toward AI's potential involvement in improving HRM quality and effectiveness. The study supports the arguments that the HR change facilitator employment is associated with adaptable and flexible methods that take into account the divergence of the IT role from established practices, in addition to the worker support HR role's resistance to the challenges caused by technological advances. While AI-based innovations are expected to keep on emerging, providing improved services and altering sectors, HR managers and policymakers have been urged to stay knowledgeable about the influence of AI on HRM and company performance.

REFERENCES

Abraham, M., C. Niessen, C. Schnabel, K. Lorek, V. Grimm, K. Möslein, and M. Wrede (2019). "Electronic monitoring at work: the role of attitudes, functions, and perceived control for the acceptance of tracking technologies," *Human Resource Management Journal*, 29(4), 657–675.

Arslan, A., C. Cooper, Z. Khan, I. Golgeci, and I. Ali (2022). "Artificial intelligence and human workers interaction at team level: a conceptual assessment of the challenges and potential HRM strategies," *International Journal of Manpower*, 43(1), 75–88.

Cooke, F. L., M. Liu, L. A. Liu, and C. C. Chen (2019). "Human resource management and industrial relations in multinational corporations in and from China: challenges and new insights," *Human Resource Management*, 58(5), 455–471.

Coupe, T. (2019). "Automation, job characteristics, and job insecurity," *International Journal of Manpower*, 40(7), 1288–1304.

Driskell, J. E., E. Salas, and T. Driskell (2018). "Foundations of teamwork and collaboration," *American Psychologist*, 73(4), 334.

Glikson, E., and A. W. Woolley (2020). "Human trust in artificial intelligence: review of empirical research," *Academy of Management Annals*, 14(2), 627–660.

Kong, H., Y. Yuan, Y. Baruch, N. Bu, X. Jiang, and K. Wang (2021). "Influences of artificial intelligence (AI) awareness on career competency and job burnout," *International Journal of Contemporary Hospitality Management*, 33(2), 717–734.

KPMG (2019). "Rise of the humans 3: shaping the workforce of the future."

Larivière, B., D. Bowen, T. W. Andreassen, W. Kunz, N. J. Sirianni, C. Voss, N. V. Wünderlich, and A. De Keyser (2017). "'Service encounter 2.0': an investigation into the roles of technology, employees, and customers," *Journal of Business Research*, 79, 238–246.

Libert, K., E. Mosconi, and N. Cadieux (2020). "Human-machine interaction and human resource management perspective for collaborative robotics implementation and adoption."

Luo, X., S. Tong, Z. Fang, and Z. Qu (2019). "Frontiers: machines vs. humans: the impact of artificial intelligence chatbot disclosure on customer purchases," *Marketing Science*, 38(6), 937–947.

Malik, A., P. Budhwar, and B. A. Kazmi (2022). "Artificial intelligence (AI)-assisted HRM: towards an extended strategic framework," *Human Resource Management Review*, 100940.

Meister, J. (2019). "Ten HR trends in the age of artificial intelligence," *Fobes*, available at www. Forbes. com/sites/ jeannemeister/2019/01/08/ten-hr-trends-in-the-age-of-artificial-intelligence.

Jurczyk-Bunkowska M., Pawełoszek I. (2015), The Concept of Semantic System for Supporting Planning of Innovation Processes, Polish Journal of Management Studies Vol.11 No 1.

Vrontis, D., M. Christofi, V. Pereira, S. Tarba, A. Makrides, and E. Trichina (2022). "Artificial intelligence, robotics, advanced technologies and human resource management: a systematic review," *The International Journal of Human Resource Management*, 33(6), 1237–1266.

41. Assessment of Personalised Engagement Using Artificial Intelligence and Marketing Implications in Financial Services

Neha Yadav[1], Neha Tomer[2], and Shubham Gaur[3]

[1]Assistant Professor, Department of UG Management Studies, Faculty of Management Studies, MRIIRS, Faridabad

[2]Assistant Professor, ICFAI Business School, ICFAI University, Dehradun

[3]Department of Management, Delhi Institute of Advanced Studies, GGSIPU, Delhi

ABSTRACT: Recent technological advancements have been endless and have had an enormous effect on everyone in all facets of their lives over the previous decades. Artificial intelligence (AI) innovations are becoming more prevalent in the modern world, serving as the basis for novel propositions of value and unique customer encounters. A questionnaire poll of generation Z participants was analysed using structural equation modelling. According to the results, AI personalised engagement marketing implications influenced repurchase intention, brand experience and preference. The connection between AI personalised engagement marketing implications and brand preference was additionally influenced by brand experience. The research will assist financial services in designing AI marketing tasks and developing more accurate marketing plans for customer purchase and retention.

KEYWORDS: Artificial intelligence, financial services, marketing, and personalised engagement

1. INTRODUCTION

The launch of artificial intelligence (AI) in the financial assistance sector was predominantly driven by rising requirements for financial legislation, a requirement for profits, and company competition. In the meantime, it is linked to improvements in data assets and related innovations, in addition to the accessibility of required financial sector facilities (Akyüz and Mavnacıoğlu, 2021). As per insider intelligence's investigation into the application of AI in the banking sector, the vast majority of banks (80 percent) have been conscious of the possible advantages associated with AI, and numerous have executed it for risk administration or to generate income (Digalaki, 2019). Nevertheless, most of the firms are still in the acquiring stage of AI adoption, attempting to figure out the way to properly incorporate different AI structures into their business processes. This special problem investigated the connection between AI and marketing financial assistance, realising the adaptability of financial assistance suppliers in the face of an enormous quantity of accessible data, making it critical for these suppliers to more effectively comprehend their customers' requirements, views and choices, and employ these data to build pertinent financial assistance to enhance service distribution (Duan *et al.*, 2019).

Personalised engagement marketing implies autonomous machine-driven decision-making for commodities, prices, web pages and advertising information that are tailored to a client's interests (Shaul, 2016). The AI in this instance employs data–company–customer interactions, customer usage patterns of

DOI: 10.1201/9781003532026-41

products and services and customer interaction patterns about firm products and services to customers – to autonomously anticipate the kind, execution and purchase of preferred rigid goods and services (Kallioniemi, 2021). Additionally, AI may customise the firm's web pages based on customer personal tastes (such as Wix, The Grid), cost offerings that correspond with customer desire to spend and communicate client interactions across every channel and gadget in an effortless and personalised way. AI instruments additionally gain insight from customer interactions to enhance the accuracy of forecasts about client tastes, which boosts the worth provided by the company to customers throughout the interaction (Kumar *et al.*, 2019). Within the context of this premise, this research offers an implication on two crucial strategic firm resources – brands and customers – across India, as well as an investigation for comprehending the significance of AI in personalised engagement marketing in financial services.

- H1: AI interaction is proportionate to individual attention provided to customers.
- H2: Customer complaints could be handled immediately and directly by AI customisation.
- H3: AI offers information that aids the decision to buy.
- H4: AI accessibility is both convenient and effective.

2. LITERATURE REVIEW

Customer reaction and decision-making could be influenced by AI personalised engagement marketing implications. Bank staff build confidence in customers via customer direction, exchanging data and responding to customer worries during customer-employee interactions (Chen *et al.*, 2022). Such AI personalised engagement marketing implications have been especially essential to banks, which should share financial data or upgrades to offer professional customised services. By providing quick access to goods or service data, AI may decrease customers' physical as well as time separation from banks. Chung *et al.* (2020) have been among the first to investigate the impact of AI personalised engagement marketing implications on consumer conduct and customer–firm relationships, yet their emphasis has been limited to AI chatbots (Paweloszek, 2015).

While multiple investigations have investigated AI in connection with business processes in the areas of marketing and personalised engagement with customers, There is an increasing amount of investigation on the way AI is particularly being implemented for personalised engagement marketing in financial assistance and the way it is improving the associated client expertise. For instance, Abdulquadri *et al.* (2021) investigated the application of AI chatbots in the insurance sector, whereas Mogaji *et al.* (2021) investigated AI applications as a kind of electronic transition in the supply of financial assistance. Mogaji *et al.* (2021) examined the application of AI in advertising financial assistance to clients with vulnerabilities, whereas Jang *et al.*, (2021) examined managers' comprehension of AI chatbots in the Korean financial assistance sector.

3. RESEARCH METHODOLOGY

In this study, an approach of random sampling has been employed. A database from a partner professional marketing investigation firm has been employed to identify the appropriate participants. The contact details from these records have been sent the link to the online survey and informed that participation has been optional. 193 online replies have been gathered overall. In addition to demographic data, each participant's IP addresses have been collected. The demographic data for each of the accepted replies are shown in Table 1. Three parts made up the questionnaire: an assessment query, constructs-related items and demographic data. The results of this investigation demonstrated the design reliability as well as the validity of the elements since composite reliability (CR), Cronbach's alpha and average variance extracted (AVE) scores have been adequate.

Table 1: Respondents' demographic data.

Variables	Number of Participants	Percentage (%)
Female	77	39.90
Male	116	60.10
25–28 yrs.	61	31.61
28–30 yrs.	53	27.46
Above 30 yrs.	79	40.93
Types of firms		
Financial service providers	58	30.05
Financial Investment Sector	51	26.42
Banks	47	24.35
Microfinance	37	19.18
Usage		
Operation	59	30.57
Customer services	34	17.62
Sales and marketing	51	26.42
Others	49	25.39

Source: Author's compilation.

4. RESULT AND DISCUSSION

Smart-PLS 3.3.3 has been utilised to evaluate the statistics due to its rigorous paradigm evaluation and suitability for investigating minimal sample sizes. It additionally proves appropriate for analysing complicated predictive designs that evaluate the connections between latent factors that have numerous structural routes. Cronbach's alpha and CR have been used to assess the scales' dependability which is demonstrated in Table 2 and Figure 1.

Table 2: Reliability and validity of designs.

Design	Element	Loads of Standardising Factor	Cronbach Alpha	CR	AVE
Interaction	Individual attention is provided to customers by AI.	0.849	0.798	0.881	0.712
Customisation	Customer complaints could be handled immediately and directly by AI.	0.846	0.875	0.914	0.726
Information	AI offers information that aids the decision to buy.	0.851	0.793	0.878	0.707

Design	Element	Loads of Standardising Factor	Cronbach Alpha	CR	AVE
Accessibility	AI is both convenient and effective.	0.9	0.882	0.918	0.738
Brand experience	My bank's experience with AI was intriguing.	0.877	0.908	0.933	0.736
Brand preference	My financial institution's services have exceeded my expectations.	0.751	0.905	0.927	0.679
Re-purchase intention	In the future, I intend to purchase additional products or services from my bank.	0.871	0.874	0.909	0.668

Source: Author's compilation.

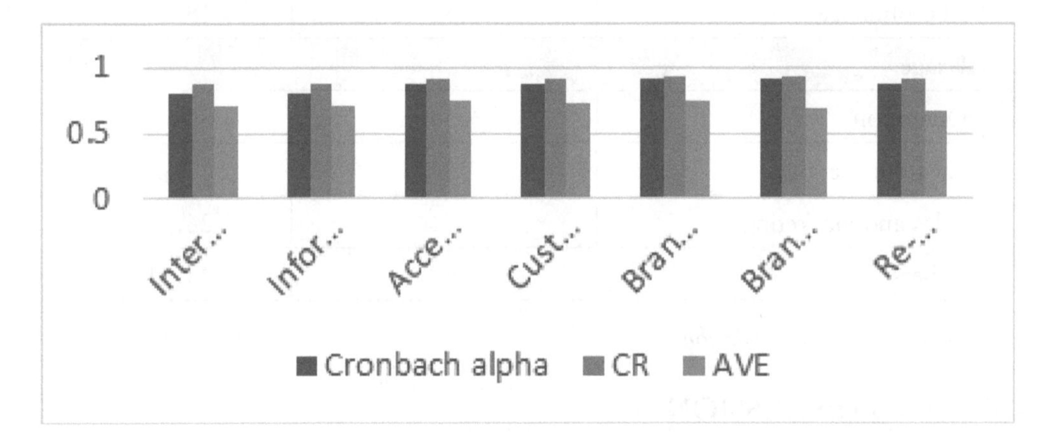

Figure 1. *Reliability and validity of designs. Source: Author's compilation.*

Cronbach's alphas covered 0.79–0.91, each above the needed 0.70 limit scale, while CR metrics spanned from 0.88 to 0.93, also over the 0.70 limits needed for reliability. Furthermore, convergent validity was determined by investigating the variable loads and average AVE. Each of the load coefficients was greater than 0.50 and statistically significant ($p<.001$). The AVEs for each of the designs have been greater than 0.5, and each of the elements demonstrated good inner consistency as well as an elevated level of convergence; therefore, the assessment scales' reliability as well as validity as convergent measures have been backed.

The outcomes indicate the implication of AI personalised engagement marketing initiatives on repurchase intention, brand experience and preference, as well as emphasise the significance of AI personalised engagement marketing efforts in customer–brand interactions. Three of the four hypothesis aspects of AI personalised engagement marketing implications information, accessibility and customisation have been discovered to be pertinent to bank customers. Using these results, AI personalised engagement marketing plans concentrate on all three of these factors. Managers can use AI to quickly offer tailored marketing interaction about services or goods and financial institution data to customers.

5. CONCLUSION

This research discovered that AI personalised engagement marketing implications possessed an important beneficial impact on brand preference, which consequently caused customer repurchase intention. As an outcome, AI personalised engagement marketing implications must be viewed not just as a way of improving customer experience, but additionally as a vital brand establishing an image tool. Banks ought to prioritise AI marketing efforts to improve brand appeal and long-term success. The validation of AI personalised engagement marketing implications is a significant measure of the worth of these expenditures, as this research indicates that customers are worth AI tasks after going through their benefits. Managers have to guarantee that AI is capable of offering efficient, dependable and precise banking services.

REFERENCES

Abdulquadri, A., E. Mogaji, T. A. Kieu, and N. P. Nguyen (2021). "Digital transformation in financial services provision: a Nigerian perspective to the adoption of chatbot," *Journal of Enterprising Communities: People and Places in the Global Economy*, 15(2), 258–281.

Akyüz, A., and K. Mavnacıoğlu (2021). "Marketing and financial services in the age of artificial intelligence," *Financial Strategies in Competitive Markets: Multidimensional Approaches to Financial Policies for Local Companies*, 327–340.

Chen, H., S. Chan-Olmsted, J. Kim, and I. Mayor Sanabria (2022). "Consumers' perception of artificial intelligence applications in marketing communication," *Qualitative Market Research: An International Journal*, 25(1), 125–142.

Chung, M., E. Ko, H. Joung, and S. J. Kim (2020). "Chatbot e-service and customer satisfaction regarding luxury brands," *Journal of Business Research*, 117, 587–595.

Digalaki, E. (2019). "The impact of artificial intelligence in the banking sector and how AI is being used in 2020," *online. Datum pristupa dokumentu*, 27(6), 2020.

Duan, Y., J. S. Edwards, and Y. K. Dwivedi (2019). "Artificial intelligence for decision making in the era of big data–evolution, challenges, and research agenda," *International Journal of Information Management*, 48, 63–71.

Jang, M., Y. Jung, and S. Kim (2021). "Investigating managers' understanding of chatbots in the Korean financial industry," *Computers in Human Behavior*, 120, 106747.

Kallioniemi, P. (2021). *The Role of Human Curation in the Age of Algorithms.*

Kumar, V., B. Rajan, R. Venkatesan, and J. Lecinski (2019). "Understanding the role of artificial intelligence in personalized engagement marketing," *California Management Review*, 61(4), 135–155.

Mogaji, E., O. Adeola, R. E. Hinson, N. P. Nguyen, A. C. Nwoba, and T. O. Soetan (2021). "Marketing bank services to financially vulnerable customers: evidence from an emerging economy," *International Journal of Bank Marketing*, 39(3), 402–428.

Paweloszek I. (2015). "Approach to Analysis and Assessment of ERP System. A Software Vendor\'s Perspective, Proceedings of the 2015 Federated Conference on Computer Science and Information Systems 2015, M. Ganzha, L. Maciaszek, M. Paprzycki, Annals of Computer Science and Information Systems\", 1415–1426, IEEE\", doi:10.15439/2015F251.

Shaul, B. (2016). "Starbucks updates mobile app to offer a personalized experience," *AdWeek Digital.*

42. AI Innovation and Implementation in Financial Services – Understanding Management and Gaining Perspective

Rahul Berry[1], Neha Tomer[1], and Sakshi Tyagi[2]

[1]Assistant Professor, ICFAI Business School, ICFAI University, Dehradun

[2]Research Scholar, School of Business Management, Noida International University

ABSTRACT: The most recent disruption caused by computer science is artificial intelligence. The dynamics of the banking and financial services business is likewise being significantly altered by AI. Combining, reconstituting and re-forming the proven and developing capabilities of AI in unanticipated ways is creating new opportunities and challenges while also bringing new risks. It is critical to consider how long technology will last since people and related individuals touch the focal point of the financial services industry, which lives on the specialty of personalisation and client fulfilment and will quite often be replaced by it. In addition to this, technology also poses ethically neutral hazards, like cybercrimes and macro-financial concerns. The current study examines the under-examined likelihood that artificial intelligence will eventually replace people in the banking and financial services industry, finishing the individualised attention and service customisation that serve as the cornerstones of customer loyalty and enhance the experience of businesses like financial and banking services, which are renowned for their trustworthy and capable nature.

KEYWORDS: Financial services, customer, artificial intelligence

1. INTRODUCTION

Natural intelligence is referred to as the ability to see, interpret, analyse and reason logically to draw valid conclusions or resolve problems, and then to grow and change as a result of experience (Sennaar, 2019). It is a quality that only people have. Both machine learning and AI are utilised to portray the very qualities that a PC has. The term 'artificial intelligence' was first used by John McCarthy, a computer and mental researcher at Stanford College in the US, to describe the most notable ability of a machine to imitate a person in thinking as a human does and selecting the best course of action from all the options available to accomplish a specific goal. Human intelligence has had a big impact on how civilisation, as we know it now, has evolved. Artificial intelligence has unlocked previously unimaginable potential for growth and advancement by augmenting and supplementing human intelligence.

In several areas, such as credit scoring, retail lending, wealth and asset management, insurance, customer support, process automation, data mining, market analysis and personal finance, the financial services sector is increasingly implementing AI to improve the client experience. The fields of robotics, remote sensing, electronic trading, healthcare, transportation, education and many other companies have also improved and undergone radical transformation. The banking sector is rapidly losing its reputation as stodgy and dogmatic due to a mentality shift towards a complete embrace of technology (Paweloszek, 2016). To communicate clearly in the future with new arrangements that will assist in enhancing connections with consumers by delivering customised, consistent and innovative banking in the digital age, banks are accelerating their development plan.

DOI: 10.1201/9781003532026-42

1.1. Objective of Research

- To comprehend management and get insight into the innovation and application of AI in financial services.
- Examining the connections between financial services, digital marketing and AI regarding vulnerable customers.

2. LITERATURE REVIEW

The implementation of AI creates a variety of challenges, according to Lui and Lamb (2018), that may undermine customers' trust and confidence. The biases and discrimination seen in the algorithms used in banking's AI applications are noted in this paper. The cost and level of execution of AI channel strategy efforts in the financial services sector are examined by Sarvady (2017). The growing application of artificial intelligence in the financial services and banking sector is brought to light by Ludwig (2018). He explains how AI helps in the analysis of client data for credit assessment and determining loanable amounts. To prevent the misuse of AI, he also recommends changing rules.

According to a recent paper by Nunn (2018), financial institutions should use existing algorithmic biases against underrepresented groups to balance off any potential AI biases. According to him, diversity in the workplace can lessen the biases present in AI applications. Satell (2016) underlines the meaning of fostering an ethical learning climate, recognising bias in algorithms, tending to ethical problems and expanding ethical norms in his exhaustive examination of the ethical issues with AI frameworks. The benefits of AI for banks are the main topic of the case study by Daks (2018) of TD Bank Group's acquisition of Layer 6 Inc. He provides examples of financial organisations adopting third-party providers' artificial intelligence models. Guy Messick (2017) examines how AI is being incorporated into financial services in the digital ecosystem in his paper. He also examines how it may be used to create new businesses and offer clients customised advice and services. The significance of artificial intelligence in forestalling cybercrime in the industry of financial services, upgrading client experience and the risk that AI might add to the acceleration of cyber-attacks is undeniably analysed by Meinert (2018).

2.1. Hypothesis

- **H1:** AI in the finance sector is advantageous to business organisations, customers and finance experts.
- **H2:** Artificial intelligence (AI) is being beneficial in a variety of finance functions.

3. RESEARCH METHODOLOGY

3.1. Theoretical Framework

The methodology is based on four processes, as shown in Figure 1. It first demonstrates how AI gathers data from various touch points, processes it and continuously learns from it. To give estimates, ideas and ends, AI collects, classifies, stores and controls the data along the way (Metcalf *et al.*, 2019). The structure additionally demonstrates the way that AI algorithms could be incorporated into digital marketing plans. Social media, messaging services or email are used to target these customers with customised communications. The third component of the framework looks at how AI and digital marketing are linked, particularly concerning financial services.

The methodology perceives the significance of the information that powers AI technologies. Financial service providers should keep up with responsibility while taking care of customer information and working with outsiders including agents, credit authorities, online entertainment, and web and mobile innovation. Algorithms may prevent vulnerable customers from accessing accounts or credit facilities if the proper precautions are not in place.

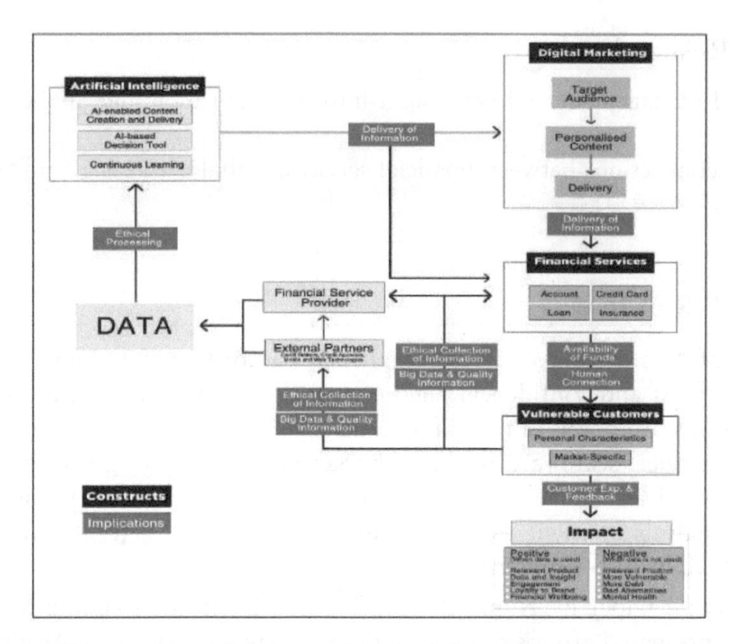

Figure 1. AI's effects on financial services' digital marketing. Source: Author's compilation.

3.2. Qualitative Research

Both primary and secondary data, the two of which are qualitative data, were obtained and further researched to make ends and proposals. The primary data were accumulated using a survey on AI in banking and financial services. A questionnaire was ready for the review, and random samples were chosen. The internet was used for the secondary data collection, including newspapers, e-books, research papers, e-magazines and the web.

4. RESULT AND DISCUSSION

AI and its ascent to notoriety are producing a major change in the financial services business as additional banks endeavour to send off developments under the standard of AI-controlled innovation. About evaluating the useful applications of AI in financial services and banking, as indicated by Table 1 and Figure 2's frequency analysis, 89% of respondents said chatbots are incredibly valuable, while 53.5% said voice assistants are not helpful, 64% said authentication and biometrics are exceptionally valuable, 91.5% said fraud prevention and detection are utilised to get the information, and 98.5% expressed uses of KYC.

Table 1: Analysis of many responses.

	YES		NO		TOTAL	
Useful applications of AI	**Frequency**	**%**	**Frequency**	**%**	**Frequency**	**%**
Chatbots	36	89	5	11	41	100
Voice assistants	17	46.5	22	53.5	41	100
Authentication and biometrics	27	64	14	36	41	100
Fraud detection and prevention	37	91.5	4	8.5	41	100
KYC/AML	38	98.5	3	1.5	41	100
Smart wallet	29	66.5	12	33.5	41	100

Source: Author's compilation

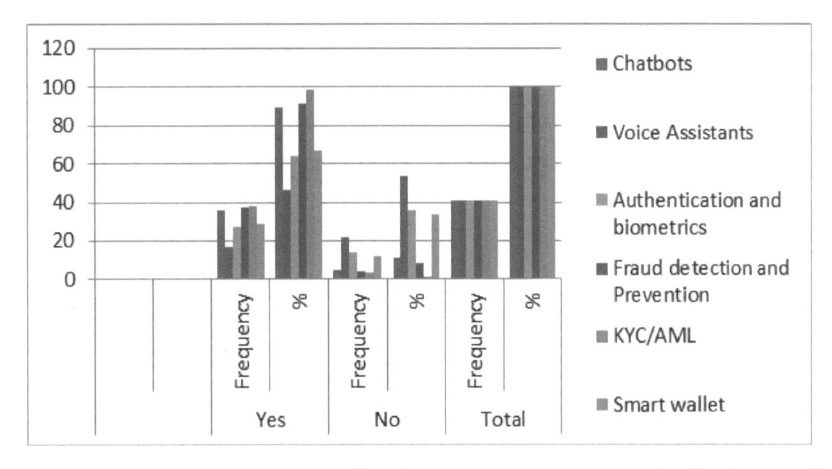

Figure 2. Graphical representation of multiple responses. Source: Author's compilation.

The literature on AI claims that AI is both a 'business capability' and a 'technological disruption'. The adoption of financial technologies like AI in marketing is required because of the changing digital environments. The independence of AI from outside inputs both during and after operations, however, is still debatable. Some contend that effective technological adoption depends on the human factor and that AI should be seen as a complex notion through the lens of human-computer interaction. According to this perspective, artificial intelligence (AI) is an ecosystem that connects human beings (marketers, customers, etc.) with technologies rather than just technology or algorithms. Before customers and AI interact, the initial human inputs must be established because they are essential for AI to learn from.

5. CONCLUSION

This article introduces another detailed structure with a flow/process map of customer input data to data processing to provide data to AI digital marketing that creates interactions and arrangements that both confirm and debunk consumers' expectations. Second, utilising the previously published framework, this study organises and evaluates the examination writing relevant to the fundamental concepts that make up the structure in the fields of financial and marketing services. The effects of digital marketing data and artificial intelligence on financial service providers and company executives more broadly.

REFERENCES

Daks, M. (2018). "Banking on technology: artificial intelligence helping banks get smarter," *Njbiz*, 31(7), 10.

Guy Messick, A. (2017). "Artificial intelligence: the ultimate disrupter," *Credit Union Times*, 28(38), 12.

Paweloszek I. (2016), Integrating Semantic Web Services into Financial Decision Support Process, [in:] Proceedings of the Federated Conference on Computer Science and Information Systems, ACSIS, Vol. 8, pp. 1189-1198. DOI: 10.15439/2016F99

Ludwig, E. (2018). "Regulators have their eye on AI," *American Banker*, 183(130), 1.

Lui, A., and G. W. Lamb (2018). "Artificial intelligence and augmented intelligence collaboration: regaining trust and confidence in the financial sector," *Information & Communications Technology Law*, 27(3), 267–283.

Meinert, M. C. (2018). "Artificial intelligence: the next frontier of cyber warfare?" *ABA Banking Journal*, 110(3), 43.

Metcalf, L., D. A. Askay, and L. B. Rosenberg (2019). "Keeping humans in the loop: pooling knowledge through artificial swarm intelligence to improve business decision making," *California Management Review*, 61(4), 84–109.

Nunn, R. (2018). "Workforce diversity can help banks mitigate AI bias," *American Banker*, 183(104), 1.

Sarvady, G. (2017). "Chatbots, Robo Advisers, & AI: Technologies presage an enhanced member experience, improved sales, and lower costs," *Credit Union Magazine*, 83(12), 18–22.

Satell, G. (2016). "Teaching an algorithm to understand right and wrong," *Harvard Business Review Digital Articles*, 2–5.

Sennaar, K. (2019). "AI in banking – an analysis of 7 Top American Banks," Available at https://www.techemergence.com/ai-inbanking-analysis/ [2018].

43. The Role Chatbots in Customer Service and Marketing

Mazharunnisa[1], Neha Chunduri[2], Nalam Lakshmi Meghana[2], Upputuri Gopichand[2], and Moussa Camara[2]

[1]Associate Professor, KL Business School, Koneru Lakshmaiah Education Foundation, KL University, Vaddeswaram, Vijayawada, Andhra Pradesh, India

[2]KL Business School, Koneru Lakshmaiah Education Foundation, KL University Vaddeswaram, Vijayawada, Andhra Pradesh, India

ABSTRACT: Businesses are using chatbots to deliver customer service more regularly. Despite this advancement, there has not been any in-depth study on user experience and motivation for this significant application area of conversational interfaces. In order to solve this study vacuum, this report identified with 24 consumers who utilised two chatbots for customer service. Particularly, service providers are increasingly using text-based chatbots as the first point of contact for customers searching for support and information. Only if they produce positive user experiences and offer value propositions that encourage customers to interact with them repeatedly will chatbots for customer service continue to be relevant and attract interest.

KEYWORDS: Chatbots, customer, marketing, service, artificial intelligence, technologies

1. INTRODUCTION

In the last few years, chatbots have created far-reaching changes in the way all the businesses interact with their customers. Chatbot, which is a shorter name for that of a chat robot, is a 'computer program', which is designed to simulate human conversations through text or voice-based communications. These are virtual assistants, which use 'artificial intelligence or AI' and 'natural language processing or NLP' techniques to understand and respond to the questions asked and the requests given to it. Chatbots are found in different platforms, including websites, messaging apps, social media platforms and in mobile applications too (Wang *et al.*, 2022). The gradual development and adoption of chatbots have been driven by several factors. Firstly, the increasing demand for instant and personalised customer service has pushed many businesses to explore new avenues to deliver timely and efficient support (Følstad and Skjuve, 2019).

1.1. Objectives and Hypotheses

The objectives of the research study are as follows:

- To investigate the role of chatbots in the context of 'customer service and marketing' and their impacts on customer loyalty and satisfaction
- To evaluate the effectiveness of these chatbots to handle customer queries, providing personalised recommendations and resolving customer issues on time.
- To identify the opportunities and challenges associated with 'chatbots for customer service' and marketing.
- To explore the perceptions of customers towards chatbots and its influence on brand image and customer engagement.

DOI: 10.1201/9781003532026-43

Some hypotheses have also developed which are described as follows:

Hypothesis 1: *In this current era, chatbots have a 'significant impact on customer satisfaction and loyalty'.*

Hypothesis 2: *Chatbots are effective in handling customers' queries, providing information and resolving issues.*

Hypothesis 3: *The perceptions of customers are positive towards chatbots to improve brand perception and customer engagement.*

2. LITERATURE REVIEW

2.1. Chatbots in Customer Service

- Research studies analysing the effectiveness of chatbots in enhancing customer experiences.
- Chatbot implementations in real-world customer support scenarios and their effect on customer satisfaction
- Chatbots have designed principles and techniques for creating user-friendly and engaging conversational experiences.

Chatbots have changed the customer service sector by providing instant and personalised support. Research studies have analysed their effectiveness in improving the customer experiences and satisfaction (Fotheringham and Wiles, 2022).

2.2. Chatbots in Education

- Exploration of chatbot-based educational tools and their impact on student engagement and learning outcomes.
- Analysis of the usage of chatbots in personalised tutoring, adaptive learning and educational support.
- Examination of the challenges and opportunities integrating chatbots into educational environments (Sarbabidya and Saha, 2020).

Integration of chatbots in education has shown promising results as well as it has helped in improving student engagement.

2.3. Chatbots in Healthcare and Mental Health

- Review of studies on the application of chatbots in healthcare, such as symptom assessment and remote patient monitoring.
- Analysis of the use of chatbots for mental health support, counselling and well-being interventions.
- Discussion on the ethical considerations and data privacy implications of chatbots in healthcare settings (Adam *et al.*, 2021).

Chatbots are being increasingly utilised in healthcare for tasks such as symptom assessment and remote patient monitoring.

2.4. Chatbots in Business and Marketing:

- Examination of chatbot applications in sales, lead generation and customer relationship management.
- Analysis of the impact of chatbots on marketing campaigns, including personalised recommendations and targeted messaging.
- Exploration of chatbot integration with e-commerce platforms and the role of chatbots in enhancing customer experiences.

2.5. Chatbot in Human-Computer Interaction and User Experience

- Analysis of the usability and user experience aspects of chatbot interactions.
- Examinations of conversational agents' natural language understanding and response generation capabilities (Ho, 2021).
- Discussion on the challenges and opportunities in designing and evaluating chatbot interfaces for improved user satisfaction.

2.6. Chatbot Limitations, Challenges and the Future Directions:

- Identification and analysis of the limitations and challenges associated with current chatbot technologies.
- Exploration of potential improvements in chatbot intelligence, contextual understanding and dialogue management.
- Discussion of future trends, such as voice-based chatbots, multilingual capabilities and integration with emerging technologies.

This review has provided an extensive overview of the research conducted on chatbots, exploring their applications and implications across various domains. The findings highlight the effectiveness of chatbots in customer service, education, healthcare, business and human-computer interaction (Cheng and Jiang, 2022).

3. RESEARCH METHODOLOGY

Secondary data collection methods are often used in the exploratory phase of research to gain a better understanding of the topic under investigation. By reviewing existing secondary data, researchers gathered knowledge, explored different variables and refined their research questions. Additionally, secondary data often include information collected over a significant period, enabling researchers to conduct longitudinal studies. By examining trends and changes over time, researchers understood the evolution of a phenomenon, identify patterns and make more accurate predictions of the research. Some secondary data sources provide access to specialised data that may be difficult or expensive to collect through primary data collection methods. Examples include government databases, industry reports or academic publications that contain specific information relevant to the research topic. The researchers have obtained material using the secondary qualitative research approach from a variety of sources, including academic journals, research publications, prior research papers, internet articles, websites, books and reviews. These have considerably aided the researchers' efforts to compile sufficient data concerning the use of chatbots in marketing and customer support. Additionally, the researchers used a thematic analysis technique to analyse the data by looking for themes, patterns and significant results.

4. ANALYSIS AND DISCUSSION

Businesses are continually looking for novel solutions to improve the consumer experience and streamline their marketing initiatives in today's fast-paced digital environment. The usage of chatbots is one such approach that has seen great success. With the development of artificial intelligence and natural language processing, chatbots are now able to serve clients in a personalised and effective manner while also collecting crucial data for every business (Kaczorowska-Spychalska, 2019).

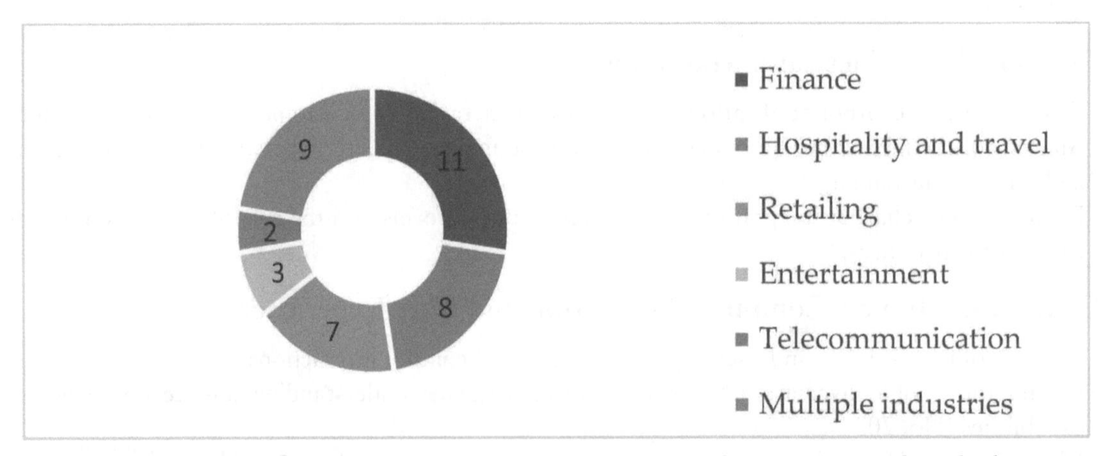

Figure 1. *'Human-Computer interaction in customer service: the experience with ai chatbots'.*

Chatbots can also provide 24x7 support, allowing customers to get assistance outside of regular business hours. This can be especially helpful for global companies that have customers in different time zones. By providing round the clock support, chatbots can help businesses meet the needs of their customers more effectively (Misischia *et al.*, 2022).

From the discussion, it can be analysed that the latest advancements in digital technologies such as AI have significantly impacted the labour market by automating tasks that require human personnel. Over the years, the maturity of chatbots for customer services has increased significantly. However, some crucial factors have been anticipated such as reputation and complementarity.

Table 1: Chatbot metrics and analysis.

Chatbots Metrics	Analysis
Response time	It is used to measure the time taken by chatbots to respond to customer queries
Customer satisfaction	It usually gathers potential feedback from customers to evaluate their satisfaction rate using surveys or other methods
Contact resolution rate	It tracks the issues of customer queries solved during the first interaction for better customer satisfaction and lowering escalation
Engagement rate	It tracks the percentage of potential customers who are actively engaged with the chatbots during the campaign to promote effective messaging
Click through rate	Chatbots usually provide clickable links to analyse the conversation rate to assess the conversion rate (Trivedi, 2019)
Lead generation	It measures the total number of leads generated by engaging chatbots. It helps analyse conversion rates to assess the effectiveness of chatbot

In addition to providing support, they can also collect valuable data regarding customers. By analysing chat logs, businesses can identify common customer concerns and areas for improvement.

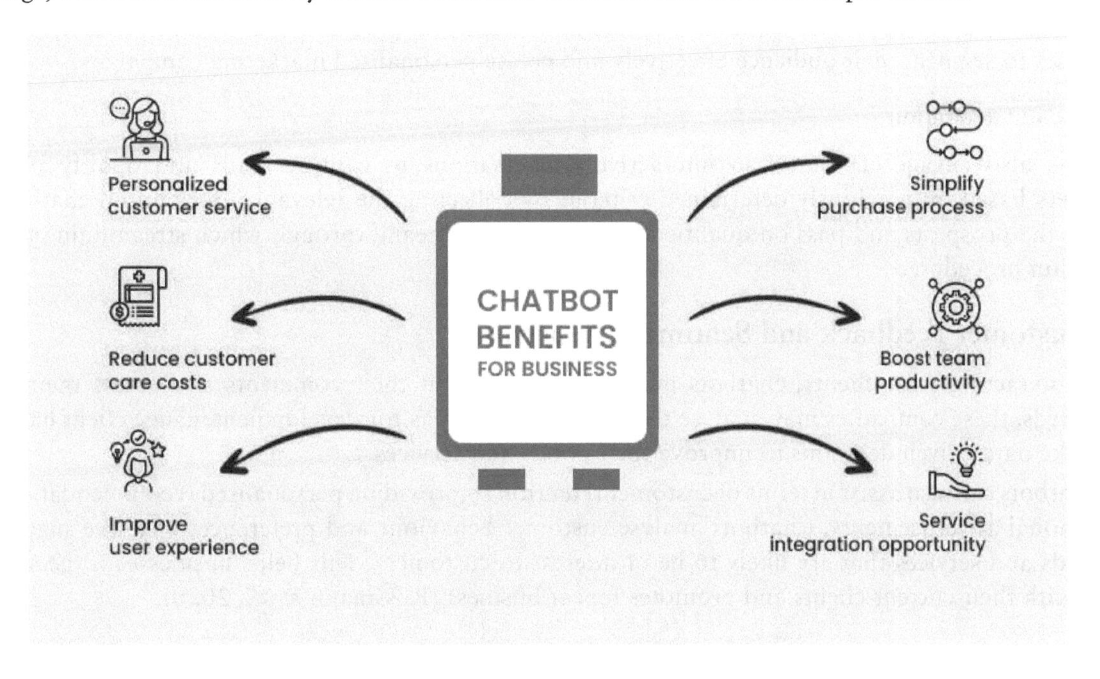

Figure 2. *Role of chatbots. Source: Trivedi, 2019*

4.1. Customer Service

4.1.1. 24x7 Availability

Chatbots are providing 24x7 availability which allows the customers to seek any assistance any time. They can efficiently handle the routine queries and also can provide information and guide customers through various processes (Følstad *et al.*, 2019).

4.1.2. Instant Responses

Customer inquiries are quickly and consistently answered by chatbots, ensuring that clients receive rapid service. Natural language processing skills allow it to comprehend consumer intentions and give correct and pertinent information regarding those objectives, thereby boosting customer happiness and experience.

4.1.3. Personalisation

Advanced chatbots can leverage data from customer profiles and previous interactions to offer personalised recommendations and suggestions. By analysing customer preferences and purchase history, it can understand and provide customised solutions which later enhance the overall customer experience.

4.1.4. Scalability and Cost Effectiveness

Chatbots are extremely scalable since they can manage several consumer conversations at once. Because of this scalability, organisations may handle a high volume of client inquiries without having to significantly increase their human resource capacity (Moriuchi *et al.*, 2021).

4.2. Marketing Analysis

4.2.1. Data Collection

Chatbots act as data collection tools which eventually gather valuable insights about the customers and also their preferences and their behaviours. Businesses may obtain a deeper insight of their target audience and improve their marketing tactics by monitoring the discussions and studying the data.

4.2.2. Customer Profiling

Through the interactions with the customers, chatbots can collect demographic data, purchasing history and browsing patterns. These data can be used to build comprehensive customer profiles, enabling businesses to segment their audience effectively and create personalised marketing campaigns.

4.2.3. Lead Generation

Chatbots also engage customers in interactive conversations to capture leads and qualify potential customers based on previously determined criteria. By collecting the relevant information, chatbots can identify the prospects and pass on qualified leads to the sales team, through which streamlining the lead generation procedure.

4.3. Customer Feedback and Sentiment Analysis

While interacting with clients, chatbots may actively seek out their comments and assess their mood. Afterwards, these companies may analyse their data to find areas for development, gauge client happiness and make data-driven decisions to improve their goods and services.

Chatbots can also assist in terms of customer retention by providing personalised recommendations and promotional advertisements. Chatbots analyse customer behaviour and preferences to make suggestions for goods and services that are likely to be of interest to customers. This helps businesses forge stronger bonds with their current clients and promotes repeat business (Rossmann *et al.*, 2020).

5. CONCLUSION

Nowadays, chatbots have become indispensable tools for businesses in customer service sector and also in the sector of marketing analysis. As the technology continues to evolve, chatbots will likely become even more advanced and valuable tools for businesses. By using the data that chatbots are providing, many companies can improve their operational efficiency and also can make data-driven marketing decision which will ultimately help these companies not only to survive but also in some cases to perform eventually better than their respective competitors.

REFERENCES

Adam, M., M. Wessel, and A. Benlian (2021). "AI-based chatbots in customer service and their effects on user compliance," *Electronic Markets*, 31(2), 427–445.

Cheng, Y., and H. Jiang (2022). "Customer-brand relationship in the era of artificial intelligence: understanding the role of chatbot marketing efforts," *Journal of Product & Brand Management*, 31(2), 252–264.

Følstad, A., and M. Skjuve (2019). "August. Chatbots for customer service: user experience and motivation," *Proceedings of the 1st International Conference on Conversational User Interfaces* (pp. 1–9).

Følstad, A., M. Skjuve, and P. B. Brandtzaeg (2019). "Different chatbots for different purposes: towards a typology of chatbots to understand interaction design," *Internet Science: INSCI 2018 International Workshops, St. Petersburg, Russia, October 24–26, 2018, Revised Selected Papers 5* (pp. 145–156). Springer International Publishing.

Fotheringham, D., and M. A. Wiles (2022). "The effect of implementing chatbot customer service on stock returns: an event study analysis," *Journal of the Academy of Marketing Science*, 1–21.

Ho, R. C. (2021). "Chatbot for online customer service: customer engagement in the era of artificial intelligence," *Impact of Globalization and Advanced Technologies on Online Business Models* (16–31). IGI Global.

Kaczorowska-Spychalska, D. (2019). "How chatbots influence marketing," *Management*, 23(1), 251–270.

Misischia, C. V., F. Poecze, and C. Strauss (2022). "Chatbots in customer service: their relevance and impact on service quality," *Procedia Computer Science*, 201, 421–428.

Moriuchi, E., V. M. Landers, D. Colton, and N. Hair (2021). "Engagement with chatbots versus augmented reality interactive technology in e-commerce," *Journal of Strategic Marketing*, 29(5), 375–389.

Rossmann, A., A. Zimmermann, and D. Hertweck (2020). "The impact of chatbots on customer service performance," *Advances in the Human Side of Service Engineering: Proceedings of the AHFE 2020 Virtual Conference on The Human Side of Service Engineering, July 16–20, 2020, USA* (pp. 237–243). Springer International Publishing.

Sarbabidya, S., and T. Saha (2020). "Role of chatbot in customer service: a study from the perspectives of the banking industry of Bangladesh," *International Review of Business Research Papers*, 16(1), 231–248.

Trivedi, J. (2019). "Examining the customer experience of using banking chatbots and its impact on brand love: the moderating role of perceived risk," *Journal of Internet Commerce*, 18(1), 91–111.

Wang, X., X. Lin, and B. Shao (2022). "How does artificial intelligence create business agility? Evidence from chatbots," *International Journal of Information Management*, 66, 102535.

44. Analysis of Social Media Marketing Impact on Consumer Behaviour

Rajesh vemula[1], Mahabub Basha S[2], Jalaja V[3], K. V. Nagaraj[4], Venkateswarlu Karumuri[5], and Manyam Ketha[5]

[1]Assistant Professor, (A), Gayatri Vidya Parishad College for Degree and PG Courses (A), Rushikonda, Visakhapatnam

[2]Assistant Professor, Department of Commerce, International Institute of Business Studies, Bengaluru

[3]Assistant Professor, School of Management Studies, Reva University, Bangalore

[4]Assistant Professor, Department of Management Studies, Gayatri Vidya Parishad College for Degree and PG Courses (A) Rushikonda, Visakhapatnam

[5]Associate Professor, Department of Management Studies, International Institute of Business Studies, Bangalore, India

ABSTRACT: Products can be investigated by the consumers, and they can able to label and can provide additional feedbacks to the products. Therefore, most of the companies have sites on social media the computer for complementing the product descriptions as well as the testimonials of the clients. Social media usage has become essential for implementing marketing strategies. Businesses have the ability to use social media likely social networking sites (SNS) to build trust and have direct conversations with their customers. To enhance their customer relationships, right marketing materials should be used by the firms that uses SNS. This will in turn affect customers' behaviour to produce sustainable performance for enterprises. This study considered social media marketing (SMM) and consumer behaviour to evaluate the quality of the customer connection, that affects the customer behavioural outcomes of participation intention, loyalty intention and purchase intention. This study gives organisations the advice that SNS marketing content must comply with SMM and consumer behaviour dimensions to achieve their marketing goals and success for long term.

KEYWORDS: Social networking, customer behaviour, Social media marketing, relationship quality

1. INTRODUCTION

The internet has made a variety of cutting-edge marketing and brand-building tools available. Additionally, it has made it possible for customers and marketers to interact with a global audience. The ability to make comments, participate in online chat rooms, use instant messaging services and exchange information through social media websites has expanded consumer connectivity (Riaz *et al.*, 2021). In recent years, online social communication has grown significantly. It is more likely that communication conducted through online social networking will reach larger groups. For both customers and businesses, this new kind of social communication seems to be a methodical and reliable source of information about products. By posting reviews, recommendations and opinions on the product and brand on their profile pages on social media platforms, consumers were given the power to influence the decisions on purchasing of their friends, coworkers or other potential customers Pawełoszek and Wieczorkowski, 2023.

SMM is the use of marketing techniques for branding, advertising and promotion on social media platforms. The most popular platform for marketers to engage with their target market and include them in advertising efforts is social media websites (Yang, 2017). SMM is primarily focused on the efficient use

DOI: 10.1201/9781003532026-44

of word-of-mouth (WOM) marketing, which is targeted to one-on-one interactions for the dissemination of information online. Electronic word of mouth (E-WOM) is the term for the use of the internet to spread and generate WOM, and it is nothing more than a form of viral marketing. Customers frequently look for reviews of the product from genuine consumers previously buying any products (Daugherty and Hoffman, 2014). E-WOM is an informal communication tool that may be found online that informs customers on the features, qualities and advantages of a product (Litvin *et al.*, 2008).

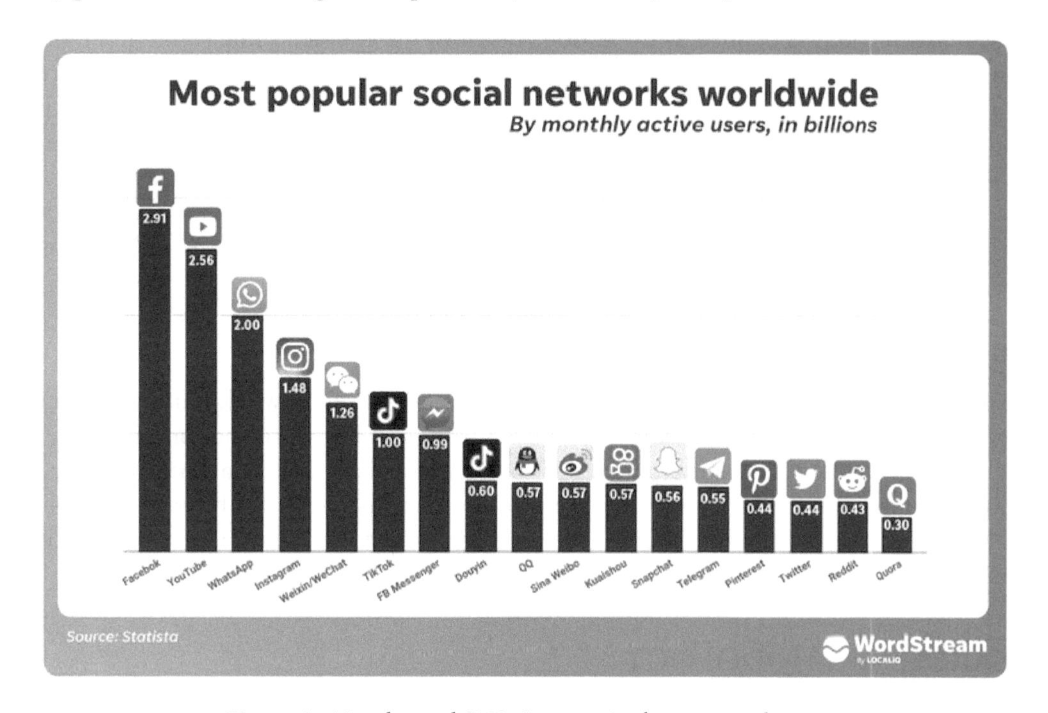

Figure 1. Mostly used SNS. Source: Author's compilation.

1.1. Research Gap and Problem Statement

Consumers frequently express dissatisfaction with the information they obtain through conventional channels. Social media increased consumer awareness since it gave them access to readily available information online. Numerous factors contribute significantly to the growth and influence of WOM developed by consumers. The majority of them are related to the wide range of investigational contexts, product types and categories, and various cultural contexts. To help consumers with their decision-making around product purchases, social media offer a variety of information linked to the product. The consumer feels satisfied after searching for and analysing information on social media, which aids them in coming to a conclusion about whether to make a purchase. By describing a model that captures the factors influencing a consumer's inclination to join in and contribute to product-related discussions on social media websites, the current study aims to close the knowledge gap.

1.2. Objective of Study

- To understand how consumers perceive their use of social media.
- To investigate how social media marketing affects consumers' attitudes.

2. LITERATURE REVIEW

Marketers use social media for their advertising initiatives and view it as a reliable, reasonably priced advertising instrument. Social media are widely accessible to the target market, allowing brands to communicate with the general public in a planned way. It is also employed as a tool to assist brands in communicating with customers effectively (Vinerean *et al.*, 2013). On social media networks, people make purchases based on the recommendations of their peers. Marketers view social media as a strategy for influencing consumer purchasing behaviour because it affects consumers' purchase decisions through online social relationships (Dabija *et al.*, 2018).

2.1. Consumer Behaviour in Social Networking Site

Online consumer behaviour is defined as the actions taken by a customer while seeking out and browsing information on a product or when purchasing goods or services online to satisfy their requirements and wants. Through social media, customers may interact with others to share their thoughts on products or services and express how they feel about them. It serves as a communication tool to get feedback and opinions from customers at each and every stage of the process of decision-making. For its users, social media have made decision-making simpler and more pleasant. Communication on platforms of social media has the capacity to affect consumers' online behaviours, from product search to post-purchase discussions (Huete-Alcocer, 2017).

2.2. Electronic-WOM

WOM marketing is described as user-to-user communication to share knowledge and experiences about a company, product or service with others. It is employed to disseminate information about the goods and services found in a conventional marketplace from person to person (Bhandari and Bansal, 2019). The development of the internet has increased the opportunity for consumers and businesses to be associated with E-WOM conduct by bringing up social media platforms, web-based social forums or communities (Lenhart and Amanda, 2009).

3. RESEARCH METHODOLOGY

Through an online poll, information was gathered to help with the current study's goal. Social media users are thought to be well-educated, and a big portion of the population is young, with college students making up the majority. Data from both primary and secondary sources are used in this inquiry.

3.1. Primary Data

In order to achieve his or her goal(s), the researcher directly collects primary data from the respondents via experiments, market research and questionnaires. Usually, it begins after the researcher has gained some insight into the issues through the evaluation of secondary data or by the study of primary data that have already been obtained and examined by another researcher.

Online survey questionnaires have been created using 'Google Docs Forms'. Students at Allahabad University and its related colleges have received links to online survey questionnaires via WhatsApp and Facebook groups for their specific classes, along with instructions on how to complete the forms through their specific lectures. For the study, students voluntarily provided their responses. The results listed below are a study of the responses using graphs and SPSS software, along with a statistical interpretation of the findings. 390 participants' responses from the study's target geographic area of Uttar Pradesh were collected as secondary data.

4. RESULT AND DISCUSSION

390 respondents provided data, which was gathered and analysed. The distribution of the sample is shown in the table 1 below based on the respondents' gender, age, level of education and profession.

Figure 2 shows responses from a sample of about 30% male and 70% female social media users. The graph below demonstrates how frequently women use SNS.

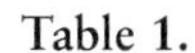

Table 1.

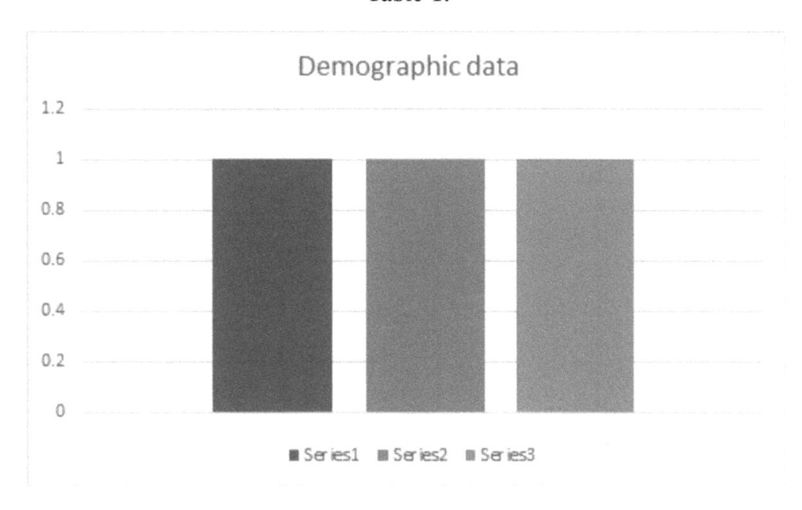

Gender

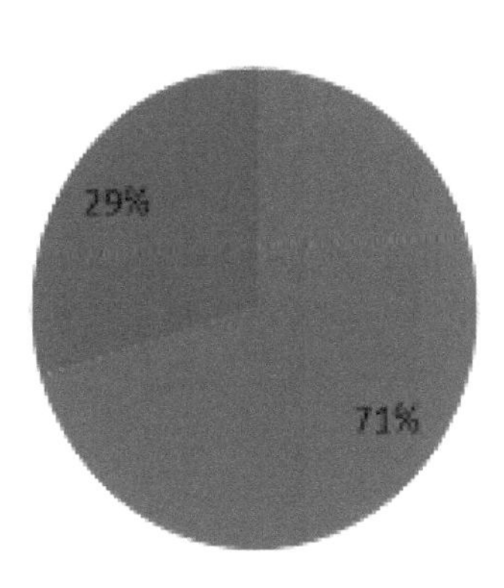

Figure 2. Gender description. Source: Author's compilation.

Figure 3 shows that roughly 69 per cent of participants are between 15 and 25 years of age, 23 per cent are between the ages of 25 and 35 years, and the remaining 8 per cent are over the age of 35. Figure 3 shows that the main age range for social media use is 15 to 25.

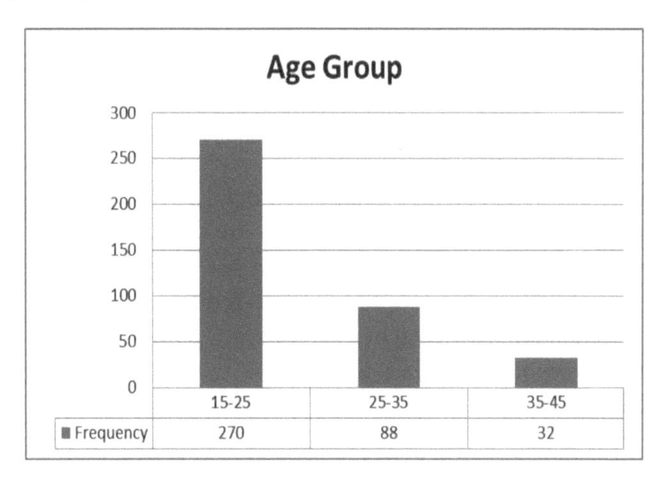

Figure 3. Age group description. Source: Author's compilation.

According to Figure 4, the majority of participants (71%) were students, 22% were in the service class and 7% were self-employed.

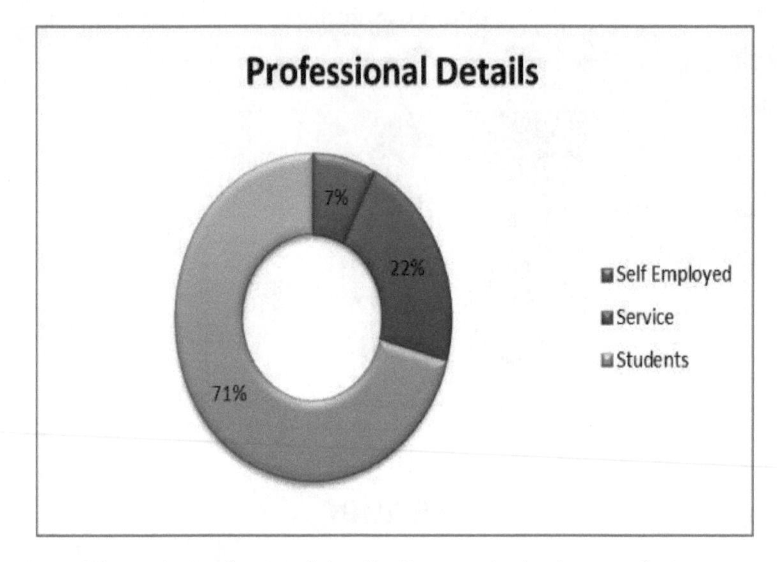

Figure 4. Professional details. Source: Author's compilation.

According to the survey's interpretation, research shows that a person's membership in a social group significantly affects their activity on social media as a voice of authority or a leader for others. This shows that when users of social media sense a connection to a group or its members, they do not hold back when offering their opinion, criticism or recommendations on social media networks.

The graph below demonstrates that for their web-based social networking requirements, respondents most frequently utilise Facebook and WhatsApp. The majority of people do not, however, use twitter frequently.

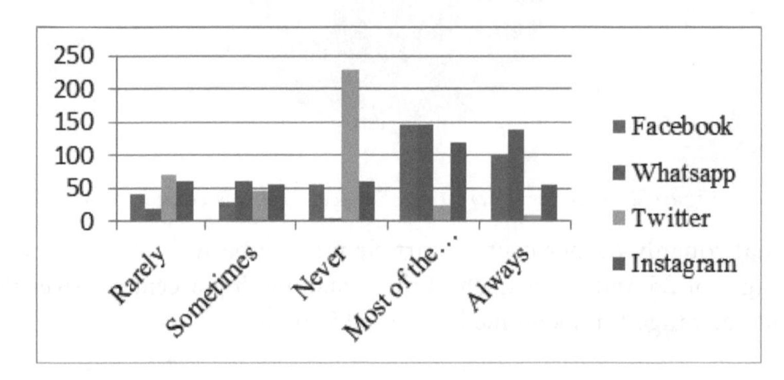

Figure 5. *Frequency of using social networking site. Source: Author's compilation.*

5. CONCLUSION

The present study was conducted to evaluate the impact of social media marketing on customer purchasing patterns. It has been noted that social media marketing focuses on developing active social relationships with the target market and generating favourable EWOM for their brand or product. According to a study of the literature, users' E-WOM behaviour can be inferred from how they act in the roles of opinion provider and seeker on social media platforms. The current study's findings will help academics understand how different factors affect how social media users behave while discussing products in E-WOM. The research's findings were anticipated to be taken into account by marketers for developing solid brand relationships with customers online.

REFERENCES

Bhandari, R. S., and A. Bansal (2019). "A study of characteristics of e-WoM affecting consumer's social media opinion behavior," *JIMS8M: The Journal of Indian Management & Strategy*, 24(1), 33–43.

Dabija, D. C., B. M. Bejan, and N. Tipi (2018). "Generation X versus millennials communication behaviour on social media when purchasing food versus tourist services," *E+ M Ekonomie a Management*, 21(1), 191–205.

Daugherty, T., and E. Hoffman (2014). "eWOM and the importance of capturing consumer attention within social media," *Journal of Marketing & Communication*, 20, 82–102.

Forbes, L. P., and E. M. Vespoli (2013). "Does social media influence consumer buying behaviour – an investigation of recommendations and purchases," *Journal of Business & Economics Research*, 11(2), 107–111.

Huete-Alcocer, N. (2017). "A literature review of word of mouth and electronic word of mouth: implications for consumer behaviour," *Frontiers in Psychology*, 8, 1256.

Lenhart, and Amanda. (2009). Adults and Social Network Websites. Pew Internet & American Life Project. January, 2009.

Litvin, S. W., R. E. Goldsmith, and B. Pan (2008), "Electronic word-of-mouth in hospitality and tourism management," *Tourism Management*, 29(3), 458–468.

Pawełoszek, I., Wieczorkowski, J. (2023), Trip planning mobile application: a perspective case study of user experience. Engineering Management in Production and Services, 15(2), 55-71. doi: 10.2478/emj-2023-0012.

Riaz, M. U., L. X. Guang, M. Zafar, F. Shahzad, M. Shahbaz, and M. Lateef (2021). "Consumers' purchase intention and decision-making process through social networking sites: a social commerce construct," *Behaviour & Information Technology*, 40(1), 99–115.

Vinerean, S., I. Cetina, and M. Tichindelean (2013). "The effects of social media marketing on online consumer behaviour," *International Journal of Business & Management*, 8(14), 66–93.

Yang, F. X. (2017). "Effects of restaurant satisfaction and knowledge sharing motivation on eWOM intentions: the moderating role of technology acceptance factors," *Journal for Hospitality & Tourism Research*, 41, 93–127.

45. Empirical Research on New Business Strategy in a Rising Economy: Managing Strategic Innovation of Business Model

Manju Dahiya[1], Simran Agarwal[2], Akash Bag[3], Mohammad Salameh Almahairah[4], Shilpa Chaudhary[5], and Col B. S. Rao[6]

[1]Associate Professor, Division of Economics, School of Liberal Education, Galgotias University

[2]Student, Division of Economics, School of Liberal Education, Galgotias University

[3]Assistant Professor, School of Law and Justice, Adamas University, Kolkata

[4]Asra University Amman, Jordan

[5]Assistant Professor, Mittal School of Business, Lovely Professional University, Phagwara

[6]School of Business, SR University, Warangal, Telangana

ABSTRACT: According to recent research, a business model's innovation is one of the most crucial success criteria for businesses and has a favourable impact on how well such organisations operate. How this link develops across an organisation's life cycle is still unclear. We contend that while business model innovation eventually becomes less significant, it nevertheless significantly impacts firm performance in earlier stages. As a result, we gathered information on 255 Indian companies and ran analytical tests using structural equation modelling. Concerning the latter, our findings demonstrate that, even though it has decreased over time, business model innovation remains a crucial path for organisations, particularly in their early years. All in all, this study opens new exploration areas by broadening and consolidating clarifications for the existence cycle theory and plan of action advancement.

KEYWORDS: Business model, strategic innovation, life cycle.

1. INTRODUCTION

Novel business models seem to assume a significant part in upsetting whole industry elements and having an impact on 'the manner in which individuals live, work, consume and cooperate with each other'. As per a few examinations, a startup called Uber, for example, offered an area-based application that permitted people to recruit a confidential on-request driver as opposed to going through the conventional taxi organisation permitting process. De Jong and van Dijk report that for a seriously prolonged stretch of time, bitcoin-based strategies really resentful how standard financial associations led business. Episodic proof shows that beneficial plans of action do not really involve a superior or more creative item, yet change the round of the business. As a result, it is unsurprising that the development of successful novel business models has become a crucial requirement for directors of various businesses (Massa *et al.*, 2017) (Figure 1).

DOI: 10.1201/9781003532026-45

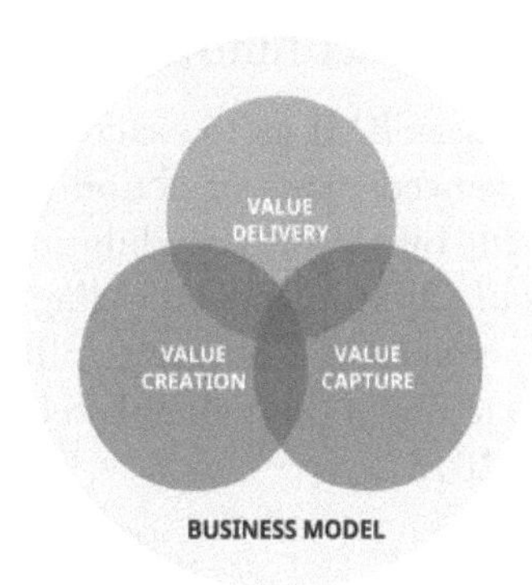

Figure 1. *Business model innovation. Source: Suh et al., 2020.*

Supervisors of occupant firms and business people are progressively utilising the business model idea to comprehend and to reconsider novel ways on the most proficient method to accomplish their organisation's objectives. Business models, on the other hand, are a topic that is extensively discussed across almost all economics disciplines, including technology and innovation management and strategy and supportability (Suh *et al.*, 2020). Since the idea has first and foremost been brought to the scholarly world, plan of action development (BMI) is considered as a wellspring of upper hand that eventually prompts financial execution (Foss and Saebi, 2017). This unmistakable connection is to some degree the essence, yet in addition, the foundation of supervisors of occupant firms and business people is progressively utilising the business model idea to comprehend and to reconsider novel ways on the most proficient method to accomplish their organisation's objectives. Business models, on the other hand, are a topic that is extensively discussed across almost all economics disciplines, including technology and innovation management and strategy and supportability. Since the idea has first and foremost been brought to the scholarly world, plan of action development (BMI) is considered as a wellspring of upper hand that eventually prompts financial execution [3]. This unmistakable connection is to some degree the essence, yet in addition the foundation of business model research.

1.1. Objectives

- Providing a comprehensive review of empirical studies examining this relationship in the literature.
- Introducing further experimental proof on the gainful person of BMI.

2. LITERATURE REVIEW

2.1. Strategic Business Model

Companies must acquire the ability to investigate, plan and construct new business areas in order to remain competitive over the long term. Towards this end, a plan of action is a portrayal of an association and how it capabilities in accomplishing its objectives, like productivity, development, and social effect (Betz, 2018). As indicated by studies, it has shown to be an important instrument for putting up novel thoughts and advances for sale to the public as well as a driver of development to open mechanical potential (Pucihar *et al.*, 2019). Researchers perceived vigour as the capacity of a plan of action to remain practical and useful in a changing business climate. Steady with this idea, an essential plan of action demonstrates that essential reasoning must be continually supported in the creation and development of an association's business exercises. The strategic capacity to quickly and successfully enter new business models, including SME, is an important source of competitive advantage and leverage that improves organisations' sustainability performance. As a direct consequence of this, research on the sustainability of businesses has begun to concentrate on the strategic business model as an essential component of innovation competitive advantage.

2.2. Life Cycle Stages and Business Model Innovation

While earlier examination frequently stresses BMI as the sacred goal for accomplishing firm execution, later examination demonstrates that advanced plans of action are not generally essentially better than existing plans of action, to such an extent that positive exhibition suggestions frequently unequivocally rely upon possibility factors (Futterer *et al.*, 2020). Understanding the possibility systems that unfurl BMI into positive firm execution suggestions is of most extreme significance for some organisations (Kranich and Wald, 2018). However, contingency factors of this valuable relationship were not thoroughly discussed in effect-side BMI research (Paweloszek, 2015).

3. METHODOLOGY

3.1. Data Collection

Using a cross-sectional research design, we gathered data from ventures in Hindi- and English-speaking nations to answer our research question. When looking into the relationship between BMI and venture performance, it has been demonstrated that cross-sectional designs are an effective method. Our example should include the key leaders of their respective businesses, also known as the top management team or the founders of the business. This is necessary because the major decision-makers shape the business model and strategic orientation of the company. From March to April, we used a self-administered survey, including one reminder email and the first approach, to gather data. In addition to the information that needed to be sent to the key decision-maker, we sent an email to 3773 individual entrepreneurs with the link to our online survey or, if there was no direct contact information, the email address of the venture. Subsequently, 270 polls were gotten back to us. In total, 19 returned surveys had critical missing qualities and straight liners that we erased, in this manner bringing about 252 respondents and a general reaction pace of 7%.

3.2. Data Analysis

We drew on laid-out measures and applied seven-point Likert-type scales with the exception of where generally expressed. We likewise pre-tried the poll with a gathering of twelve specialists, in particular PhD scientists working in the financial matter's office at college, accordingly guaranteeing validity and clearness. A few factors, which we included as control factors, are necessary for the connection between BMI and performance: in addition to the company's size, which was determined by the number of employees, the key respondents' age, sex and education level were also taken into consideration. Our research model was tested using structural equation modelling (SEM), a statistical method that allows for simultaneous evaluation of complex models with multiple relationships.

4. RESULTS AND DISCUSSION

Descriptive statistics and zero-request connections between all of the factors used in the tests are presented in Table 1.

Table 1: Descriptive analysis.

	Mean	SD	1	2	3	4	5	6	7
Performance of firm	5.45	1.135	1						
BMI	5.25	1.025	0.369	1					
Stage of lifecycle	2.03	0.456	0.330	0.060	1				
Company size	2.15	1.361	0.195	0.089	0.337	1			
Education	6.72	0.918	0.062	–0.035	–0.055	0.045	1		
Age	35.65	8.779	0.035	0.078	0.061	0.119	0.085	1	
Gender	2.19	0.346	0.034	–0.068	0.070	0.030	–0.050	–0.068	1

The essential assessment objective of this study was to examine the relationship precisely among BMI and firm execution. As a result, we gathered crucial data and tested our hypotheses by examining the underlying model's coefficients and their meanings with SmartPLS 3. Figure 2 lays out the consequences of the fundamental model.

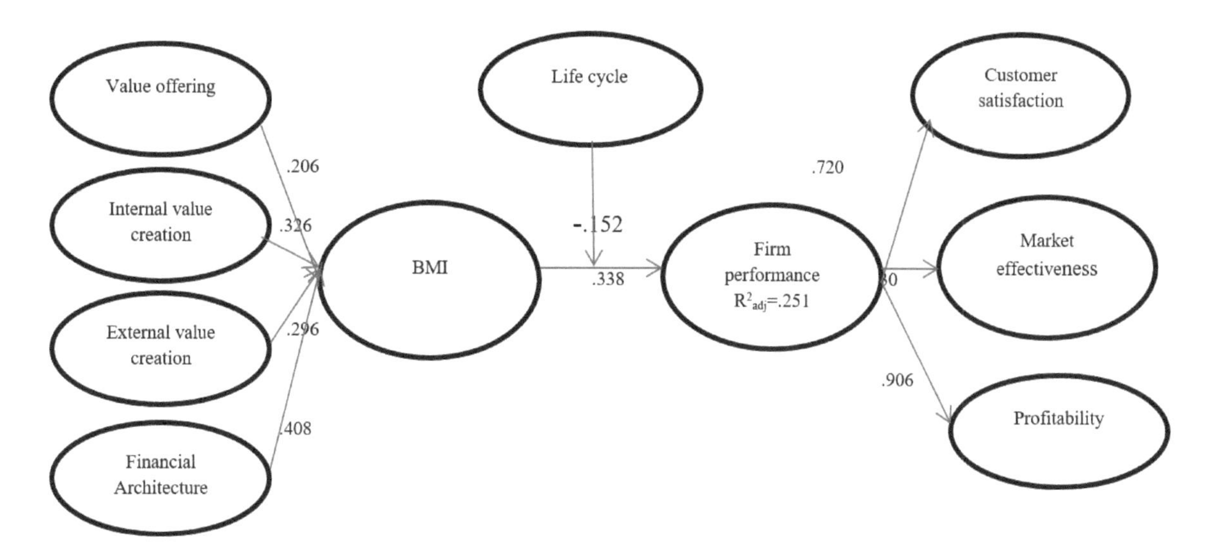

Figure 2. SEM model.

Figure 3 and Table 2 display the analysis findings. BMI and life cycle stage have a negative and significant two-way interaction (= 0.485, p = 0.020).

Table 2: An overview of the stages of life cycle moderating effects.

Main result	Term Path Coefficient	T-value
BMI ——► Firm performance	–0.485	–2.350

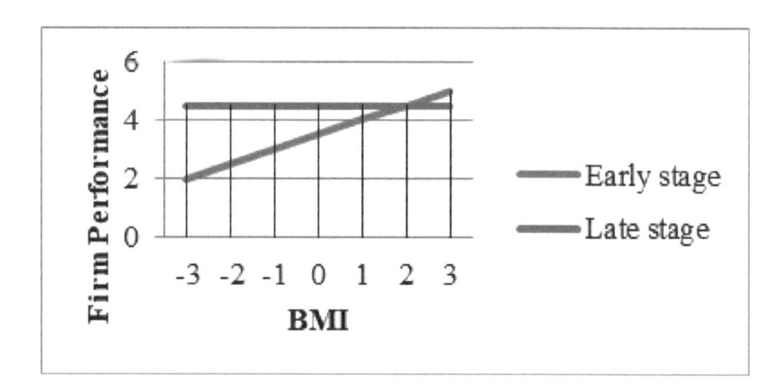

Figure 3. Directing impact of life cycle stages.

5. CONCLUSION

Managers and business owners may gain a better understanding of how to use a innovative business model to achieve success from these findings. According to previous research, BMI is a significant predictor of organisational performance implications. According to our findings, a business model that is more innovative is more beneficial to organisational performance than one that is less innovative. As a result, managers ought to make developing and putting into practice a novel business model a top policy priority. Second, our findings demonstrate that a creative action plan with a remarkable selling point is an essential component of an efficient development process, particularly in the early stages of an organisation's life cycle. An innovative business model becomes less important as the project progresses because other aspects become more important. This study demonstrates that BMI performance outcomes are significantly influenced by an organisation's individual life cycle stage, calling for careful consideration in this context.

REFERENCES

Betz, F. (2018). "Strategic idealism and reality – the case of the international financial grid," *Strategic Business Models: Idealism and Realism in Strategy* (pp. 145–164). Emerald Publishing Limited.

Foss, N. J., and T. Saebi (2017). "Fifteen years of research on business model innovation: how far have we come, and where should we go?" *Journal of Management*, 43(1), 200–227.

Futterer, F., S. Heidenreich, and P. Spieth (2020). "Is new always better? How business model innovation affects consumers' adoption behaviour," *IEEE Transactions on Engineering Management*, 69(5), 2374–2385.

Kranich, P., and A. Wald (2018). "Does model consistency in business model innovation matter? A contingency-based approach," *Creativity and Innovation Management*, 27(2), 209–220.

Massa, L., C. L. Tucci, and A. Afuah (2017). "A critical assessment of business model research," *Academy of Management Annals*, 11(1), 73–104.

Paweloszek I., Approach to Analysis and Assessment of ERP System. A Software Vendor\'s Perspective, Proceedings of the 2015 Federated Conference on Computer Science and Information Systems 2015, M. Ganzha, L. Maciaszek, M. Paprzycki, Annals of Computer Science and Information Systems\", 1415–1426, IEEE\", DOI:10.15439/2015F251.

Pucihar, A., G. Lenart, M. KljajičBorštnar, D. Vidmar, and M. Marolt (2019). "Drivers and outcomes of business model innovation – micro, small and medium-sized enterprises perspective," *Sustainability*, 11(2), 344.

Suh, T., O. J. Khan, B. Schnellbächer, and S. Heidenreich (2020). "Strategic accord and tension for business model innovation: examining different tacit knowledge types and open action strategies," *International Journal of Innovation Management*, 24(04), 2050039.

46. An Empirical Investigation on Reputation Risk Management Determinants, Value and Awareness from the Banking and Insurance Sector

Manju Dahiya[1], Vaishali Jain[2], Alim Al Ayub Ahmed[3], Souvik Roy[4], Mohammad Salameh Almahairah[5], and Seema Sharma[6]

[1]Associate Professor, Division of Economics, School of Liberal Education, Galgotias University

[2]Student, Division of Economics, School of Liberal Education, Galgotias University

[3]Associate Professor, Department of Business Administration, Fareast International University, Dhaka, Bangladesh

[4]Professor, Adamas University

[5]Asra University Amman, Jordan

[6]Department Management, Assam Down Town University, Guwahati, Assam

ABSTRACT: This research seeks to objectively investigate reputation risk management, which has grown in significance over the past few years in the Asian banking and insurance sectors. We start by utilising a text mining procedure and discover that yearly reports exhibit an expansion in familiarity with reputation risk the board and its importance in contrast with different dangers. In addition, we present the first empirical analysis of reputation risk management's advantages and disadvantages. Our discoveries show that bigger organisations are considerably bound to execute a standing gamble the board programme than are organisations that are situated in Asia and have more grounded information on their standing. Last but not least, we discover preliminary proof of the significance of reputation risk management.

KEYWORDS: Banking, reputation risk management, value, insurance sector.

1. INTRODUCTION

A regulatory definition of reputational risk was credited to the Board of Governors of the Federal Reserve System in 1995: 'Reputational risk is the possibility that unfavourable press about an institution's business practices, whether true or false, will result in a loss of clients, expensive litigation or decreased revenue'.

Reputational risk is no different from other main risks in that it eventually affects an institution's economic worth, just like operational, market and credit risks do. The perception that stakeholders have of the institution in the public eye may be impacted by a particular incident. If stakeholders ultimately decide to alter their behavior, it may have an effect on future revenues, losses and/or greater costs, and consequently the institution's final market value.

Reputation risk management is difficult because it is frequently thought of as a risk of risks, with a variety of sources (Gatzert and Schmit, 2016). At the same time, a number of factors and developments have increased the need for reputation risk management, particularly the popularity of various platforms in social media, where information spreads rapidly and unfiltered and it has allowed stakeholders to interact dynamically with one another and create reputation risks. The banking and protection areas, whose plans of action are reliant upon trust, especially need creating capacities with regard to overseeing notoriety gambles (Eckert and Gatzert, 2017).

DOI: 10.1201/9781003532026-46

The Allianz Risk Barometer has also identified this issue where loss of brand reputation or brand value ranked among in the top ten and top five risks associated with business in financial service sector. Reputational gambles even hold the best position among key dangers, as indicated by a 2014 Deloitte report. Therefore, the purpose of this paper is to conduct the very first empirical study using a sample of Asian banks and insurers on awareness of reputation risk management, its determinants and its value.

Using a sample of Asian banking and insurance organisations as a basis, this essay seeks to answer these research issues. To start with, we use text mining to dissect 830 gathering yearly reports from 80 firms north of a ten-year term to research the consciousness of notoriety risk (the executives) over the long haul. Based on the frequency with which the terms 'reputation', 'reputational risk' and 'reputational risk management' are used, the level of awareness can be identified. By allowing a correlation between ventures and districts regarding the detail and, as a result, the familiarity with notoriety risk and its development in recent years, our paper expands on the research study done by Mukherjee *et al.* (2014), who use a shorter period of time by particularly focusing on Asian Banks (Mukherjee *et al.*, 2014).

By distributing the main concentrate on the firm qualities and elements that influence the execution of a standing gamble the board programme and by breaking down the worth importance of notoriety risk the executives comparable to the effect of general ERM, we further add to the group of writing. This is finished utilising a catchphrase search and extra true measures to find organisations that oversee notoriety risk, which can likewise be utilised in future examination. It is tried for contrasts among gatherings and depends on regression and correlation examinations.

2. LITERATURE REVIEW

Since corporate reputation is a tricky idea in monetary writing, there is certainly not an exceptional meaning of 'reputation' yet rather a huge number of translations. Reputation can be for the most part characterised as 'a definitive immaterial' implying that it is just the association insight by various individuals (Eisfeldt *et al.*, 2020). Reputation finds opportunity to make can't be purchased and is effectively harmed. Specifically, corporate reputation can be characterised as a perceptual.

When compared to other powerful competitors, any depiction of an organisation's past operations and potential futures that show the firm as generally enticing for its core members (Fombrun, 2015) is valid. Dissecting corporate standing expects one to think about a firm inborn nature: a lawful fiction addressing a nexus of a bunch of contracting connections, partner collaborations with the foundation. In this manner, it is absolutely critical to think about the differentiation between the partner gatherings. In such manner, reputation can be then portrayed as an aggregate evaluation of an organisation engaging quality to a particular gathering of partners (Money *et al.*, 2017). Thusly, while characterising corporate reputation, one of the significant inquiries is whether corporate reputation is particular or plural. In this worry, there is an agreement that the corporate reputation is a complex develop, implying that various gatherings of partners can have various insights.

There is a lot of research on corporate standing, especially how to define and evaluate it and how reputation affects money-related execution (Baruah and Panda, 2020). In relationship, the sensible composition on standing gamble is respectably sparse. Exact examinations of notoriety risk typically rely on market responses to beneficial setbacks and find that, most of the time, they (by far) outperform the main hardship, demonstrating outrageous money-related reputational incident. The factors that determine banks' reputation risk are further investigated in some studies. One more strand of the writing manages ways to deal with overseeing reputation risk. A few scientists address the subject of introducing notoriety risk in a comprehensive endeavour risk the board (ERM) structure, while Gatzert *et al.* (2016) look at free security deals with any consequences regarding notoriety risk as one gamble the leader's action (Gatzert *et al.*, 2016). In addition, Mukherjee *et al.* (2014) examine the frequency of related words and the disclosures made by 21 Asian banks regarding reputation risk (Mukherjee *et al.*, 2014). Generally speaking, the results of past observational evaluation underline that it is critical to screen enthusiastically the degree of corporate standing and to control expected standing risks, since reputation and reputation hurting occasions can amazingly (ominously) influence accomplice's direct and (thus) finance-related execution for an outline of preliminary affirmation.

3. METHODOLOGY

In our exact examinations, we expect to analyse two industry areas and one region. For this reason, we start with every Asian bank and guarantors with accessible market capitalisation for 2018 in Data stream and afterwards apply a few screening models.

We centre on enormous cap firms for a time of a decade (2009-2018), since an all-encompassing standing risk the executive's framework with its significant expenses is commonly more applicable for enormous firms, which are substantially more presented to media and partner consideration. All organisations in the example had to be good to go during the whole-time frame.

We don't know about any enormous monetary administrations firm that failed because of an unadulterated standing gamble occasion during the thought about decade, furthermore, consequently potential survivor predisposition doesn't represent an issue. After the prohibition of explicit industry subsectors because of their unique status, like stock trades, and furthermore barring firms with deficient information for the example time frame, our example covers 830 firm-year perceptions.

The associations in the model are made from 50 Asian banks and 30 Asian well-being net suppliers and address more than 70% of the hard and fast capitalisation of industry market. The associated financial information in a large amount of USD was gathered from Datastream of Thomson Reuters, and the text mining analysis was done based on the event annual reports (counting changes if necessary). The annual reports were downloaded from association websites due to Asian businesses.

In our initial interpretation of the text mining approach to management, we gain insight into the leaders and mindfulness of notoriety risk as depicted in the annual reports of the associations. More specifically, we examine how the terms 'reputation', 'reputational risk' and 'reputational risk the board' have changed over time. To look at the determinants of the execution of a standing gamble the executive's programme, we utilise a Cox corresponding peril model, with regard to ERM determinants.

We also hope to precisely examine the value of the chiefs' reputational risk, as the board's goal is to acquire and further develop reputation as a significant asset. In this sense, hypothetical and exact proof exhibits that partner conduct and an organisation's financial performance are influenced by (changes in) reputation.

4. RESULTS AND DISCUSSION

Utilising a text mining examination of their yearly reports, we initially look at the ascent in consciousness of reputation risk and chance among leaders of the significant banks and insurance agency throughout the course of recent years. Predictable based on our perceptions and assumptions regarding the revelations of 19 Asian banks between 2008 and 2013, Figure 1 shows that the amount of the three analysed terms is consistently expanding and that it dramatically multiplied from 545 to 1,921 somewhere in the range of 2009 and 2018.

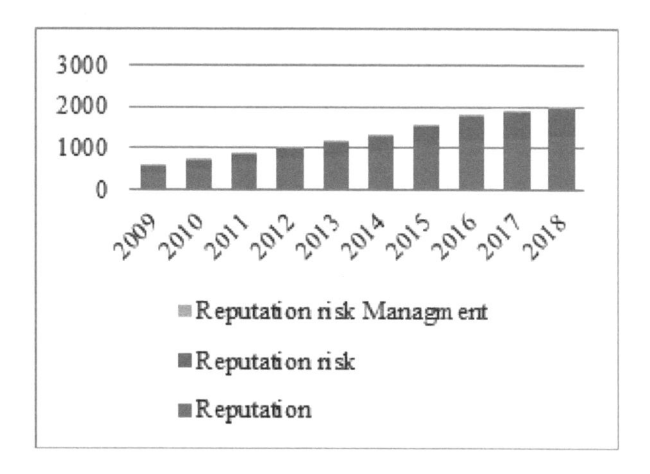

Figure 1. Increasing awareness of reputation risk.

We initially look at the importance of reputation faces a challenge concerning all risks (Table 1) to explore the overall meaning of the inspected terms. We find that the executives are the only ones in the reports who use the term risk.

According to the results presented in Table 1, an increase in the overall frequencies of the three terms under investigation is not solely attributable to an overall increase in the number of words that are included in annual reports; rather, it is also a direct consequence of a significant rise in the organisations' awareness of the significance of reputation and the risk, as reflected in annual reports.

Table 1: The level of the expression 'reputational risk (management)' on 'risk (the executives)' after some time, as well as the level of the three terms under assessment on the report's general word count for Asian banks and safety net providers.

Year	Share of Reputational Risk on Risk Management (%)	Share of the Examined Term (%)
2009	0.4	0.0066
2010	0.42	0.0073
2011	0.43	0.007
2012	0.48	0.0087
2013	0.56	0.0098
2014	0.61	0.0111
2015	0.69	0.0128
2016	0.73	0.014
2017	0.86	0.0152
2018	0.84	0.0155

As the disparities in means and medians in Table 2 indicate, businesses without a reputation risk management programme have a significantly higher Tobin's Q regarding the value of reputation risk management.

Table 2: A single variable distinguishes groups with and without a reputation risk management (RRM) programme.

	RRM (245 firm observation)		No RRM (569 firm observation)	
	Mean	Median	Mean	Median
Size	14.344	14.689	13.365	13.535
Leverage	0.921	0.942	0.860	0.922
RoA	0.730	0.533	1.567	0.845
Bank	0.730	1	0.598	1
Asia	0.815	1	0.398	0
Reputation awareness	29.315	23	10.782	8
Risk awareness	1165.342	960	589.746	491

A Cox proportional hazard model is used to first examine the factors that influence the adoption of a reputation risk management programme in the context of a multivariate, as shown in Table 3.

Table 3: The Cox corresponding danger model's results about the variables that impact reputation risk management.

	Parameter Estimate	Std. Parameter Estimate	Ratio of Hazard
Size	.497	2.862	1.538
Leverage	–.032	0.022	0.960
RoA	0.063	0.853	1.059
Bank	0.160	0.321	1.172
Asia	2.159	4.356	9.518
Awareness of reputation	0.063	2.450	1.067
Awareness of risk	–0.003	–3.347	0.989

5. CONCLUSION

Our findings highlight the growing importance of reputation risk management, a difficult field that is still mostly under development. Future study is also required, particularly with regard to empirical investigations, qualitative work and quantitative measurements of reputation risk. For example, extra exploration is expected to decide whether reputation risk management can influence how the market answers catastrophic operational loss occasions or, all the more by and large, whether it might constrict the impacts of crisis events. From the perspective of other stakeholder gatherings, for example, clients, more exploration is additionally required. Overall, we have uncovered some crucial preliminary findings into understudied areas that could be the foundation for further investigation.

REFERENCES

Baruah, L., and N. M. Panda (2020). "Measuring corporate reputation: a comprehensive model with enhanced objectivity," *Asia-Pacific Journal of Business Administration*, 12(2), 139–161.

Eckert, C., and N. Gatzert (2017). "Modeling operational risk incorporating reputation risk: an integrated analysis for financial firms," *Insurance: Mathematics and Economics*, 72, 122–137.

Eisfeldt, A. L., E. Kim, and D. Papanikolaou (2020). Intangible value (No. w28056). National Bureau of Economic Research.

Fombrun, C. (2015). "Reputation," *Wiley Encyclopedia of Management*, 1–3.

Gatzert, N., and J. Schmit (2016). "Supporting strategic success through enterprise-wide reputation risk management," *The Journal of Risk Finance*, 17(1), 26–45.

Gatzert, N., J. T. Schmit, and A. Kolb (2016). "Assessing the risks of insuring reputation risk," *Journal of Risk and Insurance*, 83(3), 641–679.

Money, K., A. Saraeva, I. Garnelo-Gomez, S. Pain, and C. Hillenbrand (2017). "Corporate reputation past and future: a review and integration of existing literature and a framework for future research," *Corporate Reputation Review*, 20, 193–211.

Mukherjee, N., S. Zambon, and H. Lucius (2014). "Do banks manage reputational risk," *IAFEI Quarterly*, 27, 22–36.

47. The Influence of Strategic Leadership and Organisational Invention on Strategic Management: The Mediational Function of IT Potential

Ashok Kumar Maurya[1], Chinki Nagar[2], Mounika Mandapuram[3], Mohammad Salameh Almahairah[4], Rajchandar K[5], and Nimisha Beri[6]

[1]Associate Professor, Division of Economics, School of Liberal Education, Galgotias University

[2]Student, Division of Economics, School of Liberal Education, Galgotias University

[3]Senior Software Engineer, Technopathz Inc, Texas, 75067, USA

[4]Asra University Amman, Jordan

[5]Assistant Professor, Department of CS & AI, SR University, Warangal, Telangana, India

[6]Professor and Head, School of Education, Block 18, Lovely Professional University, Phagwara – 144401, Punjab

ABSTRACT: Human resource is a significant considers accomplishing organisational objectives. An organisational accomplishment cannot be isolated from the job of top leaders in figuring out the right procedure. The current review looks at how organisational innovation and strategic leadership affect strategic management. It provides an additional perspective on the effects of strategic leadership and organisational innovation by focusing on the IT limit's role as a mediator. It fixates on how a connection can do persuading dire relationship by utilising the strategic leadership and organisational innovation. In this way, the data were gathered from the workers of Indian universities. Focus on overviews was used to gather the data. The review's findings revealed that organisational innovation, IT proficiency and strategic leadership are significant precursors of key administration. Moreover, it was also asserted to intervening effect of IT limit. Policymakers and academics can benefit from the review's findings.

KEYWORDS: Strategic management, leadership, IT potential, organisational invention.

1. INTRODUCTION

Leadership is a management idea that has an essential situation in doing all organisational activities to make organisational progress. This is in accordance with Bass' assertion expressing that leadership is a vital aspect for deciding an association (Bass, 1990). Despite time too as position, leaders can contribute basically in how the manner by which a firm capacity its activities. A leader's successful systems can be the main reason for using the strategic management process in a productive way. Leaders who adhere to techniques not only determine how to put those techniques into practice, but also help to facilitate the creation of appropriate vital exercises. Leaders employed by a variety of foundations are well aware of the need to develop a decisive plan for the organisation's future and participate in the appropriate sanctioning of these comprehensive plans.

Through essential administration with the motivation to reduce costs and increase advantage, the advantage and value creation are further developed. The administration provides the association's employees with strong motivations in order to accomplish the legitimate goal (Özigci, 2020). It is regarded as the foundation of management control and evaluation. Additionally, the essential administration ensures that the affiliation heads receive a comparative evaluation of activities and issues. Eventually, organisations

DOI: 10.1201/9781003532026-47

take on strategic management strategies to acquire an upper hand over their competitors (Hilman and Abubakar, 2019).

To expand the performance of the organisation, information technology capacities assume an indispensable part (Oh *et al.*, 2016). IT abilities are the heaps of IT assets. By utilising these groups, organisations can undoubtedly facilitate the exercises of the business. As a result, assume a crucial role in improving organisational performance. Associations can encourage IT limits using IT-related resources.

In the past, management composition typically uses the term 'strategic leadership'. It addresses management issues that are typically dealt with by upper management of the company. It is essential to gain feeling of that the headway of cutoff points concerning key authority is exceptionally not exactly equivalent to other association contemplations. The depiction of the leader's constraint to imagine, anticipate, engage and remain mindful of adaptability to energise key change as required is called strategic management. Enhancing an organisation's capacity to coordinate and adjust both the internal organisational climate and the external climate is the primary goal of strategic leadership.

The effects of strategic leadership and organisational innovation on essential administration are examined in this research. By evaluating the IT limit's mediational occupation, it provides yet another instrument for the relationship that was previously discussed. Accordingly, the continuous audit upgrades understanding of strategic management's parts and the techniques by which an association can actually deal with its business through strategic leadership and organisational innovation. As a result, the primary objective of this review is to investigate the connection between strategic leadership, strategic management, IT skills and organisational innovation in Indian colleges.

2. LITERATURE REVIEW

2.1. Strategic Leadership

As per Ireland and Hitt (1999), strategic leadership is a leader's capacity to expect, imagine, keep up with flexibility, think in a calculated way and lay out work joint effort to start changes that make a good future for the association. Specialists have portrayed vital authority as the limit of the strategic leader to make sensible choices concerning the targets and objectives of the affiliation. Besides, significant moves are initiated by the pioneer to eliminate the vulnerability (Jabbar and Hussein, 2017). Additionally, specialists have defined strategic leadership as the individual's capacity to envision, anticipate, foster adaptability, foster a strategic point of view and collaborate with others to initiate a significant organisational change. Scientists have in like manner conveyed it as the limit of the individual or gathering to effect, act and consider others in a way an association can get an upper hand (Alayoubi *et al.*, 2020).

2.2. Organisational Innovation

A few past researchers have introduced various meanings of organisational innovation. Hence, there are a few assortments of definitions accessible in regard to hierarchical development. These different definitions show that the impression of the association concerning authoritative development is interesting. No matter what this, portions of creativity are certified by the for the most part open composition. The analysts use the term 'organisational innovation' to describe the reception or production of novel behaviour or ideas. As a result, 'organisational innovation' refers to a new way of thinking or acting within the company. In addition, this idea significantly influences the organisation (Camisón and Villar-López, 2014).

2.3. IT Capabilities

The term capacity is depicted by some researchers as the limits, information, legitimate cycle, cutoff points and qualities, connecting with a relationship to accomplish better execution and develop a high ground in the market (Erkmen *et al.*, 2020). Along these lines, there are two potential ways of characterising IT abilities. Right off the bat, it very well may be characterised as the inside abilities and the arrangement of the organisation. Also, making IT empower to use, arrangement, and reconciliation, collect and select IT assets that team up with the abilities of the organisation.

3. METHODOLOGY

The procedure of the ongoing audit depends on positivism. Thusly, a quantitative study strategy was picked. The subsections under portray the combination of data, respondents, testing framework and evaluation plan. The ongoing survey's assessment plan is based on the quantitative methodology. In order to gather data from college employees in India, this survey employed a fundamental irregular examining method. The objective respondents of this review were school laborers. Mail frameworks and individual visits were used by the researchers to collect responses' examinations.

515 questionnaires were given to the respondents to complete. An overview was created to collect the respondents' review data. This overview was a 7-point Likert scale. The questionnaires were created in view of past writing. The 5 things of ITC were adjusted, from certain examinations, for example, our IT capability is great at overseeing gets and our IT capability can keep a productive work and so on; five things of strategic leadership were taken on, from certain investigations, for example, our chiefs rouse individuals and make a culture of greatness and our chiefs guarantee the organisation that is light-footed and adaptable enough to confront change really and so on.

The things of organisational innovation were adjusted from certain investigations, for example, 'Innovation obtaining and double-dealing is dissected and a suggestion complete intermittently'. 'The essential administration process suggested is direct', the statements went, 'The cycle steps are in a group that grants to play out the continuous step with the information from previous projects'. Preceding going through the principal data collection process, a Pilot test was facilitated to take a gander at the reliability of the survey, for sure. Smart PLS and SPSS were used later to guide the information analysis. Because the PLS-SEM method was accepted for the examination process, Smart PLS was used.

4. RESULTS AND DISCUSSION

Descriptive examination was gotten from a survey dispersed to 515 respondents who were the respondents from universities of India. Figure 1 depicts the demographics of the respondents, which show that 28% were female and 72% were male.

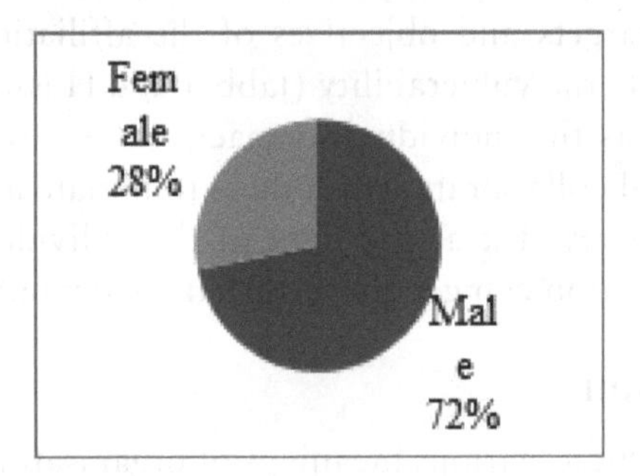

Figure 1. *Gender discrimination.*

Figure 2 also shows that 35% of respondents were over the age of 40, 45% were between the ages of 30 and 40, and 20% were between the ages of 20 and 30.

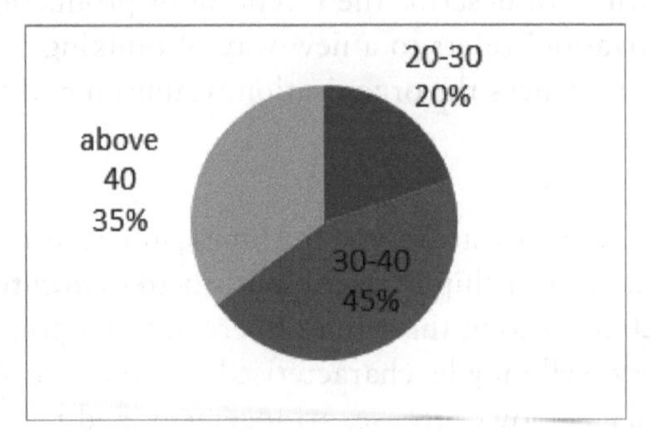

Figure 2. Age discrimination.

Figure 3 shows that 44% of respondents were qualified for a graduate degree, while 37% were qualified for a PHD. Of course, 19% of respondents held the bachelor's degree certificate.

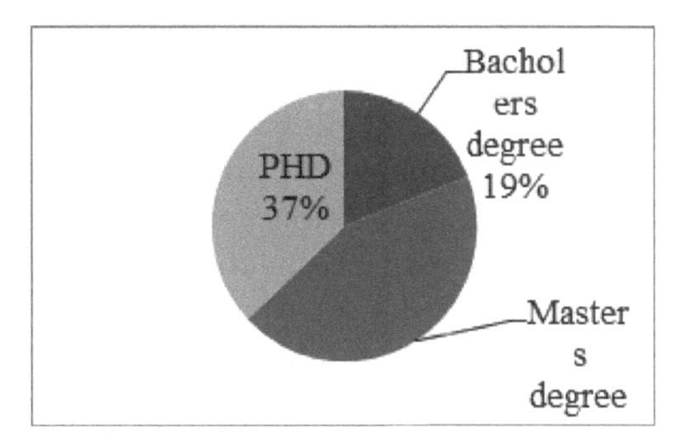

Figure 3. *Educational qualification.*

From that point onward, the information was inspected for legitimacy utilising Cronbach's alpha and composite dependability, with an adequate worth of 0.70. All CR and Cronbach alpha characteristics are greater than .70, as shown in Table 1.

Table 1: Validity and reliability.

	CR	Cronbach alpha	rho_A	AVE
OI	.909	.876	.879	.669
SM	.924	.920	.930	.615
ITC	.890	.841	.852	.619
SL	.889	.839	.868	.620

Discriminant validity is exhibited by the way that the characteristics at the corner to corner are more prominent than the excess values of the matrix, demonstrating that there is no connection issue, as shown by the upsides of AVE that are referred to in Table 2.

Table 2: Discriminant validity.

	ITC	OI	SL	SM
ITC	.781			
OI	.581	.818		
SL	.619	.620	.785	
SM	.749	.709	.720	.782

This study inspected the potential gains of R^2 to figure out the impact of independent components on outcome factors. As indicated by the outcomes accumulated as R square, ITC is made sense of 46.6% by the autonomous factors. Simultaneously, these factors make sense of SM by 53.4%.

Table 3: Values of R^2.

	Original values
ITC	.466
SM	.534

5. CONCLUSION

In this exceptionally merciless period, instructive organisations need to confront contest from adjacent and worldwide college. In light of the current circumstances, educational establishments ought to concentrate on implementing a strategic management strategy for making ends meet in this significant market. By taking on this framework, the affiliation can accomplish and maintain an advantage. In such manner, the continuous review analysed the impact of IT limit (ITC), strategic leaders and organisational innovation on strategic management. In addition, Indian universities experienced a breakdown in the ITC's mediation effect. The survey's findings demonstrate a positive correlation between these variables. These outcomes show that the colleges ought to rely on leader who can think totally in this season of genuineness. In like manner, reliance on IT factors is what's more key for definitive achievement. IT capacities will probably assume a critical part in cost reserve funds and growing organisational innovation. Eventually, progression is likewise basic for the perseverance of affiliations. The colleges ought to be creative so the nonstop understudies can be held and new understudies can be drawn in.

REFERENCES

Alayoubi, M. M., M. J. Al Shobaki, and S. S. Abu-Naser (2020). "Strategic leadership practices and their relationship to improving the quality of educational service in Palestinian Universities," *International Journal of Business Marketing and Management (IJBMM)*, 5(3), 11–26.

Camisón, C., and A. Villar-López (2014). "Organizational innovation as an enabler of technological innovation capabilities and firm performance," *Journal of Business Research*, 67(1), 2891–2902.

Erkmen, T., A. Günsel, and E. Altındağ (2020). "The role of innovative climate in the relationship between sustainable IT capability and firm performance," *Sustainability*, 12(10), 4058.

Hilman, H., and A. Abubakar (2019). "Establishing a highly competitive university: a strategic management perspective," *Journal for Global Business Advancement*, 12(1), 3–50.

Jabbar, A. A., and A. M. Hussein (2017). "The role of leadership in strategic management," *International Journal of Research-Granthaalayah*, 5(5), 99–106.

Oh, S., H. Baek, and S. Lee (2016). "Revisiting the relationship between information technology capability and firm performance: focusing on the impact of the adoption of enterprise resource planning systems," *The Journal of Information Systems*, 25(1), 49–73.

Özigci, Y. E. (2020). "Crimea as Saguntum? A phenomenological approach to the Ukrainian crisis within the framework of a transforming post-bipolar structure," *Croatian International Relations Review*, 26(86), 42–70.

48. Employing Big Data Analytics to Compare Conventional Strategies for Financial Brand Communication Assessment

Manju Dahiya[1], Pavitra[2], Parihar Suresh Dahake[3], M. Ravichand[4], Sai Srujan Gutlapalli[5], and Seema Sharma[6]

[1]Associate Professor, Division of Economics, School of Liberal Education, Galgotias University

[2]Student, Division of Economics, School of Liberal Education, Galgotias University

[3]Assistant Professor, Department of Management Technology, Shri Ramdeobaba College of Engineering and Management, Nagpur

[4]Professor of English, Mohan Babu University, Tirupathi, (Erstwhile Sree Vidyanikethan Engineering College)

[5]Data Engineer, Techno Vision Solutions LLC, Farmington Hills, MI 48335, USA

[6]Department Management, Assam Down Town University, Guwahati, Assam

ABSTRACT: Although tremendous measures of information are currently accessible to organisations, simple ownership of this information is deficient, and thorough information investigation is expected for better business choices. Data providers like social networking services (SNS) are now important. In view of the fast ascent of social networking services, they are presently broadly utilised in an assortment of examination points in the sociologies. The objective of this study is to work on existing comprehension of the amazing open doors given by friendly information for brand image in the financial firms. To that explanation, traditional procedures and digital strategies are used to inspect the brand image of financial associations. The Periodic Evaluation of the Image is a typical technique. Sentiment analysis, an AI device for huge information investigation in sociologies, is the computerised technique. The data are examined using both methods, and the outcomes are compared. While the outcomes of the big data approach and those of conventional strategies are compatible, the former facilitates faster and simpler data processing. The imperatives of this work are connected with the example size, the area broke down, and the degree of the assessed literature.

KEYWORDS: Financial brand, big data analytics, SNS, PEI.

1. INTRODUCTION

The amount of data available on the Internet regarding firms has increased dramatically in recent years. SNS, a huge wellspring of such information, have become generally used in new examination patterns in sociologies. Additionally, an innovation that benefits human prosperity and assists with improving the climate has had a significant impact on the development of our general public. Besides, due to the expanded worldwide admittance to the Web, we can discuss a democratisation of data access.

Over 90% of the total populace approaches a cell phone, and cell phone use is speeding up. In 2019, it is projected that mobile Internet penetration will be 71%, with 71% of the population using a device. This accessibility to cell phones, which can now do works previously saved for work areas and PCs, is fundamentally altering the manner in which individuals convey today (Saura *et al.*, 2019). Web-based shopping, financial administrations, and admittance to data and news are a portion of the capabilities performed with cell phones these days.

DOI: 10.1201/9781003532026-48

These modifications should be considered in marketing research (Marjani *et al.*, 2017). Clients overall use both traditional and digital media, and organisations communication with their clients by means of both conventional and digital media channels. In any case, present-day advances offer up new chances for organisations as far as creating purchaser communication and creating a brand picture. In this environment, we ought to work on our current awareness and understanding to the correspondence among traditional and digital media, as well as what the two mean for client impressions. To that reason, new systems, for example, those in light of Big Data analytics, are required. Big Data analytics definitive object is to find patterns, relationships, examples, bits of knowledge or buyer inclinations to further develop business decision-production (Philipp *et al.*, 2014). The evaluation of Big Data analytics or unstructured data is known as 'big data analytics.' To do this review, it is expected to initially find a data set, organise it utilising information text mining strategies, and afterward continue with its examination to put together decision-making with respect to the information. Nonetheless, on the grounds that the sources will generally be unique, this three-step strategy deals with different issues, strikingly in the underlying period of getting the information. Notably, given the current interplay of the traditional and digital channels, it must be explored whether the same results concerning a company's reputation image can be reached using alternative approaches.

1.1. Objective of Study

- The focus of this study is the current understanding of the opportunities social data presents for brand communication analysis in the financial sector.
- The study's objective is to investigate conventional methods for evaluating financial brand communication using big data analytics.

2. LITERATURE REVIEW

2.1. SNS

Companies no longer had complete ownership of their brands, according to research. Instead, interactions with customers are what create impressions of a brand. Thus, clients become prosumers on the Web. Twitter is the most famous social mechanism for examining organisation brands (Philipp *et al.*, 2014). Simoes and co. made and applied an opinion examination technique in a concentrated on bank pictures in light of an examination of their Twitter accounts (Sergio *et al.*, 2017).

2.2. Brand Image

While choosing and choosing an item, administration, or brand, public image is an indispensable thought. Buyers dissect the properties of different things, administrations, or brands as a component of their dynamic cycle (Paweloszek, 2015). Considering that the media is evolving these days, bringing about an intermingling of informative commitment and on-line and disconnected correspondence, it is important that brand image management consolidates the ongoing reality. According to Perez Pernas, brands are valuable resources for businesses because they can provide them with a controllable long-term competitive advantage (Perez Pernas, 2014). In this manner, financial establishments attempt to sustain their brands. One system for fostering a brand is sponsorship. For example, Banco Santander stands separated as a successful brand that has commonly used sponsorship to build up the organisation's image.

2.3. Periodic Evaluation of the Image

Utilising projective philosophies and top-to-bottom meetings, the Periodic Evaluation of the Image (PEI) strategy was developed to determine the connections between the various qualities of the brand and to gather the important components of the brand's image. This picture intermittent examination can be performed for various endeavours, including finance and the tourism business (Peters *et al.*, 2019).

3. METHODOLOGY

3.1. Sentiment Analysis

Sentiment analysis is a technique for estimating and distinguishing sentiments related with a particular occasion. This philosophy has filled lately, with the essential objective of catching and sorting the sentiments expressed in a specific example by classifying them as certain, negative or unbiased mentalities. In this review, we utilised a Python-based calculation to do the feeling examination. Different potential outcomes incorporate calculation preparing with information mining strategies, which are processes used to work on a calculation's likelihood of accomplishment with AI in light of the rightness of the outcomes. Support Vector Machines (SVMs) and most extreme Entropy (MaxEnt) have likewise been broadly utilised in social network analysis with ML to recognise huge elements in an assortment of exploration fields.

3.2. Data collection

A total of 6,689 tweets with the company hashtags (#) (#Helaba, #Deutsche, #Commerz, #DZ, #Bayern LB, and #Triodos) were downloaded for the purpose of data collection. The database was filtered, and 193 tweets were removed, leaving 6,496 tweets in the dataset. The period during which the information were gathered was from October 1 to 7, as there were no huge occasions that might have impacted the public view of the organisations being scrutinized or the brand images of SNS clients. Additionally, considering the way that the Twitter Programming connection point doesn't engage tweets to be recuperated following 7 days, the tweets were collected on October 7.

3.3. PEI Method

The PEI is an exploration approach with a few applications in the field of promoting. As a rule, the PEI is utilised to evaluate the image of a firm or brand. The gathering of data is the first step in the PEI strategy. What are gathered are insights that can be measured and consequently communicated mathematically? To gather information, every interviewee is introduced to a progression of 'stimuli' (like brands, products, and elements) that are utilised to project their thought. Every stimulus is addressed on a card with the interviewer's name.

With the right information handling, this strategy makes it conceivable to separate different kinds of data, for example, general image credits and relationship between stimuli, relative images, individual images, essential images, and worldwide situating of stimuli.

4. RESULTS AND DISCUSSION

4.1. The PEI Method's Outcomes

The data were first acquired using the PEI approach. Following that, we acquired broad picture attributes, i.e., various attributes reported by interviewers for each stimulus. The traits that emerge the most frequently are those that are most prominent in the minds of the interviewers (see Table 1).

Table 1: The characteristics of the investigated financial companies.

Characteristics	Frequency	C.F
Give more security	21,5	21,5
proximity	16,5	38,0
Advertising	12,3	50,3
Online service	8,2	58,5
Resource management	5,1	63,6
Personal service	5,1	68,7

Characteristics	Frequency	C.F
Adapted to young people	2,6	71,1
Less confidence	2,4	73,5
Social work	4,3	77,8
Variety of product	3,6	81,4
Solvency	1,1	82,5
Less commission	1,1	83,6
Custom product	3,0	86,6
More innovative	2,8	89,4
Nit been inversed	0,8	90,2
Other feature	9,8	100,0

The degree of affiliation is higher in the PEI strategy when more interviewees partner any sets of qualities a more noteworthy number of times, as well as the other way around. Among the examined elements, Helaba and Deutsche were seen as more equivalent among themselves by interviewees, with an affiliation level of 45.0 percent. On the contrary, Helaba and Triodoswere viewed as the most dissimilar. Table 2 depicts the degree of association between the financial companies under consideration.

Table 2: The relationship between variables.

Entities	Frequency
Helaba/Deutsche	46,1
Helaba/Commerz	31,4
Helaba/DZ	32,8
Bayern LB/Commerz	29,0
Deutsche/ DZ	30,1
Deutsche/Bayern LB	24,7
Deutsche/Commerz	25,3
Helaba/ Bayern LB	19,1
Triodos /Bayern LB	21,0
Bayern LB/ DZ	19,2
Triodos/Commerz	16,9
Commerz/DZ	14,0
Triodos/ DZ	9,8
Deutsche/ Triodos	9,4
Helaba/ Triodos	5,6

The vertical location of each stimulus is determined by computing a value between zero and one hundred proportionally to the score for stimuli with lower evaluation and one hundred for stimuli with higher evaluation (see Table 3).

Table 3: View of the trademark ''give more certainty/give greater security among the different elements.

Entities	Percent
Helaba	28,8
Deutsche	24,3
Triodos	5,6
Bayern LB	13,5
Commerz	10,5
DZ	17,3
Total	100,0

4.2. Sentiment Analysis Results

UGC has frequently been used to determine the most crucial criteria for a given issue. The fundamental characteristics of the investigated financial entity profiles are shown in Table 4.

Table 4: The total number of followers and extracted tweets.

#	Tweets	Followers
# Helaba	2010	56.7k
# Deutsche	2502	94k
# Commerz	1335	40.9k
# DZ	29	39.8k
# Bayern LB	899	26.2k
# Triodos	692	44k

The sentiment analysis algorithm was used, which categorises each tweet as negative, neutral, or positive. As recently expressed, a sum of 1,694 positive tweets, 3,205 neutral tweets, and 1,364 negative tweets were obtained and portioned further by monetary substance and exactness rate (see Table 5).

Table 5: According to the company, the sentiment of tweets.

Entities	Positive	Negative	Neutral	Avg. Accuracy
Helaba	545	163	701	0.686
Deutsche	637	311	938	0.707
Commerz	195	427	633	0.656
DZ	5	9	19	0.815
Bayern LB	29	559	538	0.743
Triodos	47	332	387	0.619

These findings indicate that individuals on Twitter are more likely to voice unfavourable rather than positive sentiments regarding financial institutions.

5. CONCLUSION

The growing number of social stages and Web associated gadgets produce a plenty of social information. Organizations have huge snags with regards to examining social information. With the progression of new advancements and computerized correspondence processes, it is basic to work on existing attention to the open doors managed the cost of by friendly information for brand correspondence examination in the financial business. As indicated by our discoveries, the two methodologies viable produce results that are significantly tantamount. On the other hand, the Big Data analytics approach that places an emphasis on machine learning technologies enables us to evaluate the data more quickly than the traditional approach.

REFERENCES

Marjani, M., F. Nasaruddin, A. Gani, A. Karim, I. A. T. Hashem, A. Siddiqa, and I. Yaqoob (2017). "Big IoT data analytics: architecture, opportunities, and open research challenges," *IEEE Access*, 5, 5247–5261.

Paweloszek I., Approach to Analysis and Assessment of ERP System. A Software Vendor\'s Perspective, Proceedings of the 2015 Federated Conference on Computer Science and Information Systems 2015, M. Ganzha, L. Maciaszek, M. Paprzycki, Annals of Computer Science and Information Systems\", 1415–1426, IEEE\", DOI:10.15439/2015F251.

Perez Pernas, A. (2014). Strategic management of the corporate brand in financial markets: reference to the Santander group.

Peters, M., B. Pikkemaat, and J. Frehse (2019). "The future of destination image analyses: implications of a city image research," *17th Biennial International Congress: Proceedings of the Tourism and Hospitality Industry*, New Trends in Tourism and Hospitality Management.

Philipp, E. W., M. S. Kelm, and J. O. Sausen (2014). Bibliometric study of scientific publications on competitive strategies from 2002 to 2013. Salão do Conhecimento.

Saura, J. R., P. Palos-Sanchez, and A. Grilo (2019). "Detecting indicators for startup business success: sentiment analysis using text data mining," *Sustainability*, 11(3), 917.

Sergio, R. S., T. P. Christopoulos, and E. P. Vasques Prado (2017). "Behaviour of banks on twitter and its effects on the brand image," *REGE-Revista de Gestao*, 24(1), 2–12.

49. An Empirical Investigation of the Organisational Behaviour of Citizens and Practices of Green Human Resource Management

Ashok Kumar Maurya[1], Zeenat Parveen[2], Priya J[3], Franklin John Selvaraj[4], and Melanie Lourens[5]

[1]Associate Professor, Division of Economics, School of Liberal Education, Galgotias University

[2]Student, Division of Economics, School of Liberal Education, Galgotias University

[3]Assistant Professor, Department of Professional Studies, CHRIST (Deemed to be University), Bengaluru, Karnataka, India – 560029

[4]Department of Management Studies, Dr. N.G.P. Institute of Technology, Kalappatti Road, Coimbatore, Tamil Nadu

[5]Deputy Dean Faculty of Management Sciences, Durban University of Technology, South Africa

ABSTRACT: Sustainability is a new field that gives businesses a competitive advantage. The success of companies that produce fast-moving consumer items can be attributed in part to their concern for environmentally friendly human resource management. However, there hasn't been much research on OCBE in this area. Thus, the important goal of this assessment was to make a model that could assist with enabling environmentally beneficial behaviour. Therefore, utilising a sample size of 965 hotel sector personnel in India and structural equation modelling with AMOS, the analysis was conducted, depending on the social identity theory. The factors were shown to be positively correlated by the results. Additionally, this link is moderated by organisational identification and employee environmental commitment. While the interaction between Green HRM and OCBE is moderated by Individual Green Value. The study provides HR managers and students of sustainability with knowledge that will aid in fostering OCBE among employees.

KEYWORDS: OCBC, green human resource management, green value.

1. INTRODUCTION

The arising "behavioural viewpoint" on green HRM rehearses has gotten a ton of late consideration with respect with the impacts of reasonable organisational practices on the environmental responsibility and prosperity of representatives (Shen *et al.*, 2018). As indicated by this perspective, there is a "rational cost-benefit analysis" that exists among social and environmental results. According to the "shared benefit" rationale (Pinzone *et al.*, 2019), the "human aspect" is really important for the successful implementation of reasonable organisational practices, particularly supportable HRM works on.

The issue of climate change has gained critical importance for all people, organisations, and communities. For every person who lives on Earth, environmental degradation, globalisation, resource depletion, and industrialisation pose severe concerns and problems. To overcome these challenges, it is necessary to concentrate on the most important aspects that can assist in achieving sustainable objectives, such as green human resource management (GHRM), enthusiasm for environmental protection, and organisational citizenship behaviour within an organisation. As per the exploration, GHRM exercises emphatically affect the environment performance shows in Figure 1. The desire of a company's human resources to address environmental problems and gain financial advantages for the businesses would improve with the implementation of green training and rewards (Kim *et al.*, 2019). India is a developing nation that is now working to promote environmental management and eliminate barriers to sustainability, making it an appropriate setting for the current study.

DOI: 10.1201/9781003532026-49

The firms may now turn threats into opportunities and gain a competitive edge thanks to GHRM. The motivation behind this study is to look at what GHRM means for organizational citizenship behaviour for the environment (OCBE) in the lodging business, both in a roundabout way and straightforwardly, considering the meaning of the green climate for people, associations, and society at large.

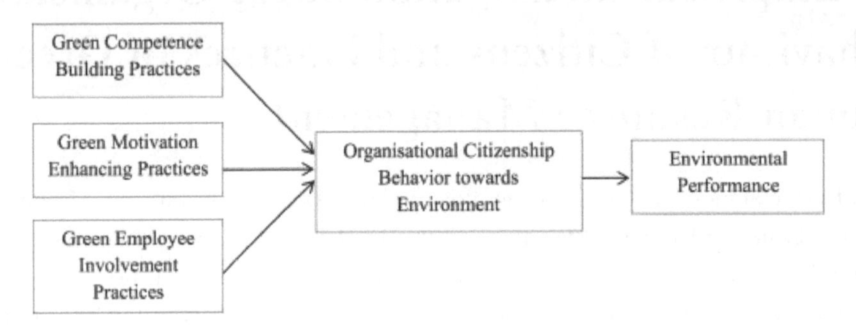

Figure 1. *GHRM on OCBE.*

1.1. Objective of Study

There is a need to investigate both direct and indirect effects through the contribution of HR green initiatives on the organisation, as it has been noted through a literature review that employee behaviour for the environment is a grey area in environmental studies. The essential objective of the examination was to look at how workers act progressively for sustainability in different settings. Therefore, in light of the foregoing discussion and literature analysis on HRM and employee ecological behavior, this research would be a step to further our understanding of how to influence organizational and individual attitudes and behaviors toward going green through the interaction between Green HRM and OCBE.

2. LITERATURE REVIEW

2.1. The OCBE and the GHRM

GHRM procedures improve OCBE (Pinzone *et al.*, 2019). Some researchers discovered that GHRM activities like as training and development, employee involvement, and performance management are strongly associated to OCBE in their study. Green training improves employees' expected environmental capacities, abilities, and information to accomplish organisation environmental objectives. Green performance management approaches support natural execution by associating worker conduct with environmental objectives and rousing representatives to take part in environmental exercises. Representatives may be encouraged to embrace green, intentional ways of behaving by surveying and assessing their environmental performance.

H1: Green HRM has a +ve relationship with OCBE.

The Mediating Role of Identification of Organisation and Environmental Responsibility in the Connection between Green HRM and OCB

As per the literature, workers construct a mind-boggling sensation of recognisable proof with the business when it has a profoundly esteemed and ideal picture, which is likewise related with proactive environmental practices. Employee appreciates environmentally cognizant ventures as their own stakeholder (Afsar *et al.*, 2018). As a result, the acceptance of green HRM practices may enhance businesses' good standing and enhance their image. In addition, research has shown that when employees' jobs enable them to fulfil a lifelong dream, they are more likely to identify with ethical businesses. Shen et al. discovered that through the role of organisational identity as a mediator, green HRM positively impacted OCB (Shen *et al.*, 2018). Shen suggest that green HRM may be very much connected with specialist organisational ID and, in this way, with positive work space brings about light of the social character hypothesis. The fact of the matter is that positive employee attitudes and behaviours are influenced by human resource management practices. As a result, one hypothesis asserts that organisational identifications have an effect on OCB as a result of green HRM practices.

H2: GHRM and OCBE are mediated by organisational identity and environmental commitment.

2.2. Personal Environmental Values

Employees are responsible for implementing green practices, but their engagement may vary for OCBE implementation. OCBE is a type of extra-role behaviour that is neither necessary nor compulsory and is not related with extrinsic benefits. They cannot, however, be forced to engage in environmentally conscious conduct. Employees participate in supplementary role activities solely because of a sense of civic duty. Representatives that have solid environmental qualities might have a higher feeling of harmony with their association as well as a more grounded feeling of consistence at work, which supports their interest for joint commitment to OCBE (Zientara and Zamojska, 2018).

H3: Employees' individual Green Value moderates the link between GHRM and OCBE.

3. METHODOLOGY

3.1. Research Design

The study is based on a positivist approach and quantitative research methodology. The idea was tested using objective measurement and statistical methodologies. The study concentrated on the most recent literature in order to assess the function of employee attitude as Organisational Identification (OI) and Environmental commitment between OCBE and GHRM. The study predicted how GHRM, OI and OCBE would change and looked at their relationship to one another. SPSS was used to analyse the quantitative data.

3.2. Data Collection

A self-reported questionnaire was used to collect the data for the study from five-star hotels in India. The questionnaire, together with the associated request letter, was delivered to employees by management. The adopted questionnaire was used, with responses ranging from strongly agreed to strongly disagree on a Likert scale of 1 to 5.

3.3. Data Analysis

The SPSS statistical tool and AMOS were used to analyse the data. All accepted measures lie between reliability scales of 0.7 to 0.8 and satisfy Hair et al. guidelines of.60 to.93. Furthermore, according to the study, the descriptive analysis included: linearity, normality, autocorrelation, heteroskedasticity, detection of abnormal values and outliers and multi-collinearity tests on for detailed analysis using SPSS to ensure the normality and suitability of data using AMOS 23.

4. RESULTS AND DISCUSSION

Correlation was used to examine the degree and direction of the association between the variables, and all of them were found to be positively associated to each other, as shown in Table 1.

Table 1: Individual-level correlation of variable.

Variables	1	2	3	4	5
OI	1				
GHRM	.43	1			
Individual green value	.38	.55	1		
Environment commitment	.33	.51	.41	1	
OCBE	.37	.59	.48	.40	1

Table 2 shows the descriptive statistics for each variable, including skewness and kurtosis.

Table 2: Kurtosis and skewness.

Variables	Kurtosis	Skewness
OI	3.2	1.5
GHRM	.35	.43
Individual green value	.72	.37
Environment commitment	.54	.15
OCBC	.05	.29

The Harman single-factor test was used to calculate the variance of the common method (CMV) in this study. The exploratory study used research structures to investigate the non-rotated factor solution. The variance of a factor is depicted in Table 3 as 29.34%, indicating that the data base did not suffer from frequent technique-biased difficulties because the variation explained by a single factor is less than the benchmark value of 51%. As a result, there were no alarming CMV concerns in the survey data, and the findings are unlikely to constitute a severe problem.

Table 3: Total variance detail.

Initial Eigenvalues of Components		
Total	Variance (%)	Cumulative (%)
8.81	29.34	29.34

After that, the measurement model was estimated using confirmatory factor analysis (CFA), and the results were found to be adequate with CMIN/DF values of 6.3, GFI values of 0.93, AGFI values of 0.86, CFI values of 0.93, and RMSEA values of 0.07.

SEM analysis produced hypothesis results. With P values less than 0.05 and a beta value of 0.55, hypothesis 1 GHRM is positively associated to an employee's OCBE was accepted. To test hypothesis 2, Organisational Identity and Environmental Commitment mediate between GHRM and OCBE, a bootstrapping analysis was undertaken. The outcomes showed that the P values were under 0.05, and the speculation was acknowledged. While underlying model gauge was additionally observed to be acceptable, with CMIN/DF = 2.4, GFI = 0.95, AGFI = 0.95, CFI = 0.97, and RMSEA = 0.04.

Finally, the hypothesis 3, that employees' Individual Green Value moderates the link between GHRM and OCBE, was examined using the interaction term of Individual Green Value and Green HRM. The result was significant when the P value was less than 0.05. As per the results, high Green HRM rehearses constantly bring about high OCBE, while personnel with strength for with values relieve the relationship between Green HRM and OCBE.

5. CONCLUSION

The current study has sought to address the research problem by arguing that, while being OCBE optional in nature cannot be compared to formal job tasks, employee volunteers for the environment can compensate for one's limitations. Furthermore, while this behaviour is not universal and differs among employees, it can assist organisations in achieving long-term sustainable goals. As a result, this research adds to the body of knowledge on the stakeholder and social identity theories-based multidimensional mediated moderated model. The results validate the speculations and show that Green HRM fundamentally affects OCBE, those environmental commitment mediators between Green HRM and OCBE, and that singular green worth is likewise a mediator for Green HRM and OCBE.

However, this study, like many research studies, has limitations that must be addressed. Despite being one of the most important industries, the management seemed unconcerned about participating in the survey, which required a lengthy time to collect data. Following that, the research was limited to five-star hotels that were participating in the quality assurance process in some way. Future study may be undertaken in SME's in developing nations with the involvement of Personnel Management because they are more numerous in offering services and can bring about a significant shift in the attitude of ordinary people toward a sustainable environment. Finally, longitudinal research is required to confirm environmental observations.

REFERENCES

Afsar, B., S. Cheema, and F. Javed (2018). "Activating employee's pro-environmental behaviors: the role of CSR, organizational identification, and environmentally specific servant leadership," Corporate Social Responsibility and Environmental Management, 25(5), 904–911.

Kim, Y. J., W. G. Kim, H. M. Choi, and K. Phetvaroon (2019). "The effect of green human resource management on hotel employees' eco-friendly behavior and environmental performance," *International Journal of Hospitality Management*, 76, 83–93.

Pinzone, M., M. Guerci, E. Lettieri, and D. Huisingh (2019). "Effects of 'green' training on pro-environmental behaviors and job satisfaction: evidence from the Italian healthcare sector," *Journal of Cleaner Production*, 226, 221–232.

Shen, J., J. Dumont, and X. Deng (2018). "Retracted: employees' perceptions of green HRM and non-green employee work outcomes: the social identity and stakeholder perspectives," *Group & Organization Management*, 43(4), 594–622.

Zientara, P., and A. Zamojska (2018). "Green organizational climates and employee pro-environmental behaviour in the hotel industry," *Journal of Sustainable Tourism*, 26(7), 1142–1159.

50. Security and Privacy Challenges in Distributed Database Management Systems

Lakshmi Prasanna Bolem[1] and Kama Sai SVM[2]

[1]Ph.D Research Scholar – College of Law, Koneru Lakshmaiah Education Foundation, KLEF (Deemed to be) University, Green Fields, Vaddeswaram, Guntur, Andhra Pradesh, 522302

[2]Associate Professor, KLEF College of Law, KLEF (Deemed to be) University, Guntur, Andhra Pradesh

ABSTRACT: The objective of this research is to investigate the challenge of recognising and evaluating distributed database management system (DDBMS) security and privacy challenges and also to create a description of important danger indicators that may lead to possible data security difficulties in distributing resources, both tangible and functioning. An investigation of the primary DDBMS security and privacy challenges that can be found while developing and managing a standard DDBMS meant for supporting data procedures and offering data services was conducted as a component of this research. The data assessment was produced depending on the findings of surveys and questionnaires administered to DDBS security and privacy professionals with varying degrees of training and emphases in their operations within this expertise field. The findings of the research include a list of primary risk variables based on their significance and frequency in practice and also a spotlight on the most important security and privacy measures.

KEYWORDS: Distributed database system, privacy, security, challenges, and distributed database management system

1. INTRODUCTION

Nowadays, DDBMS security and privacy are critical to the operations of practically every organisation that collects, procedures, and stores data using modern technology (Bouganim *et al.*, 2023). The approach depends on a periodic evaluation of data challenges, which enables one to recognise novel challenges and flaws in real time, apply suitable countermeasures and continuously track the system's current level of data security while taking previous knowledge and novel variables into account (Gupta and Agrawal, 2017).

Organisations must continually evaluate their level of risk, make suggested modifications, and actively enhance their security DDBS in preparation for possible assaults and challenges of this environment, and also avoid challenges to business procedures and activities, reputational impairment, or data loss. The foundations of any security policy are risk assessment and challenge management (Rai and Singh, 2015). Cybersecurity threats ought to be prioritised in the strategic planning of corporate procedures. As a result, every firm must establish a risk evaluation approach that best meets the firm's priorities and commercial objectives. Hence, this research aims to investigate the security and privacy challenge factors and controls of DDBMS (Curtis *et al.*, 2012).

2. LITERATURE REVIEW

DDBS are sophisticated technological structures made up of several architectural pieces that work together to offer a variety of data procedures and data services. These structures frequently function beneath random factors, the existence of numerous types of negative effects, active contact with the outside world, and the expensive nature of potential violations or failures. All of this leads to a slew of issues, the most serious of which is the security of data (Kou *et al.*, 2022). Handling cybersecurity and privacy challenge

DOI: 10.1201/9781003532026-50

evaluation in DDBS necessitates addressing several issues relating to operational dispersion and hierarchy, an elevated level of resource parallelism, & a near-complete absence of centralised management (Maurya *et al.*, 2020). As a result, implementing an innovative method for DDBMS security challenges entails the adoption of an extensive approach that incorporates data acquired from many sources as well as a broad range of instruments for their assessment.

3. RESEARCH METHODOLOGY

The investigation included two data gathering and assessment methods. The initial step will be a pilot study to test the tools used for the study and fine-tune the queries, followed by a large-scale study of the intended audience employing the finalised versions of the survey forms. To acquire a collection of examination statistics for examination, a questionnaire technique was employed. Numerous dozen data security engineers of different skill stages, penetration assessment and accounting professionals, and prominent experts in the area of security and privacy challenges of DDBMS were among those who responded. The participants have been selected at random and have prior expertise offering security and privacy for DDBS architectures of varying dimensions and levels.

The pilot investigation was conducted before the primary survey to determine if the suggested questionnaire framework is appropriate for evaluating the final measures. The primary survey included 23 experts (Table 1). Respondents in the poll ranged in age from 24–47 years old, with a mean of 34.2. Because the IBM SPSS Statistics software package is extensively used for statistical analysis, it was employed for data evaluation and modelling. The subsequent parts examine and discuss the study's results.

Table 1: Respondents' demographic data.

Variables	Number of Participants	Percentage (%)
Gender		
Female	6	26.08
Male	17	73.91
Total	23	100
Role		
Database security and privacy auditor	2	8.69
Malware examiner	2	8.69
Project manager	3	13.04
Penetration examiner	5	21.73
Infrastructure Engineer	5	21.73
Database security and privacy engineer	6	26.08
Total	23	100

4. RESULT AND DISCUSSION

Depending on an investigation by professional security and privacy specialists, Table 2 and Figure 1 depict a list of the risk factors for DDBMS. Table 3 and Figure 2 display the mean and standard deviation (SD) for every security and privacy DDBMS category. According to the findings of this research, most security and privacy measures are employed regularly and are key methods for preventing and mitigating possible challenges.

Table 2: Security and privacy challenge factors of DDBMS.

Factors	Mean	SD	Percentage (%)
Software risks	2.78	0.79	55.6
Challenges of life safety	4.17	0.65	83.4
Non-compliance legislation challenges	4.21	0.73	84.3
Security and privacy challenges	2.69	0.76	53.9
Strategic planning challenges	4.13	0.62	82.6

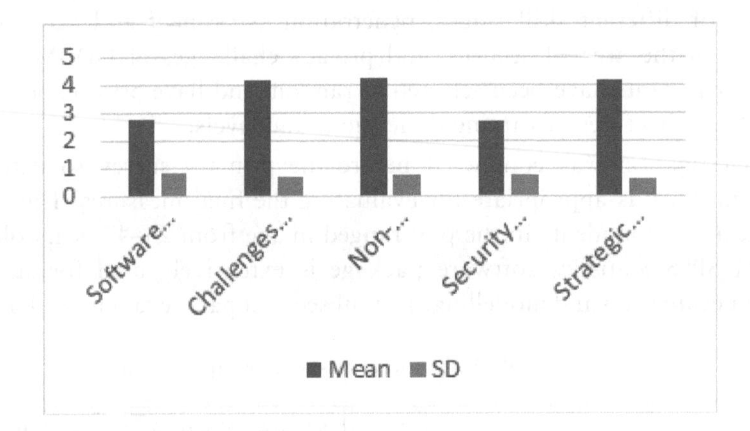

Figure 1. Security and privacy challenge factors of DDBMS.

Table 3: Security and privacy challenge factor controls of DDBMS.

Controls	Mean	SD	Percentage (%)
Backup and restore the system	4.26	0.61	85.2
Identity and access management	4.47	0.59	89.5
Prevention system for data leak	4.13	0.62	82.6
Ant-virus protection system	2.13	0.75	42.6
Network security solutions	2.08	0.90	41.7

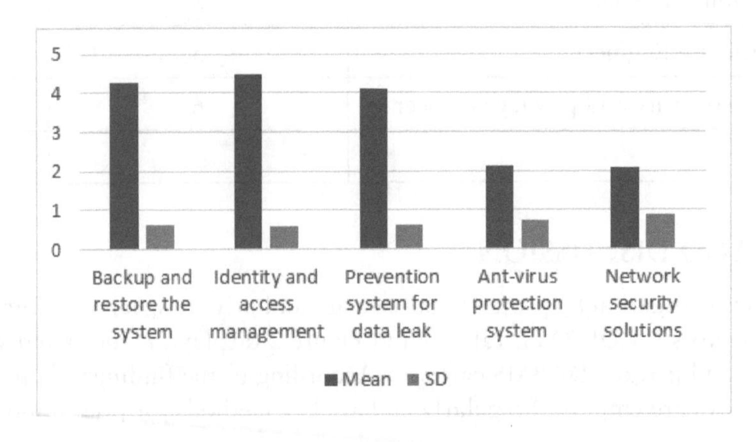

Figure 2. Security and privacy challenge factor controls of DDBMS.

5. CONCLUSION

The research investigates the challenge of recognising and evaluating DDBMS security and privacy challenges and also creates a description of important danger indicators that may lead to possible data security difficulties in distributing resources, both tangible and functioning. The study investigates the major DDBMS security and privacy challenges that can be found throughout the planning and execution of a standard DDBS meant for offering a variety of data procedures and data assistance. As an outcome, the key risk variables were ranked due to their significance and frequency in practice, and also the most important security measures were highlighted. The majority of protective measures are employed often and are essential components for avoiding and minimising possible challenges, according to an assessment of the most prevalent types of security and privacy challenges in DDBMS methods and techniques utilised in current DDBS.

REFERENCES

Bouganim, L., J. Loudet, and I. Sandu Popa (2023). "Highly distributed and privacy-preserving queries on personal data management systems," *The VLDB Journal*, 32(2), 415–445.

Curtis, L. H., M. G. Weiner, D. M. Boudreau, W. O. Cooper, G. W. Daniel, V. P. Nair, ... and J. S. Brown (2012). "Design considerations, architecture, and use of the Mini-Sentinel distributed data system," *Pharmacoepidemiology and Drug Safety*, 21, 23–31.

Gupta, N., and R. Agrawal (2017). "Challenges and security issues of distributed databases," *NoSQL: Database for Storage and Retrieval of Data in Cloud*, 251–270.

Kou, G., K. Yi, H. Xiao, and R. Peng (2022). "Reliability of a distributed data storage system considering the external impacts," *IEEE Transactions on Reliability*, 72(1), 3–14.

Maurya, A. K., A. Singh, U. N. Tripathi, S. Pandey, and D. Singh (2020). "Security in a distributed database system: a survey," *Journal of Computers in Mathematics and Science*, 11(7), 43–51.

Rai, P., and P. Singh (2015). "An overview of different database security approaches for a distributed environment," *IJISET-International Journal of Innovative Science, Engineering & Technology*, 2(6).

51. The Influence of Power and Politics on Human Resource Management

Roop Raj[1], B. E. George Dimitrov[2], Balbhagvan Acharya[3], Melanie Lourens[4], Sunitha Purushottam Ashtikar[5], and Geetha Manoharan[5]

[1]Assistant Professor in Economics, Economics, Kurukshetra, Haryana, India

[2]Assistant Professor, Department of Political Science Mount Carmel College Autonomous Bengaluru, Karnataka, India

[3]Research Scholar, Department of Management, Eklavya University, India

[4]Deputy Dean Faculty of Management Sciences, Durban University of Technology, South Africa

[5]School of Business, SR University, Warangal, Telangana

ABSTRACT: The objective of this investigation was to look into the influence of politics and power on human resources management (HRM) and the performance of staff. To study the influence of nepotism and favouritism as a kind of politics and power on HRM and the performance of staff, an explanatory study design has been used. Depending on their availability, the secondary data acquisition technique has been employed among staff functioning in various public-sector healthcare facilities. A random selection of 150 staff members has been employed for this research. The adopted questionnaire has been employed for gathering statistics. According to the correlation evaluation, there is a significant connection between favouritism, staff performance, and HRM, while nepotism has a significant connection to staff performance, yet an insignificant connection with HRM. The findings of the research revealed that nepotism encompasses a considerably negative influence on staff performance and HRM, whereas favouritism encompasses a considerably positive influence on staff performance and HRM.

KEYWORDS: Power, politics, nepotism, favouritism, and human resource management.

1. INTRODUCTION

Presented volatile surroundings and the change of flattering hierarchies, team buildings and staff empowerment, organisations nowadays use power and politics variously than in earlier times. Power, or the capacity to influence someone else, is not restricted to supervisors. Staff members at all stages, as well as strangers like consumers, can influence the acts and opinions of others (Buchanan and Badham, 2020). An individual does not have to possess power to influence someone else, and those with power might not possess influence. Furthermore, power may not be identical to control. The power invested in a specific status, like the power of the safeguarding director, is referred to as control.

Organisational Politics is defined as an individual's conduct that involves deliberate acts that influence particular choices to protect their interests. There is a link between organisational politics and staff survival because when the staff sees other individuals helping from organisational politics, they are more probable to engage in such conduct themselves (Hochwarter *et al.*, 2020). As a result, the existence of nepotism and favouritism in the context of organisational politics and power leads to disputes between staff and employer relationships in all organisations. The main obstacle during HRM adoption is the aspect of organisational politics, i.e., favouritism and nepotism, which affects the entire procedure of implementing

DOI: 10.1201/9781003532026-51

HR, like recruitment and promotions, and also staff performance, satisfaction and commitment in the government industry (AL-shawawreh, 2016). Hence, this study aims to investigate the influence of politics and power on HRM.

2. LITERATURE REVIEW

Organisational politics are influenced by the surroundings, which leads to the development of political strategies like nepotism or favouritism within the organisation (Shaukat *et al.*, 2015). According to (Ombanda, 2018), when the HRM procedure is connected to looking at nepotism (relationships of the family) or favouritism (condescending staff for close relationships) instead of the standards defining a staff member's experience and capacity, staff might view it adversely. Onthe contrary, social associations, if in the shape of nepotism or favouritism, are prioritised over organisational passions to protect their passions (Schneider, 2016).

The application of power, influence, and personal gain among staff and the leadership of an organisation give development to the perspective on organisational politics, which has consistently existed in the setting of work, like when competing for limited assets, disputes during significant choices, and the existence of self-interest among categories or people in an organisation.

3. RESEARCH METHODOLOGY

The explanatory study design has been utilised in this investigation to figure out the influence of nepotism and favouritism on HRM and the performance of staff. The secondary data-collecting methodology has been employed to gather statistics from employees at various public-sector healthcare facilities, depending on their availability. Just 150 of the 250 survey responses sent for this research have been returned correctly for additional investigation. The adopted poll has been employed for gathering information from participants. In this investigation, questionnaire items from multiple investigators were used. Statistical instruments like statistical description, dependability, connection and regression evaluation have been used to examine the statistics employing SPSS.

H1:Nepotism encompasses a negative influence on staff performance.

H2:Nepotism encompasses a negative influence on HRM.

H3:Favouritism encompasses a positive influence on staff performance

H4:Favouritism encompasses a positive influence on HRM.

4. RESULT AND DISCUSSION

The statistically descriptive results for staff performance, HRM, nepotism, and favouritism were revealed in Table 1 and Figure 1.

Table 1: Descriptives outcomes.

Factors	Mean	Standard Deviation
Favouritism	3.7	1.16
Nepotism	3.67	1.15
Staff performance	3.5	0.8
HRM	3.62	1.0

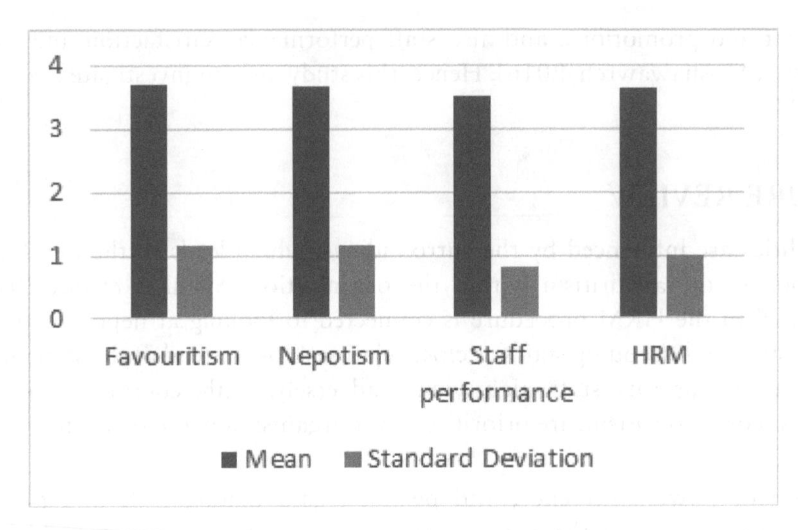

Figure 1. Descriptives outcomes.

Table 2 and Figure 2 showed the variables' Cronbach alpha values. In this research, an assessment of reliability has been carried out to evaluate the internal coherence of factors. As a result, all of the factors alpha values are inside the appropriate array of values.

Table 2: Reliability outcomes.

Factors	Cronbach Alpha	No. of Elements
Favouritism	0.81	6
Nepotism	0.84	5
Staff performance	0.75	8
HRM	0.86	6

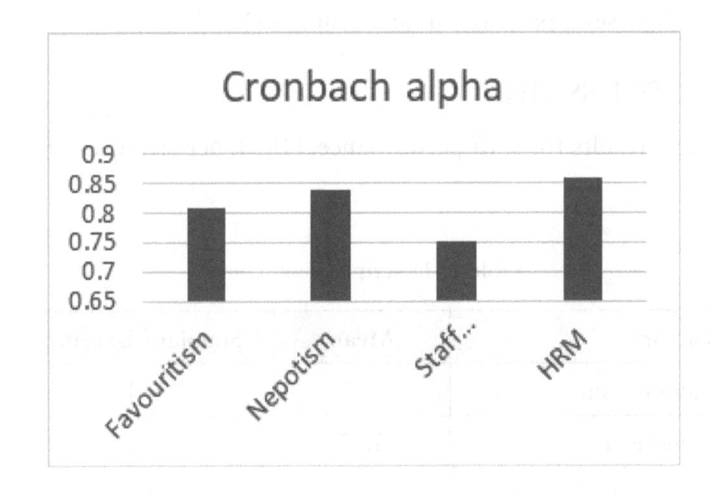

Figure 2. Reliability outcomes.

Table 3 shows the correlation number between every factor. The outcomes of the research revealed a positive connection between nepotism and staff performance.

Table 3: Correlation assessment (compensation & staff performance, coaching and mentoring recruitment and selection).

Factors	1	2	3	4
Nepotism	1			
Favouritism	0.93	1		
Staff performance	0.56	0.66	1	
HRM	0.70	0.28	0.82	1

Table 4 demonstrates the results of a multiple regression study that reveals the unstandardised beta of nepotism and favouritism. This study found that nepotism and favouritism account for 47 percent of the variance in staff performance in public-sector healthcare facilities.

Table 4: Regression assessment of nepotism and favouritism with staff performance.

Factors	β	Standard Error	t-value	Significance
Favouritism	0.75	0.11	6.26	0.00
Nepotism	–0.24	0.09	–2.61	0.01
R = 0.68, R2 = 0.47, F (2,14) = 65.09				

The multiple regression evaluation of beta numbers is shown in Table 5. Therefore, nepotism and favouritism account for 37 percent of the variance in HRM in public-sector healthcare facilities.

Table 5: Regression assessment of nepotism and favouritism with HRM.

Factors	β	Standard Error	t-value	Significance
Favouritism	1.48	0.15	9.36	0.00
Nepotism	–1.03	0.12	–8.38	0.00
R = 0.61, R2 = 0.37, F (2,14) = 44.41				

5. CONCLUSION

The Government-run healthcare facilities have to concentrate on the political and power part of the firm in terms of nepotism and favouritism, as this influences the productivity of staff HRM in its operation. According to the findings of this research, nepotism has a negative influence on staff performance and HRM. Favouritism, on the contrary, has a positive influence on staff performance and HRM. As a result, organisations must implement successful strategies to avoid the power and political element of the firm in the type of nepotism and favouritism. Because this type of politics and power influences both staff and organisational performance. As a result, HRM must be executed with excellence, without regard for friends and family members. Future studies ought to inquire into the influence of nepotism and favouritism as an aspect of organisational power and politics on HRM and staff performance in other industries. Furthermore, in the future, scientists may examine the company's power and politics in terms of nepotism and favouritism as factors that mediate between HRM and staff performance, or with other facilitating and controlling factors. Additionally, research should be conducted in the future to compare nepotism and favouritism to employment-related conduct like absences, workplace anxiety, and employment contentment.

REFERENCES

AL-shawawreh, T. B. (2016). "Economic effects of using nepotism and cronyism in the employment process in the public sector institutions," *Research in Applied Economics*, 8(1), 58–67.

Buchanan, D., and R. Badham (2020). *Power, Politics, and Organisational Change.* Sage.

Hochwarter, W. A., C. C. Rosen, S. L. Jordan, G. R. Ferris, A. Ejaz, and L. P. (2020). "Perceptions of organisational politics research: past, present, and future," *Journal of Management*, 46(6), 879–907.

Ombanda, P. O. (2018). "Nepotism and job performance in the private and public organisations in Kenya," *International Journal of Scientific and Research Publications*, 8(5), 474–494.

Schneider, R. C. (2016). "Understanding and managing organisational politics."

Shaukat, H., N. Ashraf, and S. Ghafoor (2015). "Impact of human resource management practices on employee's performance," *Middle-East Journal of Scientific Research*, 23(2), 329–338.